SCATTERED TO THE FOUR WINDS

GENERAL ORDER NO. 11 AND MARTIAL LAW IN JACKSON COUNTY, MISSOURI, 1863

by Ralph A. Monaco, II

in collaboration with the
Jackson County Historical Society

Civil War Sesquicentennial Series

Monaco Publishing, LLC
Kansas City, Missouri

Monaco, Ralph A., II, 1956-
Scattered to the Four Winds: General Order No. 11 and Martial Law in Jackson County, Missouri, 1863/
438 p. cm.
Maps and Illustrations. Includes index.

ISBN-13: 978-0974136592
ISBN-10: 097413659X

1. United States--History--Civil War, 1861-1865. 2. Missouri—History Civil War, 1861-1865. 3. Missouri—Biography. 4. Jackson County (Mo.)—History—Civil War, 1861-1865. 5. Order Number 11. I. Monaco, Ralph A. (1956-) author. II. Title.

Revised Edition, February 2014.

Published in the United States of America by:
Monaco Publishing, LLC
monacopublishing.net

TABLE OF CONTENTS

Introduction 5

Preface 9

Chapter I Union and Confederacy at the Dawn of 1863 15

Chapter II The Kansas Reign of Jim Lane 21

Chapter III Bingham: Painter, Politician and Propagandist 49

Chapter IV Quantrill Is Back! 59

Chapter V General Schofield and District of Frontier 67

Chapter VI "A Little God" Provost Marshal Comingo 71

Chapter VII General Thomas Ewing Returns to Kansas 81

Chapter VIII Death and Calamity in Kansas City 93

Chapter IX Black Friday 109

Chapter X Martial Law: General Order No. 11 119

Chapter XI Responses to Order No. 11 129

Chapter XII Smoke in the Sky 139

Saga of the Luttrell Family 144
Impact upon Family of a Future President 146
Tales of Laura (Harris) Flanery and Family 149
Family Ties near Lone Jack 154
Tragedy and Loyalty: September 6 158
Post-script of Order No. 11 163

Chapter XIII Bingham's War against Ewing 167

Chapter XIV The Fates of Major Players:
Lane, Bingham, Quantrill, Schofield, Comingo and Ewing 183

Appendix A The Emancipation Proclamation 213

Appendix B List of Known Killed at Lawrence 215

Appendix C General Order Number 9 219

Appendix D General Order Number 10 221

Appendix E General Order Number 11 223

Appendix F Letter of Comingo, et. al to President Lincoln 225

Appendix G Citizens of Jackson, Cass Bates and Northern Vernon Counties
Affected by General Order No. 11 229

Appendix H Personal, Documented Stories of Order No. 11 Victims 241

Illustrations 387

Bibliography 397

Index 417

Acknowledgements 435

About the Author 437

To My Daughters Lindsay and Lisa

May They Never Know
The Pain, Agony or Horror of War!

INTRODUCTION

Certain contemporary issues had relevance during the American Civil War that were directly or indirectly debated then as they are today.

> *Is the use of psychological measures proper to elicit information or a confession? What if they are implemented to curb terrorism or rebellion?*
>
> *Should modern technology, including drones, be employed against American non-combatants?*[1] *What about citizens' known to be criminals, insurrectionists or terrorists?*
>
> *Do surveillances such as wire-tapping require court ordered search warrants when the nations security is at risk? Has the Patriot Act enacted since 9-11 gone too far or should its scope be expanded?*
>
> *Can an arrest be made on mere suspicion and not upon probable cause when national security is at issue? Are military detainees entitled to the writ of habeas corpus?*[2]
>
> *When is it justified to seize private property without due process? Do such seizures apply in emergency or other exigent circumstance?*
>
> *Why should social profiling not be authorized when it is designed to protect American borders? What about profiling with respect to known or suspected criminals or enemies of the state?*
>
> *Does the right to bear arms under the Second Amendment include assault rifles or other similar weapons? What about mandating national background checks on purchases of firearms?*[3]

---

[1] In the spring of 2013, United States Senator Rand Paul of Kentucky waged a filibuster on the Senate floor until President Barrack Obama assured him that the federal government would not utilize drones against American noncombatants. Would Rand's question and Obama's answer apply to future terrorist attacks similar to the Boston Marathon bombings on Monday, April 15, 2013?

[2] As of May 23, 2013, there were approximately one hundred U.S. military detainees held at Guantanamo Bay, Cuba, who staged a hunger strike in excess of one hundred days in part because they have been held without trial or the right to obtain a writ of habeas corpus.

[3] On April 17, 2013, the United States Senate defeated a measure that would have required universal background checks on anyone purchasing firearms at gun-shows or over the internet.

Is the U.S. Government empowered to subjugate the rights of states? What about citizens of those states?[4]

Who could possibly oppose trying terrorists in military tribunals? Should civilian terrorists be tried in military or criminal courts?

Are the civil liberties of American citizens subordinate to national security? Does military necessity or exigent circumstances override civil liberties?[5]

Should the government be allowed to banish citizens from their property during exigent periods of time? What if it is to quash a riot, rebellion or unrest?

Are secret wiretaps issued under sealed court orders justified? Does the government have the right to subpoena cell or text phone numbers from cell phone companies?[6]

The legal debate on these queries may never end. However, these challenging interrogatories have corollaries found in the Civil War, and in particular General Order No. 11 issued on August 25, 1863. General Thomas Ewing, Jr. imposed the Order as a military measure following William Quantrill and his guerilla army's sacking of Lawrence, Kansas four days earlier. The Order banished all residents in a four county region of western Missouri and became part of the horror of war upon a civilian population. The Civil War in Missouri and in particular Jackson County was an "uncivil affair" that climaxed during the summer of 1863 with one of the most drastic military measures ever imposed upon a civilian population in American History to that date. The

---

[4] On April 16, 2013, Governor Sam Brownback of Kansas signed into law a bill declaring that the federal government had no power to regulate guns manufactured, sold and retained only within Kansas. Under the law any federal agent or officer who attempted to enforce any statute or regulation that would curb these rights of Kansas citizens would be guilty of a felony. This bill will undoubtedly create litigation in Federal Court over the Supremacy Clause and Interstate Commerce Clause of the U.S. Constitution. The debate over individual rights of citizens of a state (Kansans under this law), and federal authority was also a theme during the Civil War and it continues to this day. This debate has been raised likewise in states in which marijuana has been legalized.

[5] On April 19, 2013, four Republicans, United States Senators, Lindsey Graham, John McCain and Kelly Ayotte and Congressman Peter Kind issued a joint statement that nineteen-year-old Boston Marathon bombing terrorist suspect Zhokhar Tsarnaev should not be entitled to rights under either Article III or the Fifth Amendment of the United States Constitution and should be tried for the terrorist attack in Boston in the military court at Guantanamo Bay, Cuba and not in the United States Federal Court.

[6] On June 6, 2013, a fury of Congressional debate focused upon whether the government should have the right to issue subpoenas to obtain such records from cell phone companies.

historical fact of the Order retains its relevance today. If this is questioned, one only needs to ponder the foregoing points while pouring through this book.

The haunting journey culminating in the issuance of Order No. 11 began ten years earlier in 1854. From the adoption of the Kansas-Nebraska Act on May 30, 1854 through the imposition of martial law on August 25, 1863, Jackson County experienced almost persistent turmoil, toil and tribulation. This book attempts to weave the story of those who directly or indirectly issued, enforced, opposed or experienced Order No. 11. The tales of the horror of that singular and most notorious year along the border of Kansas and Missouri and the pervasive personal accounts of those impacted by it are recounted in the following pages.

My personal fascination or journey with the American Civil War and Order No. 11 is rooted in my childhood. The year 1961 was the Centennial of that War. Battlefields and historic sites held special events, parades, speeches and reenactments. Lexington, Lafayette County, Missouri was no exception. In September that year, my parents took me and my sister Cindy to the reenactment of the 1861 Battle of Lexington (often called the Battle of Hemp Bales). It was mesmerizing, like a quantum leap back in time. I became an eyewitness to the final successful assault by Missouri State Guard troops under General Sterling Price upon the earth works of Union forces under Major James A. Mulligan.[7] For a five-year old boy it was so realistic. Following the battle and as the men were marching off the "battlefield," I turned to my late father with glaze in my eyes and asked, "Are the soldiers going back to their graves now?" Someone who overheard my question must have considered it either cute or newsworthy as during the evening news that night Claude Dorsey, the anchor for KMBC, the local ABC affiliate, concluded the evening news with the summary: "Little Ralphie Monaco perhaps best summarized the events at Lexington today when he asked his father if the soldiers were going back to their graves."

From that moment on, my world became captivated by anything related to the Civil War. Hours were spent playing with my Civil War soldiers in the living room; fighting endless battles in the backyard and in the neighborhood (yea, we played outside back in the 1960s); dawning my Union Blue hat; caring my musket; firing the canister and ball from my Civil War cannon—it actually worked like a cannon (it mattered not that it was made of plastic). Most shocking to my parents was that I began reading Civil War books, beginning with Steven Crane's, "The Red Badge of Courage." One could say I had grown into a Civil War junkie.

This explains my interest in the Civil War, but my fascination with Order No. 11 is yet another story filled with irony. My mother and late grandmother were antique doll collectors some forty to fifty years ago. On one particular day in 1970 they attended a

---

[7] Following the Battle of Lexington, Mulligan was promoted to Colonel. He was later mortally wounded on July 23, 1864, at the Second Battle of Kernstown, near Winchester, Virginia. On February 20, 1865, the United States Senate posthumously bestowed the rank of brevet brigadier general to Mulligan.

"going-out-of business" sale at Vincent's, an antique "doll hospital" owned by two elderly, spinster sisters located on the east side of Main Street at about 13th Street in Kansas City. My mom and grandma were allowed to make an offer on any items they found (even being turned loose in the spacious attic of the old building). It was there in the damp, dark recesses they discovered a dusty, mildewed and cobweb covered picture inside a wooden frame that had to have been at least one hundred years old. The glass covering was severely cracked. All they knew was the image appeared to be a "black and white" photo of some Civil War scene. Mom bought the picture (frame, cracked glass and all) for about $100.

They were correct that it was a Civil War setting, but they were wrong in thinking it was just an old "black and white." They took the picture to a dealer of fine arts in Kansas City in order to replace the glass and reset the print. It was then that the stark truth was unveiled: the picture was a treasure trove. It was an original steel engraving by John Sartain of George Caleb Bingham's painting of Martial Law, General Order No. 11, still encased in a frame dating to the 1870's.

Neither my mother nor grandmother had any idea about the true historical significance of the artwork. I still remember mom giving the picture to me (and now safely framed behind an unbroken glass). I, too, had no appreciation of its historic value or importance other than I liked it. I told my mother that if I ever became an attorney it would be displayed on my law office wall. Sartain's steel engraving of Bingham's seminal, political artwork, "Martial Law," or "Order No. 11" has been hanging in my law office now for nearly 32 years.

PREFACE

This historical work principally focuses upon the ordeal confronted by ordinary citizens of Jackson County, Missouri under military martial law in the second-half of 1863—certainly the most pivotal year of the American Civil War. To appreciate the drama of that seminal year, a chronology is provided that is neither comprehensive nor exclusive. It is intended to infuse the reader into that tragic and heartbreak period as it had been experienced by residents of the borders of Missouri and Kansas in 1863.

1863

January 1:	***Washington City***: President Abraham Lincoln issues Emancipation Proclamation from Executive Mansion.
January 25:	***Fredericksburg, Virginia***: General Joseph Hooker replaces General Ambrose Burnside as commander of the Union Army of the Potomac.
March 3:	***Washington City***: President Lincoln signs into law the Federal Enrollment Act of Conscription (draft) in which every man between the ages of twenty to forty-five was subject to being drafted for military service.
April 23:	***St. Louis, Missouri***: Colonel Thomas Ewing, Jr. is promoted to Brigadier-General.
May 1-3:	***Chancellorsville, Virginia***: For three days, Confederate forces under General Robert E. Lee battled General Joseph Hooker and his Army of the Potomac resulting in yet another victory for Lee.
May 10:	***Chancellorsville, Virginia:*** General Thomas (Stonewall) Jackson dies from injuries suffered during the Battle of Chancellorsville when he had been mistakenly shot by a sentry in his own army.
May 24:	***St. Louis, Missouri***: General John McAllister Schofield resumes command of the District of Missouri.
June 9:	***Brandy Station, Virginia***: The greatest cavalry battle on American soil pitted the Federal Cavalry under General Alfred Pleasanton against Confederate Cavalry General J. E. B. Stuart—over twenty thousand cavalry forces were engaged in the battle.

June 3: ***Independence, Missouri***: Abram Comingo is appointed Provost Marshal of the 6th Congressional District.

June 16: ***Kansas City, Missouri***: Brigadier-General Thomas Ewing, Jr. assumes Command of the District of the Border.

June 28: ***Fredrick, Maryland***: Under General Order No. 67 General George Meade replaces General Joseph Hooker as commander of the Army of the Potomac.

June 30: ***Washington City***: The western part of Virginia is admitted into the Union as a separate state after having successfully seceded from Virginia: becoming West Virginia.

July 1-3: ***Gettysburg, Pennsylvania***: In the most decisive battle of the Civil War, Union forces under General Meade defeat Confederate troops under General Lee at the Battle of Gettysburg, thus ending Lee's military invasion into the north.

July 4: ***Vicksburg, Mississippi***: After months of attempting to capture Vicksburg, Mississippi on the Mississippi River General Ulysses S. Grant captures Vicksburg and negotiated surrender terms with Confederate General John C. Pemberton.

July 11-14: ***New York, New York***: Riots break out in New York in opposition to the Federal Conscription Law: as many as one hundred people were killed and over one million dollars in property was destroyed or damaged.

July 18: ***Charleston, South Carolina***: A failed, violent frontal assault is launched into the Confederate breastworks at Battery Wagner on Morris Island near Charleston, South Carolina by the Fifty-fourth Massachusetts Colored Infantry under Col. Robert Gould Shaw: the 54th suffered a casualty rate of over twenty-five percent and Shaw was killed in the battle.

August 13: ***Kansas City, Missouri***: The women's prison in Kansas City (owned by the wife of George Caleb Bingham and sequestered by the government) collapses killing five women and injuring many others.

August 18: ***Kansas City, Missouri***: General Ewing issues General Orders No. 9 and No. 10 in which certain slaves were freed and rebel citizens were banished from their properties in western Missouri.

August 21: ***Lawrence, Kansas***: Well over four hundred guerrillas under William Clarke Quantrill attacked Lawrence killing approximately two hundred and becoming the largest single bushwhacking raid of the Civil War.

August 25: ***Kansas City, Missouri***: General Ewing issues General Order No. 11 which banished all residents living in Jackson, Cass, Bates and Northern Vernon Counties, except for certain designated military posts and cities.

Sept. 19-20: ***Chickamauga, Tennessee***: The Battle of Chickamauga resulted in a Confederate tactical victory while Union forces were able to withstand persistent rebel attacks.

Oct. 6: ***Baxter Springs, Kansas***: Quantrill guerrillas adorned in Union uniforms appeared on the southeastern edge of Kansas and annihilated federal troops under Major-General James G. Blunt.

Nov. 19: ***Gettysburg, Pennsylvania***: President Lincoln at the dedication of the Gettysburg Cemetery gave his "Gettysburg Address."

Nov. 25: ***Kansas City, Missouri***: General Thomas Ewing relaxes certain regulations contained within Order No. 11.

Nov. 23-25: ***Chattanooga, Tennessee***: During the first two days of the Battle of Chattanooga, Union forces captured Orchard Knob and Lookout Mountain, and on the final day seized Missionary Ridge routing one of the two major Confederate armies and captured Chattanooga, the "Gateway to the Lower South," which would serve as the base of operations for General William Tecumseh Sherman's "March to the Sea."

There are several purposes for this study. First, this year marks the sesquicentennial of the issuance of the famous or infamous Order No. 11. The story remains shrouded in debate and controversy by the progenies of its antagonists and protagonists. It has become enshrined in the law of military necessity and the artistic fabric of the 19th Century. Most importantly, the meaning, purpose and ramification of Order No. 11 continue to be part of a modern America faced with national and international terrorism. This work considers the sensitive balance between the police powers of the United States to preserve and protect the country against the fundamental substantive civil rights of its citizens. This constitutional argument cannot be considered in a vacuum. Order No. 11 was a military war measure imposed against everyone, especially women, children and the elderly—or to use today's nomenclature, a war against non-combatants. Much of the following story is told by those victimized by the Civil War along the borders of Missouri and Kansas—a war that was anything but civil.

By 1863, the United States continued to be plagued with the horror and mounting death toll of the rebellion. In Jackson County the violence had become so pronounced that no one regardless of his or her loyalty was safe. Along the border, there had been no great battle fought or contested; no grand armies paraded or marched. Engagements were fought in wooded areas, over fence-lines, along by-ways, on farm fields and front-yards and in the streets of towns. Most fights were skirmishes between small-numbered bushwhackers and rebels against militias, volunteers and federal troops. Rebel guerrillas or Partisan Rangers frequently utilized an Indian style of fighting against an opponent who was generally ill equipped and understaffed. Much of the bushwhacker success came from aid and support they received from citizens principally cultured in an agrarian southern mindset.

When William Quantrill and his guerrillas raided Lawrence, Kansas leaving some two hundred people dead, the Federal government took drastic measures. The retaliatory sanctions imposed would prove to touch the heart and soul of everyone—loyal or disloyal. The supporters of Order No. 11 insisted it was issued out of military necessity while its antagonists considered it the greatest deprivation of civil rights ever witnessed in the United States.[8] This is the story of Order No. 11—Martial Law in Jackson County.

---

[8] Many historians consider Order No. 11 as the greatest warfare waged by the government upon Americans to the date of its issuance (second only to the removal of Japanese citizens following Pearl Harbor on December 7, 1941).

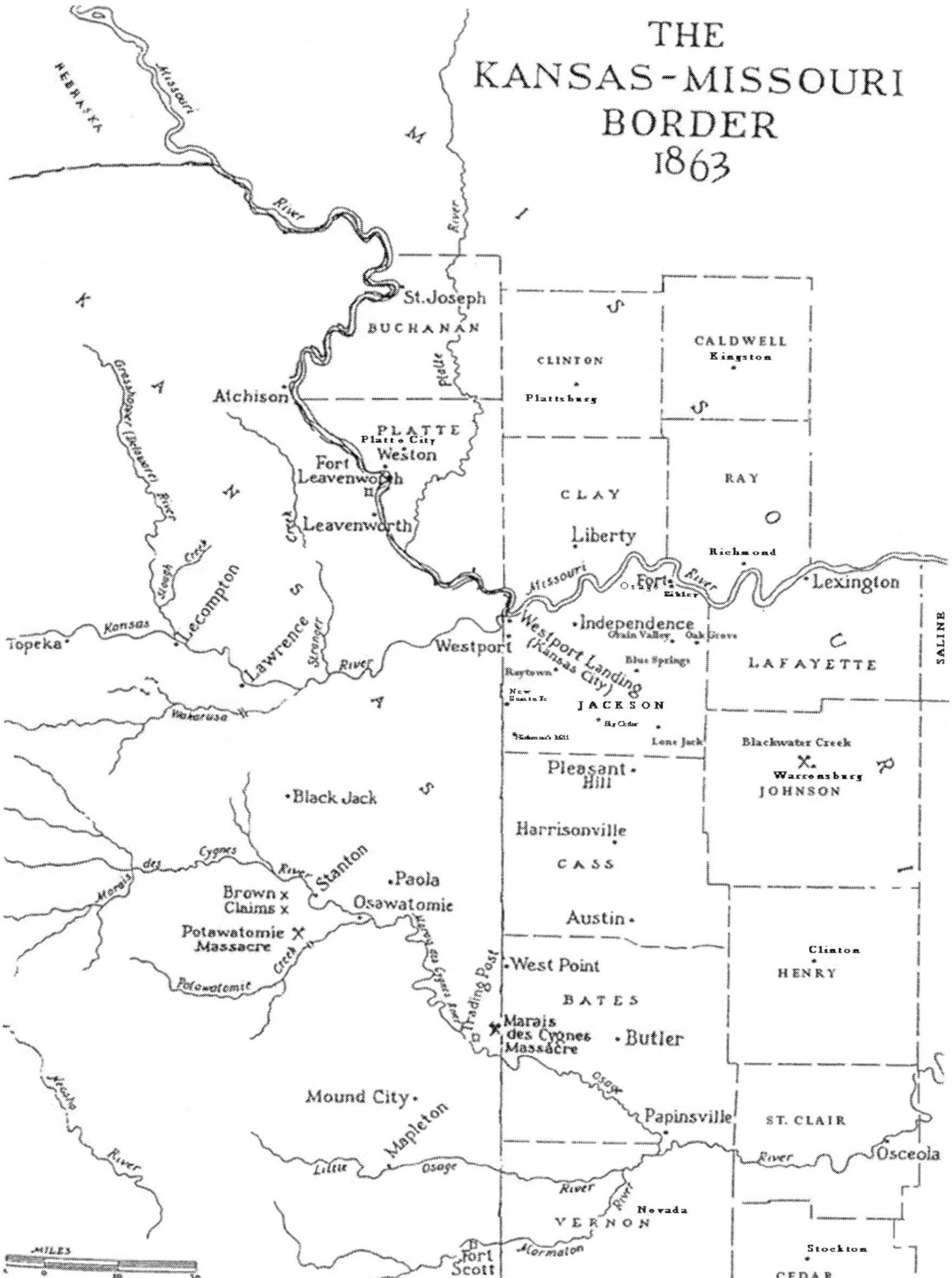

James C Malin map modified by David W. Jackson under supervision of Ralph A. Monaco, 2013

CHAPTER I

The Union and Confederacy at the Dawn of 1863

As the sun rose on New Year's Day 1863, the Civil War was entering its third year with an outcome still in the balance. To the surprise, shock and stun of the industrial north, the southern states had not been vanquished, but seemed perched upon gaining independence. For the people in New England, the entire year of 1862 had seen nothing short of repeated rebel victories and federal defeats. General Robert E. Lee and his Army of Virginia had demonstrated a resiliency that caused many in the north to doubt if any Union commander could defeat him. The last major battle of 1862 fought in December at Fredericksburg, Virginia proved to be simply a new chorus to an old song. Lee defeated Federal forces under General Ambrose Burnside (and his dashing long side hair or "sideburns"). The defeat was most horrific, as the Army of the Potomac had been nearly destroyed at Marye's Hite. This latest debacle was more disappointment to the Lincoln Administration. In Richmond the mood was elation. President Jefferson Davis predicted southern Independence was imminent. With General Lee in command, Davis was confident that when spring's warmth descended upon the land, Lee would descend upon the north and victory would soon be in hand.

General Robert E. Lee

The Lincoln Administration in Washington was gravely disturbed about the war effort and public response. Lincoln continued to deliberate on whether any Union general could ever challenge the military brilliance of Lee. Burnside would soon become another casualty of the long list of previous Union commanders. Within thirty days General Joseph Hooker, "Fighting Joe" as he had become affectionately known, would replace Burnside.

Another schism in the north was the growth of the Copperhead movement. Their philosophical agenda was for a complete cessation of hostilities. Copperheads pointed to mounting casualties and financial losses. They promoted ending the war, restoring peace and granting independence to the Confederacy.

The same uncertainty existed along the border of Jackson County, Missouri. Despite successful Confederate and guerilla recruitment efforts and victories on the streets of Independence and Lone Jack in August 1862, they failed to gain control of the region. Instead, these successes only resulted in a renewed and revitalized influx and invasion into Missouri by menacing Kansas' troops. These "invaders" once again patrolled the area with a sinister design of marauding, mayhem and murder. Only the wintry weather had brought a certain chill to the turmoil in the county. Throughout Jackson County, there was a certain sense that the coming year would not only bring Quantrill and his guerrillas back from Texas, but the blood, tears and turmoil of so many years would become an unstoppable conflagration greater than the worst summer firestorm on a Kansas prairie.

U.S. President Abraham Lincoln

Chaos, death and tribulation had ravaged America the previous two years—for those residing in Jackson County the tragedy was entering its tenth year. It was certain that the battles, violence and casualties would intensify as the new-year dawned. While the provisional state government in Missouri continued to debate the volatile issue of gradual emancipation versus immediate eradication of slavery within the state, a new public policy on slavery was announced in Washington City on January 1, 1863: The Emancipation Proclamation.

The issue of slavery had been at the very heart of the history of Missouri dating from its application for statehood in 1820. The Missouri Compromise of 1820, which admitted Maine as a free-state and Missouri in 1821 as a slave state pacified the debate for nearly thirty years. The stillness ended with America's victory over Mexico. Congress negotiated a temporary treaty to quiet the debate with the adoption of the Compromise of 1850. These resolutions proved to be at best short-term solutions to the mounting fury over slavery.

When Kansas and Nebraska were opened for settlement under the Kansas-Nebraska Act of 1854, the controversy over the expansion of the institution into the new Territories went from dialogue and debate to bloodshed and battles. Pro-slavery

Missourians waged open warfare with free-soil settlers of Kansas. There was nothing like the thunder on the border during the days known as "Bleeding Kansas."

When the Civil War erupted in 1861, Missouri found itself once again in the middle of the storm of slavery and secession. During the first two years of the war, no resolution of the vexing problem of slavery could be settled in Missouri or anywhere else in the country. Instead war continued unabated.

Under President Abraham Lincoln's new war powers policy, "all persons held as slaves within any State or designated part of a State, the people whereof shall then be in rebellion against the United States, shall be then, thenceforward, and forever free." This required every military and naval authority to "recognize and maintain the freedom of such persons." Lincoln insisted he had the authority as Commander-in-Chief of the Army and Navy under the United States Constitution to issue this edict under war powers during times of actual armed rebellion.[9]

The purpose of the Proclamation was an articulation by the Lincoln Administration of a new direction in waging war and in dealing with international relations. The Proclamation's foreign policy impact would be of short-term consequences while its affect upon domestic affairs would become the central theme of civil rights' issues that still permeates today.

The Proclamation certainly did not free any person held in slavery in the border-states, such as Missouri. It did not emancipate slaves in any area of the seceded southern states then under federal control, including Ft. Pickens in Florida, northwest Arkansas or New Orleans. Only slaves in those seceding states that remained under the authority and control of the Confederacy were freed under the Proclamation.

Lincoln realized his new policy posed a vexatious dilemma for the international community. He knew reports had been streaming across the Atlantic that Europe, especially England, considered recognizing the independence of the South. Lincoln could not allow this to occur, as it would severally compromise his ability to quash the rebellion and reunify the United States. Slavery had been abolished in Europe, and Lincoln's abolishment of the institution in those sections still under rebellion would create a diplomatic headache for the Continent. These countries would find it difficult and virtually impossible to recognize the Confederate States as a separate country since slavery remained part of its government's fundamental substantive laws.

The Proclamation's domestic success was less certain. Lincoln had spent the first two years of his presidency waging war to save the Union. This forced him to stretch the U.S. Constitution and his executive war powers' authority to their outer limits. The writ of habeas corpus was indefinitely suspended; martial law was declared; property was

---

[9] Scott, Robert N., H. M. Lazelle, George B. Davis, Leslie J. Perry, Joseph W. Kirkley, Fred C. Ainsworth, John S. Moodey, and Calvin D. Cowles. "Official Records of the War of the Rebellion: A Compilation of the Official Records." (Washington, D.C.: Government Printing Office, 1880-1901) (hereinafter "O.R."). The Emancipation Proclamation is set forth in Appendix A.

seized without any compensation; military tribunals held dominion, even where civil courts remained opened; compulsory loyalty oaths and mandatory conscription laws were implemented; searches were conducted without benefit of warrant; arrests were made on mere suspicion and not upon probable cause; and many other direct assaults on the civil liberties of Americans had been instituted.

John C. Fremont

While Lincoln's Administration usurped constitutionally protected rights, it did not interfere with the institution of slavery, which remained legal and inviolate. One explanation for non- interference with involuntary servitude was that during the first year of the War Lincoln could not run the risk of losing the slave holding border-states of Missouri, Maryland, Delaware and Kentucky to secession. These states were essential to the preservation of the Union and Lincoln was going to take whatever measures necessary to ensure they remained within the Union. This is fully illustrated by Lincoln's response to the military order issued by General John C. Fremont in Missouri during the late summer of 1861. Fremont had announced his own emancipation proclamation, and Lincoln demanded its recession; when Fremont failed to do so, Lincoln removed him from command.

Another reason slavery was not compromised was because most northerners were not abolitionists. Their goal was to save the Union—not wage war to abolish the "Peculiar Institution." They might support a program of liberating slaves through the process of gradual emancipation or colonization to Africa, Jamaica or elsewhere—but not outright eradication of it.

Lincoln knew that abolishing slavery posed significant constitutional and legal problems. Emancipation in the non-seceded border-states would require the government to compensate the slave owners for the "taking" of their "property" under the "taking clause" and "just compensation" provisions of the 5th Amendment to the United States Constitution. Lincoln recognized these legal limitations when he gave his first inaugural address. In his speech, Lincoln assured non-intervention to slavery in the areas in which it then existed.

The reasons not to emancipate certain slaves were no longer relevant as the third year of the War opened. Lincoln now saw an attack upon the institution as a valued war measure. He reasoned that the Proclamation was nothing more than confiscation of property owned by enemies of the state. This argument was bolstered by the fact that the Confederate Constitution classified slaves as "chattel property." Lincoln would use the fundamental laws of the South to impose his new war powers strategy. As Commander in

Chief, he had the authority to seize slave property as "contraband" in the same manner he could confiscate any other property without compensation to its owner to quash the rebellion. Lincoln's justification for the act of taking the "slave property" or "contraband" would be of military necessity—not human rights (that would become an issue later but not on January 1, 1863).

It was uncertain what impact the Proclamation would have on the Civil War as it went into effect on January 1, 1863. What was certain is that the Proclamation meant that the Federal government was no longer simply fighting a war to preserve the Union; it had taken on a new cause as well—a war against slavery. The Proclamation would ultimately lead to the adoption of the 13th Amendment in 1865—the total abolishment of slavery in the United States.

As the telegraph wires dispatched the announcement of the Proclamation, the citizens on the border of western Missouri remained in the throngs "of strife and bloodshed." Political bitterness and divisiveness had plagued Kansans and Missourians long before the first shot was fired at Ft. Sumter, South Carolina on April 12, 1861. For nearly ten blood-curdling years, the barbarity on both sides had been unprecedented. One atrocity had succeeded another in brutal retaliation. The situation had only worsened after the first shell was fired and fell into Ft. Sumter.

The Missouri-Kansas state line had been overrun by lawless bands of desperadoes. These gangs were composed of reckless and unrestrained soldiery who had "rioted upon the substance of the people" and had "wantonly destroyed their property and trampled upon their most sacred rights. Theft, robbery, house-burning, and other crimes" had "been perpetrated with impunity," and "plunder and vandalism" had prevailed. A large portion of Jackson County had been left impoverished and almost depopulated due to these ravages.[10] Western Missourians had been harassed, threatened, robbed or killed by Kansas Red Legs and Jayhawkers while rebels, bushwhackers and guerrillas had retaliated against Kansas border towns in similar fashion. The Proclamation simply failed to resonate any significance to the people along the border of Eastern Kansas and Western Missouri on that New Year's Day—an area in which only terrorism continued to reign at the dawn of the third year of the Civil War.

---

[10] O.R., 22.2:485-488; and, United Daughters of the Confederacy, Missouri Division. "Reminiscences of the Women of Missouri during the Sixties." (Jefferson City, Mo.: The Hugh Stephens Printing Co., 1920), 124. This book has been reprinted and is for sale by the Missouri Division (http://www.rootsweb.ancestry.com/~moudc/index.html).

CHAPTER II

The Kansas Reign of Jim Lane

It is virtually impossible to explain or appreciate the story of Territorial Kansas and the guerilla warfare that persisted throughout the Civil War without a discussion of the influence of one unique, controversial man. He was solitary, gaunt, ambiguous, and disheveled, but also gifted, influential and oratorical skilled. During those troubled years, he individually held in the tip of his tongue power, persuasion and purpose. His dominion radiated from the eastern seacoast to the setting sun. He was manipulative, intimidating, politically savvy, cunning and vindictive. To some he was a statesman to others a scoundrel. Almost single handily he had helped fuel the free-soil debate against slavery's extension into Kansas, and he held absolute command over the Republican Party from Washington City to his adopted state of Kansas throughout most of the Civil War. His power and political governance in Kansas lasted as long as his volatile reign of terror would haunt the people of western Missouri—ten years.

James "Jim" Lane

Jim Lane was a Machiavellian in the truest sense. In the end, he failed to recognize the significance of Lord Acton's famous statement: "Power tends to corrupt and absolute power corrupts absolutely."[11] When word of Lincoln's Emancipation Proclamation reached Jim Lane, the "Grim Chieftain," had reached the zenith of his political career.

---

[11] John Emerich Edward Dalberg Acton "Lord Acton" was born in Naples, Italy, in 1834. Three years after his father died in 1837, his mother married Lord George Leveson and moved the family to Great Britain. Because of Acton's firm Catholic beliefs, he was barred from attending Cambridge and went to the University of Munich. He considered himself more as an historian than a political figure. The above quote is attributed to him. Lord Acton died in Bavaria in 1902. The source of the quote and life and times of Lord Acton, see, www.newadvent.org/cathen/01114a.htm.

James Henry Lane was born in Lawrenceburg, Indiana on June 22, 1814 to Amos and Mary Lane.[12] Lane's father was a staunch, formidable and dedicated member of the Democratic Party in Indiana. He had strongly supported the leadership of President Andrew Jackson and the system known as "Jacksonian Democracy." He practiced law and ascended the Democrat ranks serving in the Indiana State Legislature and later in the United States Congress. It only seemed natural for his son to follow in the footsteps of his influential father, and he did in the early 1840s. In 1842 to further his political appetite, Lane married Mary E. Baldridge, whose own family was well known and politically connected in Indiana; three children would be born of their union.

In 1846 during the administration of President James K. Polk, the United States declared war on Mexico. Despite Lane's expanding law practice, he chose to enlist in the Indiana Volunteers. His father's political clout and his own legal talents helped garner him a captain's commission in the Dearborn Volunteer Company (later designated the Ohio River Regiment, and eventually the Third Regiment Indiana Volunteers). When the Third Regiment set off to take part in the Mexican War, Lane was elected colonel of one of the three divisions. Within less than thirty days, Lane had soared from no military experience to the rank of colonel. He would eventually serve two successful tours of duty during the Mexican War.[13]

Following America's victory over Mexico, Lane returned home to a war heroes' welcome. The overwhelming triumph ignited the American dream of "Manifest Destiny"—extending America's borders from the Atlantic to the Pacific and it extended Lane's reputation. The members of his regiment raised money and presented him with a military sword at the Indiana Democratic State Convention in January 1849 where he also received his party's nomination for lieutenant governor—an election he won. Three years later, Lane's political stock had further risen. He served as a Democratic elector at large in the 1852 presidential election where he cast his vote for Franklin Pierce, and he was elected to the United States Congress from Indiana's Fourth District.

The year 1854 marked a pivotal year in American history as it did for the future of Congressman Lane. By conquest, America had acquired significant new land from Mexico, and an outburst of public sentiment erupted favoring fulfillment of America's "Manifest Destiny." This included a clamoring for settlement into the virgin territory originally purchased from France by President Thomas Jefferson in 1803, known as the Louisiana Purchase. There was an equal insistence to extend railroad lines and telegraph

---

[12] Collins, Robert. "Jim Lane Scoundrel, Statesman, Kansan." (Gretna, La.: Pelican Publishing Company, Inc., 2007), 15-16; and, McCullough, David. "Truman." (New York: Simon & Schuster, 1992), 30. To prove that history is stranger than fiction consider the names in this paragraph: Lane's birth home was Lawrenceburg—his death home was Lawrence; his father's first name was Amos as was the first name of the namesake of Lawrence—Amos Lawrence, the great benefactor of the Massachusetts's Emigration Society whose financial wealth had established the town in 1854—it remains the city's name.

[13] Collins, 18-19.

wires to the Pacific Coast of California and to ensure the integrity of the three western trails that originated in Independence, the county seat of Jackson County: The Oregon, Santa Fe and California Trails.

Since the Louisiana Purchase, the land west of Missouri, except military posts, such as Fort Leavenworth and fur trading stations, had remained dormant and under the dominion of Native Americans. The public reason for ignoring the region was a lack of public interest. The private reason was to evade the sensitive issue of slavery. For nearly fifty years, the only interruption to this vexing question occurred in 1820 when Congress approved the Missouri Compromise, which had authorized the admission of Missouri as a slave state and Maine as a free state. The Compromise further prohibited any further extension of slavery into the remaining areas located within the Louisiana Purchase.

The Missouri Compromise had quieted the furious debate over slavery for well over a generation. Congress also found a way to avoid the issue. It adopted a steadfast but unwritten policy known as the "Gag Rule." Under this procedure no Congressman was authorized to speak or debate on the slavery issue. The Gag Rule remained ironclad until the victory over Mexico. With the gag removed, the slavery controversy raged and rocked the country. Although the Compromise of 1850 quieted matters for a brief time, it proved to be nothing more than the calm before the storm. Western expansionism created a thirsty demand for the opening of the once undeveloped territory. Congress could neither ignore the issue nor fall back upon the Gag Rule. It had to find a solution to the slavery issue to avoid an irrevocable seism from erupting between north and south.

David Rice Atchison

No resolution of this potentially volatile discussion seemed possible until Democrat Senator Stephen A. Douglas of Illinois, known as the "Little Giant" because of his diminutive height, came forward. He intended to establish a plan to quash the national dilemma. Douglas received significant support from the President Pro Temp of the Senate, David Rice Atchison of Missouri. Their height difference and background made the two men an odd pair: the Little Giant stood only about five foot four, while Atchison towered over his colleague at six feet; Douglas was from the industrially powerful City of Chicago; Atchison was from an agrarian southern culture. They each were members of the Democrat Party and held real political clout in Congress and in their home states. They united their pedigrees to propose the Kansas-Nebraska Act to Congress in 1854.

Under this plan, expansion of the "Peculiar Institution" into the new territories would be determined by the doctrine of "Popular Sovereignty." This system would leave the issue of whether the new territories would include or exclude slavery to be decided by the popular vote of the people of Kansas and Nebraska. Douglas and Atchison were confident that no one, north or south, could object to the democratic process of allowing free citizens in each territory to resolve the matter on their own. Subtlety, they were equally convinced that Nebraska would enter the Union as a free-state and Kansas as a slave state, thereby assuring that a sensitive national balance would remain in Congress between slave and free states. Privately, Douglas wanted a northern railroad line extending from Chicago through Nebraska to San Francisco, while Atchison knew that the survival of the institution of slavery depended upon its extension into the Kansas Territory.

The two senators realized adoption of the Kansas-Nebraska Act by the U.S. House would not be an easy task. They had sufficient votes in the Senate, but the House was another matter. They had to gather their support from all their Democratic allies in the House and Senate. Congressman in the New England States and Whigs openly opposed the Bill. Through diplomacy, arm-twisting and other political intrigue, the Bill finally passed the House in May 1854, with Lane voting in favor of its passage. President Pierce promptly signed it into law on May 30, 1854.

The Bill's adoption received a protest of public scorn in the north and a reluctant acceptance in the south. Northern Democrats received the bulk of the criticism. When the Congressional election of 1854 was held that fall, the Kansas-Nebraska Act proved to be a disaster for those Democrats who had cast an affirmative vote for the Bill. Only seven out of the forty-two Northern Democrats who had voted for its passage were re-elected; the Democrat majority became the minority. Lane experienced the same protestation from the electorate in Indiana and lost his Congressional seat.[14]

The Act soon produced a cry from both free-soil and abolitionist factions in the north and pro-slavery and "fire-eaters" in the south. Each side was determined to resolve the issue of Kansas by "Popular Sovereignty" or by the "knife." The leading northern organization was established in Boston, Massachusetts by Eli Thayer, known as the "Massachusetts Emigration Aid Society." Their determined goal was to guarantee the new Territory of Kansas would be established as a free state. Through donations from leading philanthropists, including Amos Lawrence, a large number of New Englanders were encouraged to immigrate to Kansas for that purpose. Thayer sent Dr. Charles Robinson known for his anti-slavery sentiments to Kansas in 1854 to locate and lay out a town for settlement—Robinson selected an area later to be known as Lawrence.

Southern communities likewise called upon settlers to garrison their resources to settle Kansas to ensure the expansion of slavery. Organizations in western Missouri were

---

[14] Collins, 29.

as determined to see that the institution extended into Kansas. Sam Ralston, a leading citizen from the Independence area of Jackson County, expressed this viewpoint in a letter he wrote to a friend on June 16, 1854:

> *We are on the eve of exciting events, we are menaced with threats, over-whelming and destructive threats, from the Abolition board of N. England, and we are organizing to meet them on the soil of Kansas - and we will burn every blade of grass sooner than their unhallowed feet shall prostitute the soil. Yes, we are organizing to meet them on the border and dispute every inch of ground before they shall have a home, or a resting-place there... you are with us in sentiment & spirit, why not give some aid and comfort? Write to some of your soil stirring Editors and arouse them on the subject, tis the cause of the south for which we are contending. But I must stop - my feelings are excited and carry me away from the proper theme of reflection.*[15]

The Act had an irreparable impact upon Lane's political fortunes in Indiana. His popularity in Indiana plummeted from the mountain top to the depths of the valley after he had cast his vote in support of the Kansas-Nebraska Bill. He had no political future in Indiana. He chose to seek his fortunes in the territory he had helped create. There are several theories as to why Lane chose Kansas. These vary from a desire to form and become the leader of the Democratic Party; to obtain a divorce from his wife Mary; to help Senator Douglas with the latter's presidential aspirations thereby garnering a political chip; to ensure that no blacks, free or slave, would enter Kansas; or to avoid judgment creditors.[16] Regardless of his motives, Lane turned his back on the east and ventured west into the new Territory of Kansas.

Lane entered the Kansas Territory in late spring 1855. He first evaluated the political climate and air. By June, he had made attempts to form the Democratic Party proclaiming he had personal knowledge of the policy plans that President Pierce and Senator Douglas had with respect to Kansas.[17] Whether this was true or not mattered

---

[15] http://genforum.genealogy.com/ralston/messages/1873.html. See, Sam Ralston family history in Appendix H of this book.

[16] Collins, 37-46 and 149 states that the divorce was his wife's idea as she had returned to Indiana where she filed for divorce on the grounds Lane had abandoned her and their children. The divorce was granted to Mary Lane, and Lane was ordered to pay her $300 per month for support with custody of the children being split. After the divorce Mary reverted to her maiden name of Baldridge until Lane returned to Indiana. They did eventually reconcile and remarry. However, Mary insisted that she would remain in Indiana until the Kansas fight had been resolved. Lane agreed, and Mary and the children stayed in Indiana while Lane returned to the Territory.

[17] Goodrich, Thomas. "War to the Knife: Bleeding Kansas 1854-1861." (Mechanicsburg, Pa.: Stackpole Books, 1998), 71.

nothing to Lane. He held no reservations in making comments to enhance his own reputation. He seemed to be inspired by his own bravado. Lane soon realized he could not successfully establish the Democrat Party and that further efforts to do so would not be in his personal best interests. The anti-Democrat sentiments in Kansas were similar to what he had experienced in Indiana following his vote for the Kansas Bill during his failed reelection bid to Congress the previous fall. This enlightenment eventually led him to abandon the Democrat Party and adopt the Republican mantra—but this would be a process and not an overnight fact.

Stephen A. Douglas

During the period from 1855 until the outbreak of the Civil War, Lane held center-stage on every issue that confronted the Territory of Kansas. Every measure he took was designed to ensure that the Territory entered the Union as a free state. Despite his late arrival in Kansas, Lane's participation in the debate as a delegate and negotiator catapulted him to the forefront of every pivotal issue during his first year in Kansas.

He was a principle participant in the formation of the free-state constitution. By the fall of 1855, he had served as a delegate to the Topeka Convention. There he helped forge and draft the free-state constitution. He vehemently denied the legality of the parchment known as the Lecompton Constitution—a pro-slavery constitution adopted by the "duly elected" Territorial Legislature of Kansas at Lecompton. Lane and other free-state politicos insisted that the Lecompton Constitution was nothing more than an act of mob rule.

Lane's demonstrative evidence of the illegal assemblage in Lecompton was the Congressional inquiry into voting fraud earlier that year. The committee investigated reports of irregularities during the March 30, 1855 territorial election that had elected the Lecompton Legislature. Their investigative report chastised the large contingency of Missourians who had crossed into Kansas and overwhelmingly voted for a pro-slavery legislature. There was no dispute that some election shenanigans had occurred, but the electorate in Kansas on March 30 was certainly more supportive of a pro-slavery legislature and in opposition to a free-state

Kansas; the committee findings may have been correct, but the truth is that the outcome would have gone the way of pro-slavery supporters even without the election fraud.[18]

As winter approached on the Kansas plains, pro-slavery advocates were determined to ensure that slavery would become the law of the land. The Lecompton Assembly viewed the Topeka Constitution as an act being prosecuted by a tyrannical government that could not be tolerated. They compared their opposition to the Topeka document to that revolutionary spirit of the country's Founding Fathers in their war against "the haughty tyrant George the Third."[19] To the men in Topeka those gathered in Lecompton were usurpers of the law; to the men in Lecompton those gathered in Topeka were illegally assembled. The fuse was lit and it would not take much to ignite it.

The explosion erupted in the early days of December 1855. The tipping point did not occur over merely the controversy of two competing constitutions. It occurred when citizens of Lawrence had apparently interfered and prevented the legal process. According to the "duly" appointed Sheriff of Douglas County Samuel J. Jones, he had been directed to serve an arrest warrant upon a free-state agitator named Jacob Branson. In compliance to his authority, Sheriff Jones arrested Branson and as the Sheriff was transferring Branson to jail, a Lawrence posse at gunpoint forced the Sheriff to release the prisoner—Jones acquiesced. When word reached across the state line about the illegal abuse of the law by "abolitionists," pro-slavery advocates marshaled their resources and marched upon Lawrence and the Wakarusa War had begun.

Lane recognized the situation and took on the ironic role as peace negotiator. To avoid open warfare, Lane actively engaged in settlement negotiations. His efforts resulted in the execution of a "peace treaty." The conflict between the largest numbers of Americans aligned in battle against one another since the days of the Revolution had subsided.[20] Lane also knew that Lawrence was outnumbered and faced certain destruction if a resolution of the dispute was not quickly achieved. Lane's success was certainly further aided by the unseasonably thirty below zero temperature and mounting snow which chilled the invaders desire to attack the town. Regardless of his reasoning, Lane appeared the statesman in quieting the situation, and the Wakarusa War came to an inglorious end.

The year 1856 saw an intensification of bloodshed and violence never seen before in America. Surprisingly, there had only been a splattering of casualties during the first

---

[18] http://lecomptonkansas.com/page/samuel-j-jones. A Congressional committee later determined that only 1,410 votes were legal and 4,908 were fraudulent. The report found that armed proslavery Missourians had forcibly overrun polling precincts throughout the Kansas Territory, and had run off the election judges and had elected themselves to the legislature.

[19] Siddali, Silvana R. "Missouri's War, The Civil War in Documents." (Athens, Oh.: Ohio University Press, 2009), 30.

[20] Harrold, Stanley. "Border War: Fighting Over Slavery Before the Civil War." (Chapel Hill, Nc.: The University of North Carolina Press, 2010), 168-169; and, Goodrich, "War to Knife," 86.

couple of years of struggles in Kansas. The ink had barely dried on the Wakarusa "peace treaty" when conflict resumed on the border on January 15. During the free-state elections in Leavenworth some thirty pro-slavery men clashed against free-state advocates in which both sides suffered casualties. Similar episodes on election-day were reported in other Kansas border districts. The election crisis marked the beginning of a period of bloodshed and open warfare.

On April 19, 1856 someone in Lawrence attempted to assassinate Sheriff Jones. This marked Jones' second encounter with "free-soilers." Jones was shot, and despite newspaper accounts to the contrary, his wounds were not mortal. Death or no death the assault upon a peace officer demanded retaliation by pro-slavery supporters.[21]

As trouble intensified along the prairies of Kansas, Lane traveled east to garner aid and support for the cause of a free Kansas. The Grim Chieftain first spoke in Chicago where he passionately and emotionally told crowds about "the crimes against Kansans." His zealous pleas convinced the audience to open their pockets and money was raised. Leaving Illinois, Lane continued his eastern trek articulating outrage over persistent and unrelenting events in Kansas and the nation's Capitol. His vivacious velocity and prose style presentations intensified as current affairs from the hallowed chambers of Congress to the plains of Kansas provided him significant fodder. The nation's misfortunes Lane laid at the feet of the pro-slavery hordes that he insisted infected the country.

The bloody conflict that had been transpiring in Kansas had remained non-violent at the national level. Other than the occasional challenge to a dual between warring Congressman on different sides of the Kansas issue, there had been no bloodshed as they continued to debate the sensitive issue of slavery's expansion into Kansas. Oratory bravado held dominion over Congress until May 19 when the attacks turned personal. The first salvo in the halls of Congress was not fired by a slave owner. They were shot by abolitionist Senator William Sumner of Massachusetts.

On May 19, the galleries of the United States Senate were filled to capacity. Doorways were blocked with ladies and gentlemen. The mass assemblage had come to hear Senator Sumner's expectant remarks on the Senate floor about the Kansas issue. At one o'clock, Sumner's long anticipated oration began; it would continue well into the following day. Sumner blistered and railed against the "scourge" of slavery. He described the slave movement in Kansas as an outright "rape" of the "virgin territory," and he exposed "The Crime against Kansas." On day two, Sumner turned his condemnation into a personal attack against the senior senator from South Carolina, Andrew Preston Butler, who was not in the chambers but was in his home state. Sumner proceeded with a scathing criticism of a man he described as the "Don Quixote" of slavery and climaxing with the following crescendo:

---

[21] Goodrich, "War to Knife," 110-111; Harrold, 169; Monaghan, Jay. "Civil War on the Western Border." (Lincoln, Ne.: University of Nebraska Press, 1955), 52; Goodrich, "War to Knife," 120; and, Collins, 109-110.

> *[He] has chosen a mistress to whom he has made his vows, and who, though ugly to others, is always lovely to him; though polluted in the sight of the world, is chaste in his sight . . . the harlot, Slavery.*[22]

Sumner's speech had not just opened but had ignited "Pandora's Box."

The violent propensities between north and south had reached Congress. For three days, Sumner's remarks festered and percolated in the mind of Congressman Preston Brooks of South Carolina, Butler's nephew. Brooks like a slow fuse to a powder keg finally erupted. With cane in hand, he approached Sumner as the Senator sat in his senate chair engaged in work. Brooks promptly began plummeting savage blows to Sumner's head, neck and body; the assault nearly killed him. Sumner claimed he never fully recuperated from his injuries; he was physically unable to return to his senate seat for nearly two years.[23]

The Brooks-Sumner episode sparked a conflagration throughout the land. The division between north and south carved along the Mason Dixon Line opened a cavern wider than the Grand Canyon. U.S. Congressmen began carrying concealed revolvers, derringers and knives fearing an eruption of a volcano.[24] Newspapers in the north and south captured the divisions in the land. The Richmond Whig applauded Brooks' act as one of chivalry while papers in the north described it as a crime committed by a "cowardly ruffian."[25] A Congressional committee was convened to investigate the matter for purposes of expelling Brooks from his seat.

The drama in Washington paralleled the saga that was contemporaneously being played out in Kansas. The day before Brooks' assaulted and battered Sumner, pro-slavery forces had marched upon Lawrence out of revenge for the shooting of Sheriff Jones. In what has become known as the "sack of Lawrence" an invasion of forces, many of whom came from Missouri, seized the town. They destroyed the two printing offices and threw the print-type into the river. The Free State Hotel, the "citadel of stolen property" became the final focal point of the Lawrence invasion. Former Missouri United States Senator David Rice Atchison and one of the architects of the Kansas-Nebraska Act held

---

[22] Donald, David Herbert. "Charles Sumner and the Coming of the Civil War." Originally published in 1960. (Naperville, Il.: Source Books, Inc., 2009), 236-239.

[23] Donald, 277; Goodrich, "War to Knife," 99, 119-120. Sumner's speech infuriated the South and Senator A. P. Butler from South Carolina, to whom Sumner's criticism had been directed. The assault nearly caused Sumner's death and the House attempted to remove Brooks, but he resigned. Brooks' Congressional District in South Carolina reelected him on August 1, 1856 to the Thirty-fourth Congress to fill the vacancy caused by his own resignation. He was also re-elected in November to the Thirty-fifth Congress, but he never took his seat as he died in Washington on January 27, 1857.

[24] Goodrich, "War to Knife," 121.

[25] *New-York Daily Tribune*, 29 May 1856, quoting *Richmond Whig.*

dominion. He fired the first cannon shot into the Free State Hotel before his followers set fire to the building.

For Lane, these episodes evidenced beyond any doubt the compelling need for eastern support. He continued to remind his audiences about the atrocious attack upon a United States Senator by a slave owner and the annihilation of an American town by vicious "slaveocrates"—the principal demon had been Atchison, a former United States Senator. Lane masterfully weaved these and other stories into his oratory repertoire. As he continued to stoke the fire of war, he raised an "army" to accompany him back to Kansas. Lane had not only returned to the political stump, he was determined to recapture his militarily reign as well. His Mexican War experience would neither be ignored by the people of Kansas nor anyone in opposition to him—especially those "pukes and ruffians" from Missouri.

John Brown

Lane sent word to Kansas that he had returned with an "Army of the North." The intended invasion abruptly stopped near the territorial line. Word reached the troops that Lane had no legal authority; the Army stationed at Leavenworth was prepared to prevent their intrusion; and Lane represented a lawless group similar to the border ruffians from Missouri. The "army" promptly departed Lane and returned home. The Grim Chieftain was not deterred and continued to proclaim he had returned with an "army" despite having only six followers. Lane and his "army," quietly slipped into Lawrence where preparations were being made to wage war against pro-slavery advocates for an invasion into Missouri.

Reports encircled western Missouri that Lane and his "army" was advancing. In Lexington, Missouri a mass unity meeting was held to assemble forces to defend against Lane and his three thousand abolitionists. Missourians may have decreed Lane an abolitionist, but he was not. He remained an outspoken leader for the free soil movement, but despised the political viewpoint of abolitionism. He felt that men such as John Brown, Osawatomie Brown, and his radical abolitionist adherents needlessly incited the pro-slavery sentiment.[26] Instead, Lane had been an ardent and insistent supporter of laws

---

[26] Monaghan, 69-84; Goodrich, "War to Knife," 120; and Collins, 109-110. John Brown had opposed Lane's negotiated treaty the at the Wakarusa in December 1855, and the "sack" of Lawrence during the early morning hours of May 25, 1856 resulted in Brown's retaliation that resulted in the murder of five pro-slavery men living along the Pottawatomie Creek. These killings did not sit well with Lane who objected to Brown and his abolitionist zeal.

that prohibited all blacks, free and slave, from the Territory. Lane believed in the sword and he was not reluctant to encourage its use, but not to abolish the "Peculiar Institution" or its cause but to establish a free white state.

The territorial and national debate on slavery evolved into violence, bloodshed and battles. Organized and disorganized militia units paraded back and forth across the border between the Territory and Missouri inflicting the first musket and cannon fire in a war that would grow in intensity and horror, only to be eclipsed by the Civil War.

Following the sack of Lawrence, John Brown retaliated the following evening. He and his sons and other confederates hacked to death five pro-slavery men who lived along the Pottawatomie River in Kansas. The bleeding in Kansas mounted during the summer of 1856. Battles were fought at Black Jack, Lecompton, Ft. Titus and Ft. Franklin. On one such engagement, Lane had personally led a force of free-state followers into Missouri. Both armies achieved no tactical gain; it only proved Lane and his boys were willing to take their cause into Missouri as Missourians had done in Kansas. The term "Bleeding Kansas" proved not to be mere words.

In March 1857 the Administration of Franklin Pierce ended, and President James Buchanan assumed office. Neither his inaugural address nor the Dred Scott decision calmed the delirium of revolution and counter insurrection between Missouri and Kansas.[27] During those summer months, Lane served as president or chairman at various free-state meetings and at the Topeka Convention. He openly protested against what he considered the usurpation of the laws by the "slavocrates" in the Lecompton Legislature. He objected to complying with their laws or paying their taxes. Lane further insisted that all free Kansans should vote at the upcoming October territorial election to elect a new legislature; he insisted it would provide "free-staters" their best opportunity to take control of Kansas—and under the protection of his own militia forces.

Lane's stalwart position surprisingly garnered the support of Dr. Charles Robinson of Lawrence. Robinson had been a central character in the fight over Kansas since its inception's in 1854 and had been politically at odds with Lane since Lane had first arrived in the Territory the following year. Robinson politically leaned towards abolitionism and Lane did not; they also differed on most political issues—specifically leadership in the Territory. However, in 1857 they had finally found common ground. In the end, the election in October went the way of the free statesmen. The Lecompton Legislature ignored the results of the free-state election and proceeded to frame a strong

[27] Scott v. Sanford, 60 U.S. 393 (1857). In the Dred Scott decision Chief Justice Roger Taney of Maryland held that the Missouri Compromise of 1820 was unconstitutional, as Congress had no authority to legislate against slavery in any of the Territories of the Union. Taney also ruled that Dred Scott was not a person recognized under the Constitution because he was a slave and thereby had no right ("standing") to present his case in federal court. The high Court's decision was handed-down two days after Buchanan had taken office.

pro-slavery constitution. To Lane and Robinson and their allies, the Lecompton Constitution had been nothing short of "a fraud upon the people."[28]

As the year 1858 opened, the 5th Territorial Governor of Kansas assumed his position. Buchanan in the waning days of December had appointed forty year old John W. Denver as the new Territorial Governor. Denver had experience in politics, war and especially Missouri. He appeared to possess the perfect pedigree to tackle the persistent intrigue in the Kansas Territory. Denver, a Democrat, had practiced law in Missouri, served in the Mexican War and later represented California in Congress. However, like his predecessors, Denver found it difficult to maintain order in Bleeding Kansas. Denver not only faced the anti-slavery faction but the abolitionist spirit and delirium of John Brown and his disciples, including Rev. James Montgomery. Both Brown and Montgomery had been born in Ohio, were ardent abolitionists and harbored moralistic certainty of the correctness of their social cause.[29] They also shared the common exploit of making frequent raids into Missouri to liberate slaves and march them off to freedom.

Denver's dilemma was shared in Washington. Congress had been presented with two conflicting constitutions—the pro-slavery one from Lecompton, and the free-state document from Topeka. Congressmen promptly rejected the Topeka instrument—it had been adopted by an illegal assembly. Consideration was only given to the Lecompton Constitution—the legally constituted body of the Territory. Rumblings, feuds and back-room deals filled the backrooms in Congress as its members pondered the vexing issues contained within the Lecompton Constitution. Finally, the Little Giant from Illinois once again came forward and brokered a compromise. The Lecompton document would be ratified by Congress contingent upon the approval of the citizens of Kansas.

On August 2, 1858, the voters in Kansas overwhelmingly rejected the Lecompton Constitution. This defeat marked the beginning of the end for the slave movement in Kansas. It also resulted in John Brown and his boys abandoning their cause in the Territory and traveling east for a destiny at the federal arsenal in Harper's Ferry, Virginia.

Although Lane reveled in the defeat of the Lecompton Constitution, he faced a serious personal problem prior to the vote. On June 3, 1858, he had shot and killed Gaius Jenkins. Lane argued he had acted in self-defense, but Lane's enemies and he certainly had many, including Dr. Robinson, called it murder. The Lane-Jenkins dispute evolved out of a land claim argument and litigation remained pending for nearly two years. The land claims court had not entered its judgment by June when Jenkins reportedly came upon Lane's property. Lane saw Jenkins approach and an exchange of gunfire erupted;

---

[28] Collins, 127-129 and 141.

[29] Dirk, Brian R. "By the Hand of God." *Kansas Territorial Reader*. (Topeka, Ks.: Kansas State Historical Society, 2005), 39.

Jenkins fell mortally wounded.[30] Lane was charged with murder, and he hired Thomas Ewing, Jr. a well- known lawyer from Leavenworth and foster brother and brother-in-law to William Tecumseh Sherman, as one of his defense attorneys.

The preliminary hearing on the murder charge was held before three judges. It lasted two long, contentious weeks. In the end, Ewing was able to convince the court that his client had acted in self defense and that the prosecution had failed to show the shooting had been willful and wanton. Although the judges vindicated Lane, the killing of Jenkins would politically plague him the rest of his life, and by the following summer, Lane morally or for his own political future took on religion.[31]

Thomas Ewing, Jr.

In 1859, the same year in which Lane found his Methodist Faith peace seemed to have finally settled upon the Kansas Territory. Governor Denver resigned from office claiming to President Buchanan that Kansas had been delivered from its days of bleeding and his job had been completed. At the same time, leaders within Kansas, including Lane, took measures to formally establish the Republican Party within the Territory. Lane once again played a pivotal role in the formation of the Party.

Former one-term Illinois Congressman, Abraham Lincoln, a former Whig turned Republican, made an early campaign trip to eastern Kansas in December, 1859. When Lincoln disembarked from a steamboat at Leavenworth, Lane not to be denied the opportunity to finally meet Lincoln was on hand for the reception. Lane considered Lincoln more of a moderate on the slavery issue and chose to support the tall, angular and gaunt figured man from Illinois at the Republican Convention scheduled for 1860. Despite their former differences over the enactment of the Kansas-Nebraska Bill—Lincoln openly opposed it; Lane had cast his Congressional vote in support of it—the two

[30] Goodrich, "War to Knife," 209; Castel, Albert. "A Frontier State at War, Kansas 1861-1865." (Lawrence, Ks.: Kansas Heritage Press, 1958), 22; Smith, Ronald D. "Thomas Ewing, Jr.: Frontier Lawyer and Civil War General." (Columbia, Mo.: University of Missouri Press, 2008), 76; Collins, 157; and Monaghan, 107.

[31] Collins, 157 and 162; Monaghan, 107; and Smith, 4-17. On May 1, 1850, Sherman married Ellen Ewing, the daughter of Thomas and Maria Ewing. The marriage was described as an opulent, political and social affair in Washington City. When the newlyweds honeymooned, her brother, Tom Ewing, Jr., accompanied them.

men united and Lincoln's visit to Kansas had been successful. A Lincoln presidency would certainly give the Grim Chieftain influence inside the Executive Mansion.[32]

By the time Lincoln visited Kansas and received Lane's backing, the Kansas fight had virtually ended. There was no doubt that it would eventually enter the Union as a free state. It also marked the end of abolitionist John Brown, and not to Lane's displeasure. The beginning of Brown's demise began on October 16, 1859, when he and his followers captured the U.S. military arsenal at Harpers Ferry, Virginia. Brown's efforts to unite a black rebellion proved disastrous to him and his raiders. U.S. Marines under Colonel Robert E. Lee had been dispatched to the federal arsenal at Harpers Ferry after Brown and his men had seized the facility and took refuge inside it. The following day Lee and his Marines, including young Lieutenant "Jeb" Stuart, stormed the building. Brown was captured and two of his sons along with other conspirators were killed. Brown's dream of ridding Kansas and ultimately the country of the curse of slavery had played out its final mission—or had it?

On December 2, 1859, at 11:30 a.m., Brown was hanged for his crime. Among the witnesses to the execution was an actor named John Wilkes Booth, Major Thomas J. Jackson (later known as "Stonewall") and many cadets from the Virginia Military Institute—an institution where Jackson served as a professor. Both Jackson and Stuart had their own personal experiences with Lane, Brown and the free-state movement during those many, troubling years as they both had been stationed at Ft. Leavenworth. Before Brown's execution, he had prophetically written a letter to his wife. He declared with certainty that "the crimes of this guilty land will never be purged away but with blood." Brown may have died, but his cause had not.

By the summer of 1860, Lane had become wholly entrenched in the Republican Party's cause for Lincoln. It also turned out to be a disastrous year for the Democratic Party. Democrats were as fractioned and as irreparably compromised as they had been in during the Congressional election of 1854 following the adoption of the Kansas-Nebraska Act. The contentious relations within the Democratic Party, primarily over the slavery question, split the party and two candidates were nominated with each representing the party's fraction: Senator Douglas, the Little Giant from Illinois and John Breckenridge of Kentucky and sitting Vice President under Buchanan. Douglas represented the northern mindset within the party while Breckenridge captured the sentiments of the South. The old Whig Party along with other disgruntled Democrats nominated Senator John Bell of Tennessee. Their party platform was ambiguous: it stood firm to the Union and Constitution.

Republican's rallied behind Abraham Lincoln and Lane soon took to the stump in Kansas and in his former state of Indiana for "Honest Abe."[33] In November, Lincoln was

---

[32] The White House was called the Executive Mansion during the period of the Civil War and did not become known as the "White House" until the Presidential Administration of Theodore Roosevelt.

[33] Collins, 168.

elected the 16th President of the United States. His election had been a polarizing one. He failed to carry any state in the South, but he dominated the polls in the North as well as Kansas. In Missouri, Lincoln received his only support in St. Louis from its strong German immigrants. In the western section of the state, he faired dreadfully; he received a splattering of votes in Jackson County and not one in Clay County. Douglas only earned the electoral votes from two states, New Jersey and Missouri, where John Bell had finished a close second in each. The Missouri Democratic gubernatorial candidate Claiborne Fox Jackson and his Lieutenant-Governor running mate Thomas Reynold, both having publically supported Douglas were elected to office. Lane gave no heed to the political outcomes in Missouri or in the South. His Republican candidate had won, and he would not let the next president forget all the support he had given him.

Despite Lincoln's victory and the certain admission of Kansas as a free state, nothing curtailed the deep-seeded and steadfast determination of some Kansans to inflict retaliatory measures upon their enemies in Missouri. During November and December of 1860 Jayhawking raiders led by Dr. Charles "Doc" Jennison, preacher James Montgomery, George Hoyt, the future founder of the "Red Legs," and even John Brown, Jr. the son of militant John Brown whose "body laid molding in the grave" converged upon western Missouri. These guerrillas many of whom had been fanatical extremists during the days of Bleeding Kansas viewed western Missouri as open for the taking. The raising of the rebel banners in the seceded states only further bolstered their sentiment and desire to attack those "Missouri pukes." Spurred by these and other evil motives, they descended upon western Missouri like locust to corn. They soon engaged in willful atrocities including brutality, thieving, pillaging, stealing and burning. Their activities caused mounting fear in Jackson County and elsewhere along the border. Many in this region were convinced that once Lincoln took office in March nothing would stop the Kansas terrorists.

Charles "Doc" Jennison

Western Missouri appeared to be at the mercy of these Kansas invaders. Outgoing Governor of Missouri Robert Marcellus Stewart fearing the situation took executive action to defend the border from Jayhawking. The Governor dispatched the Missouri state militia from St. Louis under Brigadier-General Daniel M. Frost in what became known as the Southwest Expedition. Frost and his men were garrisoned in northwest Bates County and surrounding areas and remained until the end of December.[34]

Nearly three months after Lincoln's election and while Jennison, Montgomery and others continued to terrorize western Missouri, Kansas was admitted into the Union on January 29. The territory's admission had been assured when seven southern states, with South Carolina leading the way, seceded from the Union. The seven long years over the struggle for statehood in Kansas had come to an end. Dr. Charles Robinson became the state's first governor. The new state's constitution was similar to the one Lane had proposed and championed as early as 1855—a constitution that prohibited slavery, the introduction of slavery and for that matter any blacks whatsoever.

The admission of Kansas in 1861 did not thwart the persistent struggles between Lane and the new governor. They each sought to be named as United States Senator by the Kansas Legislature. Lane and Robinson had grown into bitter, detested enemies. Their feud caused a division within the Republican Party in Kansas. Robinson even feared Lane more than personally winning the senate seat for himself. From the mouths of both men assertions and allegations of misconduct and corruption spewed. Robinson reminded his constituents that Lane had murdered Jenkins and that his conduct had demonstrated a lack of suitability to serve in the Senate. Lane did not remain silent. He accused Robinson and his agents of fraud, unfair play and secret dealing. Their power struggle resulted in the development of alliances and factions within Kansas and the new legislature. The debate over the election of the new United States Senator gave Ewing, Lane's former defense attorney, the feint hope that he would be elected as the dark-horse. In the end on April 4, 1861, Lane was able to muster his political weight and by the margin of seven votes became Kansas' first United States Senator.[35]

On March 4, 1861 Lincoln was inaugurated in Washington. He had been forced to clandestinely enter the Capitol by night and under the watchful eye of the Pinkerton Agency. During his inaugural address, the new president attempted to assuage the anxiety and fears of his "dissatisfied" friends in the South. He told them they had no reason to be concerned about his victory. He would not wage war to bring them back into the Union, and the federal government would not "assail" the south; any blood spilled would be with the South as the first "aggressors." Lincoln also attempted to soften their concerns over the slavery issue. He proclaimed:

> *I have no purpose, directly or indirectly, to interfere with the institution of slavery in the States where it exists. I believe I have no lawful right to do so, and I have no inclination to do so.*

---

[34] "Norton, Richard L., ed. "Behind Enemy Lines: The Memoirs and Writings of Brigadier General Sidney Drake Jackman." (Springfield, Mo.: Oak Hills Publishing, 1997), 28; and, Monaghan, 120.

[35] Collins, 176; and Castel, "Frontier State," 25.

Lincoln's concluding remarks were an historical, rhythmic, impassioned appeal for peace:

> *We must not be enemies. Though passion may have strained it must not break our bonds of affection. The mystic chords of memory, stretching from every battlefield and patriot grave to every living heart and hearthstone all over this broad land, will yet swell the chorus of the Union, when again touched, as surely they will, by the better angels of our nature.*[36]

Nothing Lincoln or anyone else could have declared or decreed would have mollified the secessionist spirit. The South was determined and committed to form a "more perfect confederated union" composed of a confederation of seceded states.

For over a month, Lincoln stayed true to his Inaugural Address. He did not attack or wage war upon the South. Instead he waited for the South to act first, and they did. On April 12, during the early morning hours in the Harbor of Charleston, South Carolina a barrage of battery fire opened upon Ft. Sumter. For nearly two days, Major Robert Anderson and his men withstood a relentless barrage of cannon fire. There was little that anyone within the fort could do to defend their position. Ultimately, Anderson was compelled to raise the white flag and capitulate. The Civil War had commenced, and as Lincoln promised it had been started by the Confederate South (or, had it not begun with the enactment of the Kansas Nebraska Act seven years earlier?). Regardless of the official start date of the rebellion, Civil War had come to a divided nation, and Jim Lane was prepared for the new challenge.

The day following the battery attack upon Ft. Sumter Lane was safely ensconced in the Nation's Capitol. The city had become volatile and under siege by divided sentiment, fear and turmoil. Emotions had reached a boiling point and Lane personally experienced the mounting anger within the city. Shortly after his arrival, he reportedly encountered a furious mob of southern agitators who abruptly threatened him. They did not realize the man they challenged was the Grim Chieftain. Lane was never one to run from a fight and he would not back down to the crowd. He confidently boasted to the assemblage that he had "a hundred men from Kansas in the crowd, all armed, all fighting men, just from the victorious fields of Kansas! They will shoot every damned man of you who again cries 'Mob,' 'Mob.'"[37]

Lane's boast to the agitators and their hostile rabble was no mere bluff. He and Samuel Pomeroy, the other new Kansas Senator, had journeyed to Washington accompanied by a large contingency of Kansans who had followed the two senators to

---

[36] United States House of Representatives. "Inaugural Addresses of the Presidents of the United States." Document, 93-208. (Washington, D.C.: United States Government Printing Office, 1974), 119 and 126.
[37] Collins, 178.

seek political favors. The day following Lane's threat and dismissal of the "mob," Major David Hunter requested troops to aid in the protection and defense of President Lincoln and the Executive Mansion. The Capitol simply lacked sufficient military defense. Lane, always the opportunist, knew his boys, many of whom had been fighting against Missourians for years, could fill that role. Lane quickly formed these Kansas boys into a presidential guard unit he called the "Frontier Guard." Among the members of the Frontier Guard was Dan Anthony, Thomas Ewing and many others known from those many, long, suffering years of "Bleeding Kansas."[38] The garrison took up refuge in the first floor of the Executive Mansion where Lane's army promptly spread out their blanket roles and muskets in defense of President Lincoln.

Lane and his Frontier Guard held dominion for about thirty days until reinforcements entered the Capitol. The Frontier Guard was soon "mustered" out of service. Lane had triumphed. As he and his men vacated the Executive Mansion, they were congratulated by a very pleased president. Certainly, Lane's former one-term experience in Congress had ended in defeat, but his first days after his return to Washington had been successful. Lane no longer wore the mantle of a local, sleazy, western politician. He had become the President's friend. He had arrived and returned to the national spot light and had successfully defended his country and president—perhaps, he even pondered the thought of one day making the Executive Mansion a more permanent residence for Mary, the children and himself.

Congress had adjourned and Lane and his Frontier Guard had been discharged. Lane had become a national and local hero. He returned triumphantly to Kansas in the middle of May 1861. His success in Washington had been as well received in Kansas as he had once been celebrated in Indiana after his return following the Mexican American War. He had shown his military valor in the foreign country of Mexico, and he had now displayed it in Washington. It was time for the "colonel" to resume his duties in Kansas—waging a campaign against his foes in Missouri.

Once in Kansas, Lane promptly announced that Lincoln had commissioned him to recruit Kansas regiments. These were not merely Lane's embellished remarks as on June 20 Lincoln confirmed Lane's proclamation authorizing him to raise two regiments. The President also requested Secretary of War Simon Cameron to name Lane brigadier-general. Stretching the envelope as only Lane could do, he eventually formed three regiments, known as the "Lane Brigade."[39]

Governor Robinson was overtly pleased when he received word that Lane had been granted a general's commission. By law this forfeited Lane's senate seat. Robinson immediately replaced his political foe in the Senate. The Governor saw this as the most efficient manner in which to rid him of the trouble-making Lane. The Grim Chieftain

---

[38] Collins, 179-180; Goodrich, "War to Knife," 226; and Castel, "Frontier State," 34.

[39] Collins, 185-186.

desired the dual power of U.S. Senator and commissioned general, but he recognized Robinson's argument. Lane refused to step down from the Senate. He insisted he had not taken the commission. The controversy over the claims of two competing senators became fiercely debated in Kansas and Washington. The Senate Judiciary Committee initially ruled against Lane, but when the committee learned Lane had not in fact officially accepted the commission as a brigadier-general, their ruling was reversed and Lane retained his U.S. Senate seat. The Grim Chieftain had held onto his political fiefdom, but he was not going to be deterred from assuming the rank of general or precluded from forming three Kansas volunteer regiments.

During the secessionist fervor of 1861, Missouri did not join her sister southern states but remained within the Union. Its citizenry though not overwhelmingly loyal elected a delegation of representatives to the state convention opposed to secession. Some delegates preferred neutrality while the vast majority held steadfast for the Union. The western part of the state, and in particular Jackson County, shared the sentiments of the non seceding constitutional convention. Area residents had been primarily hewn from immigrants from the south and engaged in agricultural pursuits rooted in their old familial traditions, including the institution of slavery, but saw no reason to secede.

After the fall of Ft. Sumter on April 13, Lincoln requested seventy five thousand volunteers to quash the rebellion. Governor Claiborne Fox Jackson of Missouri decreed the request as illegal. To the people of Missouri, they preferred neutrality or enlistment into the Missouri State Guard so as to keep "Yankees" and other invaders out of the state. The vast majority did not go south or join the Union Army. The only area for enlistment into the federal ranks came out of the German population in St. Louis. Missouri teetered between neutrality, loyalty and disloyalty. Governor Jackson was convinced that the President's call for each state to provide forces to save the Union would be the measure that would push Missouri into the southern column.

The federal government recognized Missouri had voted against secession and had strong political allies in St. Louis. Lincoln realized that these facts did not mean the state was soundly devoted to a successful prosecution of the war to save the Union. The Lincoln Administration had to ensure Missouri did not join the South. The value of the state was obvious. Continental trade and transportation relied upon the two great rivers: Missouri and Mississippi; western banking was controlled by St. Louis; the western trails originated in Jackson County. The Mississippi composed the eastern border of the state while the Missouri River bisected the state.

Political leadership in St. Louis firmly held to the Union. Congressman Frank Blair of St. Louis and the tyrannical, red-headed, openly abolitionist, New Englander, Captain Nathaniel Lyon took decisive action to preserve Missouri. Although they were pleased delegates at the state convention had found no basis to secede, Blair and Lyon determined that stronger military and political measures had to be imposed to ensure the state's safety. Lincoln saw the need as well. He appointed John C. Fremont, the

"Pathfinder," son-in-law of the late, long-term Missouri United States Senator Thomas Hart Benton and the first Republican candidate for President in 1856 as commanding general of the Union Army in Missouri and the western theatre. Lincoln knew Blair and Fremont had more than just experience and knowledge of Missouri; they had strong political followings on the state and national levels that would prove useful for the federal cause and for Lincoln's own political fortunes. Future events would cause Lincoln to reconsider his decision of ever having placed trust in Fremont.

General Sterling Price

Fremont following assumption of control ordered Lyon, now a Brigadier-General, to chase Governor Jackson out of the state and to destroy the Missouri State Guard. The plan to expel Jackson and General Sterling Price reached a climax along Wilson's Creek some eight miles south of Springfield on August 10. Price had been joined by Confederate forces under Ben McCullough and members of the Arkansas State Guard. Their combined forces reportedly exceeded twelve thousand men. Lyon's army equaled less than half that number. Lyon repeatedly dispatched requests to Fremont for reinforcements.

In the early morning hours of August 10, Lyon surprised the enemy by dividing his forces and launching a two front attack. While caught off guard, the combined Confederate and state guard forces were able to eventually regroup and withstand the Union challenge. Wilson's Creek (or the Battle of Oak Hills) was noted as the bloodiest battle fought during the first year of the Civil War. Lyon's decision to divide his outnumbered forces did not adhere to military protocol. Nonetheless, if his battle plan had been properly coordinated by his subordinate officers, Lyon might have produced a northern victory. The greatest part of the Union defeat at Wilson's Creek was the loss of Lyon who had been killed in action. This was the same man who during the spring of 1861 had promptly, decisively and dictatorially ensured Missouri remained in the Union. Lyon's became the first Union general to die in action. By his death, Major John McAllister Schofield, Lyon's adjutant, removed the troops to Springfield and left the field to Price and his Confederate allies.

Price's victory naively convinced him that he could liberate Missouri. He recognized the need for his southern allies to assist him. He requested McCullough and the other southern troops to march northward with him to reclaim the state for the South. Price's request was refused. McCullough reminded Price that Missouri had not seceded from the Union and his men had not been authorized to make any further advance. Instead of assisting Price, McCullough's forces turned their back and withdrew into Arkansas. The rebuke did not deter Price of his plan to liberate Missouri. He chose to march his Missouri State Guard northward.

Word soon began to spread that Price intended to attack Lexington, Missouri or Ft. Scott in Kansas. James Montgomery received this information and promptly notified Lane of the potential of an invasion into Kansas. General Fremont in St. Louis heard the same reports. The General could not allow Price to successfully advance into Kansas. He had already received complaints over his failure to reinforce Lyon at Wilson's Creek and he did not want further ridicule. Fremont dispatched an order to Lane directing him to intercept Price as his forces advanced northward. When Lane received Fremont's directive, he had been training his men at a camp named Camp Lincoln, located some four miles east of Ft. Scott and near the western Missouri border. As Lane struck camp to move into Missouri, he admonished his forces with a singular directive that had been long-standing within his mind: "Everything disloyal, from a Shanghai rooster to a Durham cow, must be cleaned out."[40]

When the Lane Brigade entered Missouri, they soon encountered state guard troops under Col. Sidney Drake Jackman. Lane's troops battled Jackman at Dry Creek on September 2 in Vernon County. Jackman, a resident of Bates County (the county to the immediate north of Vernon and to the immediate south of Cass) had personally experienced years of turmoil on the border. He had fought free-staters during the days of Bleeding Kansas, and he had served as a scout under General Foster in the Southwest Expedition the previous year. Jackman's troops pushed Lane and his men back into Kansas in what Jackman described as a "very spiritual little fight." Lane later wrote that his withdrawal was because he had been "surrounded by a superior force."[41] Regrouping his men at Camp Lincoln, Lane eventually renewed his pursuit of rebels into Missouri. However, his recent setback at the hands of Jackman may have caused him to consider the advisability of pursuing Price cautiously and leisurely.

No one, not even Lane's Brigade, was able to thwart Price's northern invasion of Missouri. When Price reached the Missouri River town of Lexington, the county seat of Lafayette, on the south side of the river, he laid siege to the city. Union commander Major James A. Mulligan, like Lyon before him, was greatly outnumbered and in desperate need of reinforcements. Mulligan requested aid, but Fremont once again failed to dispatch troops from the eastern side of Missouri. Fremont was unaware that his Kansas forces under Lane were not in any hurry to defeat Price or to engage his troops'

[40] Brownlee, Richards, S., "Gray Ghosts of the Confederacy: Guerrilla Warfare in the West, 1861-1965." (Baton Rouge, La.: Louisiana State University Press, 2011), 38; and, Monaghan, 195.
[41] Norton, 28; and Collins, 187-188.

directly. As Mulligan considered his precarious situation at Lexington, which ultimately resulted in his surrender to Price and Governor Jackson, Lane's Brigade was some one-hundred and twenty miles south of Lexington.

Lane had chosen a military strategy not to assail Price's army directly but rather indirectly. Lane struck Price's rear by inflicting punishment upon Missourians who had supported or provided aid to Price. Lane reasoned that taking the war to those disloyal citizens who had aided Price during his northern march was legal and predicated upon the doctrine of military necessity. To Lane and others, the destruction and confiscation of property that belonged to southern sympathizers would help quash the rebellion. Loss of property would deprive southern citizens of their ability to assist the enemy in the future. Lane further reasoned that this military measure was nothing more than an extension and continuation of two edicts General Fremont had previously issued.

On August 30, 1861 Fremont took drastic military action to quash the rebellion. First, he emancipated all slaves owned by disloyal citizens, which he said was intended to impede the southern war effort in Missouri. Next he ordered the execution of any "guerrillas" captured behind Union lines. Fremont and Lane had agreed upon policies of total war. Therefore, over a year before Lincoln issued the Emancipation Proclamation and Brigadier-General Thomas Ewing, Jr., had declared Order No. 11, Fremont had issued his own statewide martial law plan of emancipation coupled with a policy of total war against a civilian population.

With a war policy supported by Fremont, Lane and his brigade continued their mission to inflict harm upon the towns of Missouri. They made slow progress eastward by burning whatever stood in their way or seizing what they could carry with them. By September 21, Lane had reached the outskirts of Osceola. The Lane Brigade consisted of about two thousand infantry, cavalry and two artillery pieces.[42] The city was the county seat of St. Clair and had become a thriving port town on the Osage River with a population of about two thousand. The days of Bleeding Kansas may have ended, but Fremont's martial law orders provided Lane the legal license to bestow the wrath of God or the wrath of Lane upon the slave owners of western Missouri.

On September 22, 1861, a day that would be long remembered by the people of Missouri, the first great outrage of outlawry occurred. Osceola would mark the genesis of terrorism and criminality on the border throughout the Civil War. Lane's men laid waste to the town by killing nearly two dozen men, burning all but three buildings and houses, seizing courthouse records and stealing not only what they could carry or load into wagons, but marched off with about two hundred slaves and three hundred and fifty horses.[43] Infuriated Missourians and politicians called Lane's actions unwarranted and unscrupulous waged upon an undefended and non garrisoned town. From that moment,

---

[42] Brownlee, "Gray Ghosts," 38; Castel, "Frontier State," 53; Collins, 188; and Monaghan, 194.

[43] Goodrich, "War to Knife," 227; Collins, 189; Brownlee, "Gray Ghosts"; and Castel, "Frontier State," 54.

all subsequent guerilla and ruffian activities on both sides were said to have been christened by the blood of Osceola.

Price's Missouri war in 1861 appeared on its face to have been quite successful. He had defeated Union forces under the radical Lyon at Wilson's Creek; he had captured Lexington and seized state financial funds; his men had acquired much needed supplies and ammunition; and of most intrinsic value, he had recruited southern sympathizing men.[44] These gains failed to recapture Missouri. They proved only short term successes as Price quickly recognized he had no ability to hold Lexington or any other town in Missouri. By the end of September, he had no option but to retreat towards Arkansas. The Lane Brigade despite having been in the direct line of Price's southern route took no action to interfere and thereby failed to inflict any damage to Price and his retiring forces.

Instead of challenging Price, Lane had promptly fled back to Kansas after the destruction of Osceola. One possible reason for Lane's speedy return was because he had delayed responding to Fremont. Four days before Lane sacked Osceola, he had been ordered to march towards Kansas City near the state line. Lane preferred destroying property than obeying orders. A second explanation may have been because of the ridicule and protests Lane had received over what had been described as nothing more than criminal behavior. Even defenders of the Union protested the actions that Lane and other Kansas raiders had taken in Osceola and elsewhere. Governor Robinson of Kansas became incensed fearing retaliation and retribution. Major W. E. Prince, the federal commander at Ft. Leavenworth, joined the chorus in opposition to Lane, and he ordered the Jayhawkers out of Missouri.[45] More complaints would bellow across the land.

Western Missouri had become victimized by numerous raids led by Lane's Brigade, Jennison's 7th Kansas Jayhawkers' and Montgomery's Red Legs. The meaning of the term "Jayhawker" is uncertain. One claim to the term's origin is traced to a daredevil Irishman named Pat Devlin. In 1856, Devlin was met returning from a foraging expedition, loaded with spoils whereupon the following exchange allegedly took place:

> *"Where have you been, Pat?"*
> *"Jayhawking."*
> *"Jayhawking?" What's that?"*
> *"Well, in the old country we have a bird called Jayhawk, which kind of worries its prey. It seemed to me as I was riding home that this was what I had been doing."*

The word "Jayhawker" might be a term that does not mean anything in itself but once the word is spoken it becomes significant to the one who hears it. Others have

---

[44] Gerteis, Louis S., "The Civil War in Missouri: A Military History." (Columbia, Mo.: University of Missouri Press, 2012), 110.
[45] Brownlee, "Gray Ghosts," 39, 43.

described a "Jayhawker" to be a pirate, privateer or freebooter. Jayhawkers "banded together like other marauders and galloped over the country, appropriating horses, cattle, farming implements--in short, whatever came their way. They claimed they never robbed or hung anyone who entertained the same political opinions they held...."[46]

The initial use of the term "'Red Legs" is said to have originated out of a group of about thirty men who came from around Atchison, Kansas, and were organized by General Blunt as part of his staff. They were so called because they wore red yarn or leather leggings that covered their military boots. Some of their red leather had been seized from a mercantile store in Independence when Jennison had overtaken and looted the county seat in November of 1861. Red Legs were a paramilitary organization in which looting and murdering was part of their credentials. They were known as some of the toughest set of men that ever infested any country.[47]

The passion of the Jayhawkers and Red Legs was a by-product of their deep hatred for Missourians that had originated during the days of Bleeding Kansas. Once the activities of these Kansas guerilla organizations had begun they "gradually increased, and the horrible barbarities" they perpetrated uniformly "grew at unchecked and appalling rate."

> *Their warfare was characterized by savagery. The lives of the people and the material wealth of Jackson County had been wantonly and wickedly destroyed in a manner and had been unknown and unheard of among a*

---

[46] Hale and Eakin, 140, quoting *Missouri Historical Review*, Connelley, and *Daily Missouri Republican*, 3 Oct. 1861. William Connelley in, "Quantrill and the Border Wars," wrote about a group that crossed Death Valley in 1849 and who for unknown reasons called themselves Jawhawkers. One member claimed he coined the term upon reaching the Platte River, but no one else in the party made such a claim.

[47] Hale and Eakin, 226, quoting the 25 April 1917 *Confederate Veteran*. A list of Red Legs was published in the Sycamore Wabash [Indiana] True Republican, August 10, 1863, and sent by C. M. Chase, of Leavenworth. George E. Hoyt was captain. He was a young lawyer of Delaware, Ohio, who defended John Brown in West Virginia. The others were: Theo. Bartles of Wyandotte, Kansas, whose brother founded Bartlesville, Oklahoma; "Wild Bill" Hickok; "Red" Clark, of Emporia, Kansas; Charles R. Jennison, gambler, colonel of the 7th Kansas Volunteer Cavalry, the most notorious figure of the Missouri-Kansas border; "Jack" Harvey, brother of Fred Harvey, of the Santa Fe Railroad eating houses; Joe Guilford, Wyandotte half-breed; Al Sevier; [-?-] Also; Jim Flood; Jerry Malcolm; "Bloom" Swain, lieutenant, also captain of Company K, 15th Kansas Volunteer Cavalry; William S. Tough, "tough by name and nature" (the True Republican said, "He is pure horse thief and murder" and was at that time exiled from Fort Scott, Kansas, for killing a man. he founded Kansas City, Missouri, horse and mule market with stock stolen from Missouri.); "One-Eyed" Blunt; [-?-] Hawkins; Walt Sinclair; John Sinclair; Jack Blatchly, profligate son of a Presbyterian minister of Wyandotte; Jack Bridges, a little Englishman (after the War he sold a fake which he called the "Cardiff Giant," a huge figure said to have been a petrified woman, which he claimed to have unearthed while digging in a mine in Colorado); Newt Morrison, lynched after the War for stealing a horse from Joe Guilford; "Pickles" Wright, killed by Guildford after the War; [-?-] Gladheart, Lawrence, Kansas; "Pony" Searl; John Salthiel; Hank Grosscup and Hand Starr, Kansas City, Missouri, and one or two others. They met at the Six-Mile House, between Wyandotte and Leavenworth, Kansas."

civilized people. Lane and these evils produced a degree of consternation that language could not be described, and which none could comprehend unless they had witnessed it.[48]

Nothing ordered, directed or said in protest to these persistent atrocities and privations inflicted upon Missouri by Lane or his compatriots, Jennison or Montgomery, resonated or impeded their despicable exploits. Lane and his Kansas allies continued as was their want to inflict revenge and to loot whenever they entered western Missouri and this continued during the final two months of 1861. In November, Doc Jennison and the Independent Mounted Kansas Jayhawkers, legally the Seventh Kansas, acted wantonly as they traveled through Jackson County pillaging, burning, intimidating and murdering at will. They rode from Kansas City to Independence capturing the county seat and corralling male citizens. Jennison threatened to kill any disloyal person. Their march continued to Westport leaving in their wake burned out houses, farms and other property.[49] On November 17, 1861, Jennison's men burned and looted the town of Pleasant Hill in Cass County; Col. James Montgomery burned Butler in Bates County and on December 13, Lane's 1st Kansas burned houses as well as the Pappinsville Courthouse in Bates County; the year of pillage, fire and raids came to a vindictive end on January 1, 1862. Jennison had ordered the 7th Kansas under Dan Anthony to seek and destroy rebel supporters. On that New Years' day, Anthony marched boldly upon the town of Dayton in Cass County and proceeded to burn the entire town and surrounding areas to the ground.

Lane and other Kansas guerrillas were not the only problems Missouri faced during the first year of the Civil War. There was mounting political unrest over Fremont's command and his military orders. Fremont's emancipation proclamation had brought formal protests from Governor Hamilton Gamble, Congressman Frank Blair in St. Louis and other loyal Missourians. They equally protested Fremont's extermination

[48] O.R. 22:2:485-488.

[49] Brownlee, "Gray Ghosts," 42-43; Goodrich, "War to Knife," 227; and Wallace, William H, "Speeches and Writings of William H. Wallace, With Autobiography." (Kansas City, Mo.: The Western Baptist Publishing Co., 1908), 247-260.

order. There was grave concern and apprehension that this decree would provoke retaliatory measures by southern agitators and rebels. Fremont was also criticized for having failed to send reinforcements to Wilson's Creek and Lexington. Many Unionist thought this lapse of military leadership caused the defeat and death of Lyon at Wilson's Creek and the capture and surrender of Lexington. To these antagonists, Fremont had demonstrated stubbornness and unwillingness to work with the political leaders of Missouri. As the year 1861 began to close, these and other persistent and mounting complaints against Fremont could no longer be ignored in Washington. Lincoln and his Cabinet reached the conclusion that changes had to be made in Missouri.

General Henry Halleck

From the Capitol, orders were issued to reorganize the federal military structure in the west. The first action resulted in the removal of Fremont on November 3, 1861. He was succeeded by General David Hunter. For the politically powerful Blair Family, Fremont's removal had been long overdue. These changes were followed by the appointment of General Henry Halleck, "Old Brains," as commander of the District of the Frontier. Halleck divided Missouri and Kansas into two separate districts, the District of Missouri and the District of Kansas. The creation of the District of Kansas impacted Lane's Brigade as it was mustered into the Union Army, given new assignments and ceased to exist. As for Jennison, his misconduct culminated in an unceremonious transfer to the war's front east of the Mississippi River.

With the formation of the District of Kansas, Lane's military career took a sabbatical. He resumed focusing on politics and launched another speaking tour of eastern states. Although he spent the bulk of 1862 off the battlefield, he waived his political sword over the District of Kansas thereby shoring up his political power and dominion over the state. From Washington, the Grim Chieftain maneuvered western troops, made appointments, doled out promotions and advocated recruitment of black soldiers. He had a hand in the promotions of James Blunt to Major General and Thomas Ewing, Jr. from Chief Justice of the Kansas Supreme Court to Colonel. By September 1862, the First Regiment of Kansas Colored Volunteers was being trained.

Throughout this time, Lane waged open warfare against Governor Robinson. Questionable railroad bond transactions had caused financial problems within the State of Kansas. Soon the Governor and other statewide office holders came under the scrutiny of

Lane and his political forces. To the Grim Chieftain this afforded him the opportunity to inflict mortal injury upon his long-time political rival. Impeachment proceedings were commenced against the Governor and others by two of Lane's men in the Kansas Legislature. Robinson was acquitted of any culpability or wrongdoing, but the scandal had compromised him. Robinson would forever blame Lane for his downfall. Robinson's political demise much to the gloating satisfaction of Lane came during the fall of 1862 when Thomas Carney, a Lane candidate, was elected as the second governor of Kansas.[50]

Throughout 1861 and 1862, public and military sentiment had soured over Jayhawking by Lane, Montgomery, Anthony and Jennison. General Halleck now in full command declared that Lane and Jennsion had done more for the enemy than twenty thousand soldiers. Their atrocities had turned many men who had once been neutral or loyal Unionist into bitter enemies and bushwhackers. Provisional State Treasurer George Caleb Bingham announced from his office in Jefferson City that if Jennison was tried for war crimes, it would aid the Union cause along the western border of the state, and if Jennison were hung "Price would lose thereby the best recruiting officer he has ever had."[51]

As 1863 opened, Lane had returned to the Capitol, and a calm coolness settled along the border of Missouri and Kansas. The calmness was surreal and the cold was just the season. The outrageous conduct in which the Grim Chieftain and his Kansas raiders had waged upon Missouri had taken their toll upon the people of the western border. Throughout 1862, recruiting activities by Confederates and formation of rebel guerrilla units proved successful. Many of the men who enlisted did so out of a conviction to retaliate against the domestic affairs that the Kansans had inflicted upon their homes, families and towns. Eventually this would produce an unrestrained lawlessness never seen previously in western Missouri. As 1863 would progress everything would become as hot as hell on the borders of Missouri and Kansas and Lane would be at the center of the controversy.

James "Jim" Lane

---

[50] Collins, 210-214.

[51] Goodrich, "War to Knife," 227-228; Brownlee, "Gray Ghosts," 46-47; and Castel, "Frontier State," 58-59.

CHAPTER III

Bingham: Painter, Politician and Propagandist

George Caleb Bingham

As the New Year of 1863 opened, George Caleb Bingham had just completed his first full year as the Provisional Treasurer of the State of Missouri. On that date, Bingham was preparing his report to the Missouri General Assembly of the financial condition of the state. This would include an announcement that the state had honored $56,317.54 in warrants issued under the new Missouri law that had established the Missouri State Guard on May 14, 1861. Despite a nativity in Virginia and a slave-owning childhood in Missouri, Bingham supported implementation of the Emancipation Proclamation in those states still under rebellion. For Missouri, however, he held firm to the Whig tradition of freeing slaves only under the process of gradual emancipation.[52]

Bingham's rise to state treasurer in January 1862 had been charted by events throughout 1861. Bingham staunchly supported the Union and openly opposed secession. He considered the inaugural remarks of "Old Abe" as "reasonable," but unlike the new president Bingham was willing to fire the war's first shot as he had grown "tired of

---

[52] Letter from Bingham to the Missouri General Assembly dated January 2, 1863, from "But I Forget That I am a Painter and Not a Politician: The Letters of George Caleb Bingham." (Columbia, Mo.: The State Historical Society of Missouri and Friends of Arrow Rock, Inc., 2011), 278-279 (hereinafter "Bingham Letters"); letter from Bingham to James S. Rollins dated February 16, 1863, "Bingham Letters," 279-280; Steve Carr, *200th Birthday of George Caleb Bingham: American Artist of the American System*, June 2011, schillerinstitute.org/edu/reviews/2001/george_bingham.html. The term "gradual emancipation" was the process whereby slaves would be awarded their freedom after a specific date in the future. The debate on this issue centered not upon the emancipation of slaves but which date and time in the future would be appropriate. Bingham set forth his support of this measure in his February 16 letter to Rollins.

submission to traitors."[53] The Missouri State Constitutional Convention held to determine the relationship between the state and federal governments first convened on February 28, 1861 and then reconvened on two other occasions in 1861. During the second convention the delegates declared the offices of Governor and Lieutenant Governor vacant. They concluded Governor Jackson and Lt. Governor Reynolds had treasonously voided their offices by conspiring to lead a secessionist movement in Missouri. They filled these open seats by appointing Hamilton Gamble as provisional state governor and Willard Hall as provisional Lt. Governor. These provisional statewide offices would remain in the hands of the unelected provisional office holders throughout the war.[54]

The State Convention then expanded the scope of its legislative authority over Missouri. In July 1861, the Convention imposed an oath on all public officials, attorneys and clergy. In Boone County, six of the seven justices of the peace, among others, refused to take the oath, and were summarily removed from office. Other oaths were required to be taken by any former member of the Missouri State Guard, Confederate, guerilla or aider and abettor who desired to be paroled. During the same period in which the provisional government held control at the Capitol of Missouri, all other power was held under the military dictates of the federal government. This created a significant amount of overlapping of jurisdictional authority between the state government and the federal government. Both however shared the goal of ensuring that Missouri did not secede from the Union.

On December 6, 1861, General Halleck, the new commander of the Department of Missouri, following the lead of the Convention expanded the loyalty oath to all citizens of Missouri. Failure to submit to the oath constituted disloyalty and sedition and would result in imprisonment or exile from the state without benefit of trial. Halleck's oath read:

> *"I solemnly swear that I will bear true allegiance to the United States and support and sustain the Constitution and laws thereof; that I will maintain the national sovereignty paramount to that of all State, county or Confederate powers; that I will discourage, discountenance and forever oppose secession, rebellion and disintegration of the Federal Union; that I disclaim and denounce all faith and fellowship with the so-called Confederate States and Confederate armies and pledge my honor, my property and my life to the sacred performance of this my solemn oath of allegiance to the Government of the United States of America."*

---

[53] Bingham's letter to James S. Rollins dated March 6, 1861, "Bingham Letters," 246-247; and Nagel, Paul C. "George Caleb Bingham Missouri's Famed Painter and Forgotten Politician." (Columbia, Mo.: University of Missouri Press, 2005), 106.

[54] Hamilton Gamble died on January 31, 1864 as a result of a fall, and Lt. Gov. Willard Hall served as governor until Governor Thomas P. Fletcher succeeded him on January 2, 1865.

State Treasurer A. W. Morrison refused to submit to the recently enacted loyalty oath and was removed from office. Morrison's removal was to Bingham's good fortune. Governor Gamble appointed Bingham, whom he knew had strong political allies inside the state and in Washington, as the new state treasurer. Bingham unquestionably was one of the most noted American genre and portrait artists in the 19th Century.[55] While his artistic notoriety is well known, he is often overlooked as an elected public official, statesman and political force. Bingham assumed his new position at the State Capitol on January 2, 1862.

Bingham was born on March 20, 1811, in the Shenandoah Valley of Augusta County, Virginia. His father Henry Vest Bingham married his mother Mary Christina Amend on September 8, 1808, in Rockingham County, Virginia. They were closely aligned with the spirit and culture of the southern way of life. Within two years of their marriage, they owned six slaves. The Bingham family moved to mid-Missouri around 1819. The family consisted of Henry, Mary, their three children, Mary's father Matthias Amend and six slaves. They settled at the end of western civilization—Franklin, the "Metropolis of the West," in the Missouri Territory.[56]

The Bingham's new home in Franklin was the second largest city in the Missouri Territory, second only to St. Louis. Like the larger city Franklin was a river city, but unlike St. Louis on the Mississippi, Franklin was located on the southern banks of the unforgiving Missouri River. Henry soon opened an inn, "Square and Compass," and opened a tobacco factory. He became involved in local government and served as justice of the peace. Tragically, when Bingham was only twelve years old his father died of malaria in December 1823.[57]

The Missouri River had a pronounced affect on Bingham's life. Both his maternal grandfather and his brother Isaac Newton drowned in the River. The Bingham property and the entire city of Franklin were literally wiped off the map by a flood in 1829. Henry Bingham before his death once described the "ole Muddy River" as "the strongest and

---

[55] Nagel, 108. Today Bingham's artwork is on display in museums across the county including the Metropolitan Museum of Art, New York City, the Nelson-Atkins Museum of Art in Kansas City, Missouri ("Canvassing A Vote"), the St. Louis Art Museum, Arrow Rock Historic Site, Arrow Rock, Missouri, Washington University, St. Louis, Missouri, Missouri Historical Society, Columbia, Missouri (Order No. 11), Museum of Fine Arts, Boston, Mass., Detroit Institute of Arts, Detroit, Michigan, Cincinnati Art Museum, Cincinnati, Ohio (Order No. 11), Colby College Colby College Museum of Art, Waterville, Main, Los Angeles County Museum of Art, Mildred Lane Kemper Museum of Art, St. Louis, Missouri, and National Portrait Gallery, Washington D.C., See: http://www.artcyclopedia.com/artists/bingham_george_caleb.html on the many other locations in which Bingham's Art is displayed across the United States.

[56] 1810 Census, Harrisonburg, Rockingham County Virginia, 147; Christensen, Lawrence O., William E. Foley, Gary R. Kramer and Kenneth H. Winn, eds. "Dictionary of Missouri Biography." (Columbia, Mo.: University of Missouri Press, 1999), 70.

[57] 1810 Census, Harrisonburg, Rockingham County Virginia, 147; Christensen; Nagel, 4-7; Letters, 523.

most turbulent current in the world." Old Franklin was gone and a New Franklin formed—this time on higher ground. In the succeeding years, the River would become a central theme of Bingham's artwork.[58]

In his youth, Bingham demonstrated an interest and talent for art. Before relocating with his family to Missouri, Bingham had previously shown a proclivity in art, as he often sketched drawings on barns in his native Virginia. Bingham continued this pattern of drawing on structures after the family settled into the Missouri Territory. As early as 1820, his artistic skills began to receive recognition and encouragement. The famous early 19th Century American artist, Chester Harding, had the opportunity to see Bingham's early-aged talents. Harding at that time was in the Territory to paint the portrait of the famous trailblazer, Daniel Boone, and he stayed at the Bingham tavern, Square and Compass. Harding told Bingham and his parents of the young boy's great artistic promise. He encouraged the family that Bingham's art would benefit from learning the trades of woodworking and cabinetmaking. Acting upon Harding's recommendation, Bingham served an apprenticeship with a cabinetmaker in Boonville. Within ten years, Bingham was painting portraits ("painting heads" as he called it), and illustrating and painting the scenery and people of Missouri ("genre art").

James S. Rollins

By 1834, Bingham's artistic career and personal life began to merge with his growing political interests and ambitions. His future career in politics was propelled forward because of a friendship he made that year with a young Whig politician, James S. Rollins from Columbia, Boone County, Missouri. Their meeting proved to be the beginning of a lifelong relationship of intense personal friendship and strong political affiliation. They shared the philosophy of the Whig Party and they supported the American System of Henry Clay and President John Quincy Adams—internal improvements, protective tariffs, gradual emancipation and a strong federal banking system.[59] In a letter to Rollins dated May 6, 1837, Bingham expressed his heartfelt sentiments as to his artistic goal:

---

[58] Nagel, 7.

[59] Bingham personally knew former President John Quincy Adams when the former president was a member of the U.S. Congress, and President Adams even sat for a portrait painting by Bingham during the period in which Bingham was working in Washington City.

> *There is no honorable sacrifice which I would not make to attain eminence in the art to which I have devoted myself I am aware of the difficulties in my way . . . and by determined perseverance, I expect to be successful.*[60]

Despite his desire to become a famous artist, Bingham was evolving into a Whig politician. Although his political and philosophical support of the Whig platform included the doctrine of gradual emancipation of slavery, the Bingham family continued to own slaves even after the death of his father Henry. Throughout the 1830s and 1840s Bingham's mother, until she died in 1851, resided in the Arrow Rock Township of Saline County, Missouri and owned slaves. The 1830 census reveals that Mary owned two slaves, a male between the ages of ten and twenty and a female about twenty-four years old. Until her death, Mary still owned slaves. She is identified as having one slave in both the 1840 and 1850 census. Her last slave of record was a black female, aged forty-five years. Bingham was no different than his mother and late father. According to the 1850 slave schedules, he was the owner of two slaves, a twenty-one year old black female and a fifty-five year old male.[61]

By the mid 1830s, Bingham had developed a reputation as a gifted painter. In 1840, he turned his artistic talents to political themes. His first known opinionated artwork was painting campaign banners for local Whig candidates and Whig clubs in central Missouri. The Whig candidate for President in 1840 was General William Henry Harrison of Ohio and Bingham supported the General. He painted a large four-sided banner that promoted the candidacy of Harrison, and artistically depicted the Whig Party theme of the American System, which focused upon a system of internal improvements for the national good of the country. Harrison won the election and Whigs, including Bingham, soon sought patronage positions in Washington. Bingham desired a federal art commission and an opportunity to paint portraits of political figures in the Capitol. Four years later, Bingham painted murals in support of the Whig Presidential Candidate, Henry Clay—the founder of the American System. Bingham had developed into a pronounced disciple of Clay's political thought and agenda.

Bingham, the political painter and artist, in September 1834 married Sara Elizabeth Hutchison. Their first child Newton was born on March 26, 1837. Their second son Nathaniel was born in 1838 but died in infancy. In 1844 Sara gave birth to their first daughter, Clara. To support his family, Bingham returned to the canvas and began to focus on rural settings and genre art.

His artwork soon attracted attention outside his adopted home state. By 1846, the New York-Art Union had purchased several of Bingham's paintings, and his painting,

---

[60] "Bingham Letters," 43.

[61] 1830 census, Arrow Rock Township, Saline County, Missouri, 235; 1840 census, Arrow Rock Township, Saline County, Missouri, 97; and 1850 census Arrow Rock Township, Saline County, Missouri.

"Fur Trappers Descending the Missouri River" gave him national acclaim. Bingham was well on his way to merging his artistic talents with his political interests. The theme of many of Bingham's genre paintings during this period, especially those focusing on river scenes were designed to extend a political statement of the urgency for internal improvements—a theme that was central to Henry Clay's "American System."

Two years later in 1846 Bingham returned to campaigning, but this time not for other Whigs, but for his own candidacy for state representative of Saline County. In that election, Bingham defeated the Democrat candidate Erasmus Sappington by three votes. The Sappington family was part of the Booneslick Democratic regime in Missouri. They were a politically powerful family with significant connections at the Capitol in Jefferson City. The Sappingtons' challenged the election before the Missouri General Assembly. In the end, the legislature sided with Sappington, and Bingham's election was overturned and the seat was awarded to Sappington—a decision protested by Bingham as being contrary to the will of the people.

Bingham re-challenged the Sappington's and their "loco focus" (as Bingham and his political allies called them) Democratic power in the next election in 1848. This time Bingham received thirty-six more votes than Erasmus Sappington; there was no election-challenge and Bingham, the Whig, was seated as the state representative of Saline County. Bingham's excitement over winning the election that November was promptly overshadowed by the tragic deaths of his twenty-nine year old wife, Elizabeth, and his third son, Joseph Hutchison. His election as state representative proved to be the first and last time in which Bingham held public office by the vote of the people. Nearing the end of his first term in the Missouri House, Bingham on December 3, 1849 married his second wife, Eliza Thomas, a daughter of a Baptist minister. Within a couple of months of his second marriage, Bingham's first term in Jefferson City came to an end, and he chose not to seek re-election in 1850.

The 1850's marked the only decade in Bingham's adult lifetime where he withdrew from the political arena to focus upon art. He and his family had moved to Independence where Bingham devoted much of his energy prior to 1856 in painting his election series, which included Stump Speaking (1852), Canvassing for a Vote (1852), The County Election (1854), and the last in the series, Verdict of the People (1855). With the exception of traveling to Philadelphia in 1855 to meet with the famous steel engraver, John Sartain, Bingham spent the majority of the year engaged in portrait painting.

Although Bingham may have stayed out of public office, he remained attentive to the political wrangling over the adoption of the Kansas-Nebraska Act. He recognized that his friend, James Rollins, had labored over the territorial debate and the "hot work" over the "goose question." Bingham knew Rollins would continue to hold the "conservative" line, and Bingham was pleased to inform Rollins that other politicians in Independence, including three leading attorneys of the city Sam Woodson, Abram Comingo and

William Chrisman, for their firm position on the question as well.[62] Bingham was convinced the Kansas Bill had created "a storm" in which he prophesied would "sweep onward with a fury, which no human force" could withstand and that its enactment would destroy the Whig Party.[63]

Bingham and other members of the former Whig Party certainly held serious apprehension to the unsettling activities transpiring in Kansas. He voiced concerns over "all the violence and blustering" from pro-slavery supporters, such as former Senator David Rice Atchison, but he felt confident that "clear-headed thinking men" would be able to demonstrate that the "establishment of slavery in Kansas" was "utterly impossible." With the growing tension over the Kansas issue, Bingham informed his friend Rollins that he intended to leave Independence and travel to Paris to complete his paintings.[64]

Rollins prevailed upon the Missouri General Assembly to commission Bingham to paint two life size portraits of George Washington and Thomas Jefferson. Thirty five hundred dollars was appropriated for the paintings. This sum was unquestionably the largest amount Bingham had ever received.[65] These funds made it possible for Bingham and his family to travel to Europe so he could study art on the continent and complete his commission paintings.

From 1856 to 1859, the Bingham's lived in Europe. They temporarily resided in Paris, France before settling in Düsseldorf, Prussia (Germany today). While in Düsseldorf, Bingham completed the life-size portraits of Washington and Jefferson, and he followed-up their completion by writing to Rollins suggesting that the legislature appropriate money for him to paint similar portraits of President Jackson and Henry Clay.[66] During the family's European sojourn, Bingham's children Horace and Clara learned to speak French, German and other continental languages, and Bingham found the artistic style of Germany to his liking.[67]

In late 1858, word reached Bingham in Prussia that his wife's father had become gravely ill. He sent his family back to Missouri while he made plans to follow them. Before Bingham left Düsseldorf in June 1859, he learned that his wife had suffered

---

[62] Letter from Bingham to James S. Rollins dated June 21, 1855, Letters 146-147.

[63] Letter from Bingham to James S. Rollins dated May 27, 1854, Letters 142; and Rash, Nancy. "The Paintings and Politics of George Caleb Bingham," (Yale University, 1991), 185. The term "sound on the goose" meant being a pro-slavery supporter; it had also been utilized as a code for Missourian during the fight over Kansas. To know one's position on slavery was to inquire whether he was "sound on the goose." "Yes," meant he was a pro-slavery advocate for Kansas; the reply of "no" would signal that the man was not pro-slavery and wrath was often hurled upon him.

[64] Letter from Bingham to James S. Rollins dated June 21, 1855, "Bingham Letters," 147.

[65] Rash, 159-160.

[66] Letter from Bingham to James S. Rollins dated March 8, 1858, "Bingham Letters," 194; and Rash, 165-167.

[67] Letter from Bingham to James Rollins dated March 8, 1858: "Bingham Letters," 194; and Nagel, 96-97.

another miscarriage.[68] When Bingham arrived in Missouri, his father-in-law had already died. By his death, Bingham's wife inherited property from her father, including a building located at about 14th and Grand in Kansas City. It soon became the Bingham residence. Based upon 19th Century law, Bingham had the responsibility of managing his wife's inheritance, which entitled him to all rents and profits derived and generated from said property. Bingham soon added a third floor to the structure to provide sufficient studio space for his art gallery and artwork.

As the Civil War approached, Bingham staunchly stood strong for the Union. He had supported the candidacy of Senator John Bell in the 1860 presidential election. Bell, like Bingham, had been a former Whig and an Unconditional Unionist in opposition to secession—not in support of immediate emancipation of slaves, but rather gradual emancipation.[69] Bingham quickly completed his paintings of Jackson and Clay in his new studio in Kansas City. Like the favorable response he had garnered from his portraits of Washington and Jefferson, these two paintings received acclaim from the press and Missouri Legislature. On January 7, 1861, Bingham's paintings of Clay and Jackson were placed side by side in the Hall of Representatives at the Capitol in Jefferson City. The following evening the paintings were formally dedicated and Bingham was asked to give some remarks to the Missouri General Assembly.

Bingham used the occasion to quote from Jackson and Clay on the "Union," and to proclaim his own unwavering loyalty to the "Stars and Stripes" of the United States. His speech was a pro-union statement that defiantly declared secession to be evil and illegal. He also provided prophetic remarks about the impact the pending calamity of war would have in the country, and especially its potential devastation to his home in Kansas City, Jackson County.

> *I live within three-quarters of a mile of the extreme border of our State. Bombs from an enemy in Kansas can demolish my dwelling, and in half an hour's time, made a waste of all the property I own in the world.*
>
> *As the war which must follow dissolution will be waged in my peculiar locality with all that rancorous hate which years of crimination and recrimination have necessarily produced, I shall be forced to participate in it, devoted, as I am to the pursuits of peace, or ingloriously desert my country in the hour of her peril. My property will be destroyed, my life may be sacrificed, my wife becomes a widow and my children orphans.*

---

[68] Nagel, 100. In 1859, Bingham painted one of his most famous portraits. It was of Dr. Bentoist Troost of Kansas City (Troost Ave. in Kansas City is named for him) and who died later in the year. The painting is owned and on display at the Nelson-Atkins Museum of Art in Kansas City.

[69] Letter from Bingham to James S. Rollins dated February 16, 1863, "Bingham Letters," 280; and Nagel, 103.

> *I have only alluded above to a few of the immediate evils which must come upon my State, my neighbors, and myself, as soon as dissolution takes place.*[70]

Following the speech, Lt. Governor Thomas Reynolds (a secessionist) condemned the pro-union slant of Bingham's speech. In defending himself, Bingham declared he was not expressing his own views, rather those of Jackson and Clay. When Bingham's comments received ridicule from the press, he wrote a letter to Nathaniel Paschall on January 12 in which he defended himself. Paschall had Bingham's letter published in the *Missouri Daily Republican* in St. Louis. The letter was certainly indicative of Bingham's style, political view and aggressive nature. Vigorously and without reservation, Bingham held firm to his "Stars and Stripes" speech. He was unwilling to accept the secession movement in the South; he was uncompromisingly for the Union.[71]

Despite having no military training and being fifty years old, Bingham on June 1, 1861 enlisted as a private in the U.S. Volunteer Reserve Corps in Kansas City under Major Robert T. Van Horn's Missouri Home Guard. He was soon wearing officer's bars as the men of Company C elected him captain. In September 1861, Bingham and his company under Van Horn joined Major James A. Mulligan in defending the town and earthworks at Lexington, Missouri.

The total Union forces did not exceed three thousand men while General Sterling Price and his attacking Missouri State Guard had as many as fifteen thousand troops. Price on September 20 with men hidden behind the protection of hemp bales attacked the Union earthworks. This final assault compelled Mulligan to surrender. After the surrender at Lexington General Price paroled the Union troops, and Bingham retuned to Kansas City.[72] Bingham's military career ended on January 2, 1862 when he resigned his commission to become the Provisional State Treasurer of Missouri.[73]

Bingham's appointment as State Treasurer was certainly aided by the efforts and political connections he had with Congressman Rollins. Following Bingham's

---

[70] Letter from Bingham to Nathaniel Paschall, "Bingham Letters," 233-239; and Rash, 183-185.

[71] On the same day Bingham wrote his letter to Paschall, his bachelor brother Mathias died at his residence near Houston, Texas. Bingham traveled to Texas to settle the affairs of his late brother. In the same month, the Missouri General Assembly passed an ordinance calling for an election of delegates to convene at a state convention for the purpose of determining the relationship between Missouri and the federal government. The portraits of Washington, Jefferson, Jackson and Clay were hung in the Missouri State Capitol but they were lost in a fire that consumed and destroyed the Capitol building in 1911.

[72] The day after the Union surrender at Lexington, Bingham's wife Eliza gave birth to the only child born of their marriage: James Rollins Bingham (named in honor of Bingham's long-time friend, James Rollins).

[73] Bingham later filed a claim against the federal government in the amount of $275 for back pay for the period in which he had served as Captain of Company C from November 15, 1861 through January 4, 1862.

appointment on January 2, 1862, he moved his family from their home near 14th and Grand in Kansas City to the State Capitol in Jefferson City.

From his Capitol office, Bingham continued to follow the saga and events along the western border of Missouri. His concern was both personal and public. The Bingham residence inherited by his wife from her late father was located in Kansas City about "three-quarters of a mile" east of the Kansas state line. He feared the struggles in the area had the real potential of compromising the integrity of the property. He also objected to the persistent Kansas guerilla raids. On May 6, 1862, Bingham wrote a letter to the *Missouri Dailey Republican* in which he wanted to call to the attention of the whole country and President Lincoln his public sentiments about the persistent intrusion into Jackson County by Kansas Jayhawkers and Red Legs. Bingham was outwardly agitated by the "record of Jennison's crimes" and was extremely gratified when he learned Jennison had been arrested for war crimes. Bingham declared that Jennison's "gross abuse of power exhibited in low acts of oppression against defenseless individuals, and the plunder and desolation of entire communities has been such as should impel good citizens everywhere to unite in invoking against him the punishment due to a felon." He firmly asserted that "scoundrels" from all quarters of Kansas had "flocked" to Lane and Jennison's standards "as naturally as buzzards collect around their favorite carrion." He insisted that their followers "had no scruples in furnishing themselves with horses from the stables of our citizens without money, without price and equally without leave." He also hoped that the lessons of the first two years of the war would not be repeated but would rather "serve as a fountain of wisdom for many generations."[74]

Bingham would continue to retain office as the state treasurer throughout the war. He never ceased expressing his protests over the outrageous conduct that Lane, Jennison and their cohorts had inflicted upon Missouri. His anticipated dream that the prior war years would educate leaders so as to avoid further unrest was only a fantasy. The year 1863 was destined to become the ultimate climax of a blending of Bingham's artistic and political careers.

[74] Bingham's letter to the Missouri Republican, May 6, 1862, Letters 269-271; and Bingham's letter to James S. Rollins, dated February 16, 1863, "Bingham Letters," 280.

CHAPTER IV

Quantrill Is Back!

As the 1863 frozen tundra along the border of Missouri and Kansas surrendered to the flowers of spring, Quantrill and his guerrillas made their annual journey back to Jackson County from their winter quarters in Texas. The New Year not only would prove to be the climax of the Civil War, it would prove to be the seminal year of Quantrill's reign in Jackson County.

William C. Quantrill

Quantrill's nativity, background and mission have remained clouded in ambiguity. Some considered him a hero and knight errant while others knew him to be a brutal, psychopath and murderer. It all depends on one's perspective of the Civil War. The mystery of Quantrill's adolescences remains as shrouded today as it was during his lifetime. The scant information truly known is that he was the eldest of eight children born to the marriage of Thomas Henry Quantrill and Caroline Cornelia Clarke.

The general historical consensus is that his date of birth and birthplace was near Dover, Ohio on July 31, 1837.[75]

---

[75] Leslie, Edward E. "The Devil Knows How to Ride." (New York: Random House, 1996), 36-37; Houts, Joseph K. "Quantrill's Thieves." (Kansas City, Mo.: Truman Publishing Company, 2002), 48; Hale, Donald R., "We Rode with Quantrill: Quantrill and the Guerrilla War as Told by the Men and Women who Were With Him, and a True Sketch of Quantrill's Life." Originally published in 1975. Reprint. (Independence, Mo.: Blue and Grey Book Shoppe, 1982), 8. Quantrill's parents were married in Ohio on October 11, 1837 and four of his siblings died in infancy. Quantrill's date of birth and nativity according to both Captain Harrison Trow and W. L. Webb was July 20, 1836 and at Hagerstown, Maryland—this is actually the birthplace of Quantrill's father. See, Burch, J. P. "Charles W. Quantrell: A True History of his Guerrilla Warfare on the Missouri and Kansas Border during the Civil War of 1861-1865."(n.p.: J. P. Burch, 1923), 15; and, Webb, W. L. "Battles and Biographies of Missourians or the Civil War Period of Our State," Originally published in 1900. Indexed Edition. (Springfield, Mo.: Oak Hills Publishing, 1999), 266. The confusion over his birth can be laid at the feet of Quantrill himself who was often less than truthful to his friends and guerrilla members about his personal past.

William C. Quantrill birthplace

Even the history of Quantrill's childhood is steeped in controversy and uncertainty. Some biographers have asserted that his cruelty during the Civil War had its genesis in his youth. These authors contend that Quantrill delighted in such mischievous activities as shooting pigs through their ears to make them run and squeal; nailing snakes to trees to watch them suffer; tying cat tails together to watch them claw each other; strolling down the lane and randomly stabbing a horse or cow; or throwing snakes on girls to cause them fear and grief.[76] Other less biased or prejudiced Quantrill scholars seriously question and highly discount the veracity of such tales. Regardless of the truth or falsity of these stories, they are certainly part of the Quantrill myth that remains to this day.

Quantrill's parents

Quantrill's parents did ensure he received an education. By the age of sixteen, Quantrill had graduated from high school and began teaching school in Dover. Within the year on December 7, 1854, his father, a stern taskmaster and rigorous disciplinarian, died of consumption (called tuberculosis today). Quantrill had a close relationship with his mother but had no affection or bond with his father. By his father's death, the young man began to search for new opportunities.

The year that the elder Quantrill died marked a pivotal time in the life of his oldest son and for the United States. The Kansas-Nebraska Act had opened the land to settlement creating a hornets' nest for warring factions over the issue of slavery. This also made the Territory an inviting venue for young men such as Quantrill who wanted to seek their fame and fortune, or become embroiled in the pro-slavery or free-soil movements. Quantrill's motives for adventuring towards Kansas may be unknown, but he did slowly begin his westward trek. His official arrival date to the border is given as March 31,

[76] Houts, 54; Connelley, William E. "Quantrill and the Border Wars." Originally published in 1910. Reprint. (New York: Pagant Books Co., 1956); and, Leslie, 42.

1857. On that date Quantrill disembarked from a Missouri River steamboat at the Wayne City Landing near Independence—the county seat of Jackson County.[77]

For the next few years, Quantrill experienced much of the new Territory of Kansas and the far western regions of the United States. He taught school, worked as a teamster, gambled and panned for gold near Pike's Peak. His westward adventures brought him face-to-face with Mormons in Salt Lake City, Indian attacks on the plains, frostbite and snow-blindness from the wintry elements, and associations with jayhawkers' such as Captain John E. Stewart, the "Fighting Preacher," whose reputation was second only to Charles "Doc" Jennison and James Montgomery.[78] Qunatrill's political views during this period supported the free-soil movement. Quantrill in a letter he wrote to a friend in Dover, Ohio in January 1857 openly protested against the pro-slavery Lecompton Constitution and saluted Jim Lane as being "as good a man as we have here" in Kansas.[79]

As the new decade of the 1860's opened, Quantrill's political view on slavery and southern states' rights was experiencing a transformation. His philosophical evolution may have started during his teamster days out west when he had worked closely with southern men who were not hesitant to share their insight and perception as to the benefits of the institution of slavery. Another possible basis for Quantrill's new attitude on emancipation may have stemmed from the outrage he felt over how abolitionists had supported John Brown's raid on the Federal military arsenal at Harper's Ferry the previous fall. There is also the legendary account that his new attitude arose out of revenge. Quantrill claimed it stemmed from Kansas' abolitionists killing of his younger brother and leaving him for dead.[80] There was no dispute that Quantrill had the ability to spin a web that no spider ever knew.

Morgan Walker

Quantrill's final conversion to the ranks of the southern cause occurred at the Morgan Walker Farm in Jackson County on December 10, 1860. James Morgan Walker was a wealthy slave owner who owned approximately two thousand acres of land in eastern Jackson County. He and his wife Polly (Cox) Walker had twelve children. Their extensive asset holdings were well known throughout Jackson County and as far west as Lawrence. The 1860 census lists the value of their estate to be in excess of $35,000. Their property included twenty-five slaves, twenty-five horses, forty-eight mules, thirteen oxen and thirty-five cows. The home was a

[77] Leslie, 59.
[78] Leslie, 58-60, and 66.
[79] Leslie, 51.
[80] Burch, 15.

stately two story antebellum residence that contained nine large rooms and was located about a quarter mile south of Pink Hill Road.[81] Kansas' raiders had been eyeing the plantation for some time.

Early on the morning of August 10, Quantrill secretly conveyed information to Andy Walker, one of Morgan's sons, that their farm was going to be attacked by Kansas Jayhawkers. Their purpose Quantrill explained was to plunder Walker's property and steal his slaves. Armed with this information, Quantrill and the Walkers agreed to surprise the Kansas raiders with buckshot when they arrived later that evening. Unbeknownst to the five Kansan abolitionists, Quantrill had conspired to lead them directly into a mortal ambush.[82]

Quantrill and his five Kansas abolitionists journeyed into Missouri in route to "steal" Walker's slaves. The ambuscade was successful leaving one abolitionist killed outright. Of the other four men, two were found the next day hiding in the woods nearby and were summarily executed. The two other men made good their escape back to Kansas.

Following the clandestine decimation of the Kansas raiding party, Andy Morgan escorted Quantrill to the Jackson County Jail in Independence for Quantrill's own protection.[83] The Walker's knew that the people of Independence did not realize Quantrill had converted to the slavery cause. For this reason they felt Quantrill's safety required him to seek sanctuary in the county jail. Eventually the raid became the rallying call for Quantrill and his name

1859 Jackson County Jail. The jailor's residence is visible from the street; the cell blocks ajoin at the rear. Drawing attributed to Virginia Jennings.

---

[81] Eakin, Joanne Chiles. "Recollections of Quantrill's Guerrillas as Told by A.J. Walker of Weatherford, Texas, to Victor E. Martin in 1910." (Independence, Mo.: Two Trails Publishing, 1996), 91-92.

[82] Morgan Walker survived the war but died in 1866. Jackson County (Mo.) Historical Society Archives, Document ID 11F10; Edwards, John M. "Noted Guerrillas." Reprinted with Index. (Shawnee, Ks.: Two Trails Publishing Company, 1996), 59; Houts, 55; Hale, 35-37; and Leslie, 72-73.

[83] Jackson, David W., and Paul Kirkman, "LOCK DOWN: Outlaws, Lawmen & Frontier Justice in Jackson County, Missouri." Sixth Edition. (Independence, Mo.: Jackson County Historical Society, 2012), 37.

would soon become synonymous with the term "terror" in Kansas. By stepping upon the steps of the Morgan Walker Farm, Quantrill became the "lead player" in the drama of the guerilla warfare on the Missouri-Kansas border. Single-handedly, Quantrill would make the events of "Bleeding Kansas seem like child's play by comparison."[84]

Although Quantrill did participate in some guerilla and Confederate activities during the first year of the Civil War, his involvement took on a much more sinister role in 1862. He and other young men along the border had grown weary of the numerous forays by Kansas Jayhawkers and Red Legs and their wanton atrocities, indiscriminate murder and unwarranted destruction of property in Jackson County and other Missouri counties. These Kansans led by such radicals as Col. Charles "Doc" Jennison, James Montgomery, Col. Dan Anthony, General Jim Lane acted with a conscious intent to inflict terror and leave behind burned, ransacked and pillaged farm properties. Quantrill and his new boys had enough; it was time for retaliatory measures to be inflicted upon Kansas.

The outrageous conduct displayed by these Kansans had not only gained the attention of politicians like Blair and Bingham. It soon emboldened Quantrill and his young followers as they began to run "wild" in Jackson County. Quantrill's men in 1862 commenced a series of counter-retaliatory measures including ambushing border towns in Kansas, seizing United States Mail escorts, assailing Union patrols along the Missouri and Kansas border, assaulting steamboats on the Missouri River and burning bridges.[85]

The summer of 1862 witnessed a massive Confederate recruiting effort in eastern Jackson County. These attempts proved to be successful as numerous men who had assembled near the Frances Cowherd Farm enlisted in the Confederate Army or in a militant guerilla organization. The success of the recruitment was in part based upon a desire to seek revenge against Kansas' troops. Another reason was directly related to new regulations imposed by military and state governmental officials upon Missouri citizens. Governor Gamble and Major General Henry W. Halleck ordered every citizen to sign a loyalty oath or post bond to ensure good conduct and behavior. More disheartening was the mandate that required all men between eighteen and forty-five to aid in the preservation of the Union by joining the militia. This decree proved to be the final tonic and incentive for men to enlist in the Confederate Army or to join one of the various guerilla units, including Quantrill in Jackson County. The ranks of the guerrillas and Confederate were beginning to swell.

The swelling of men into rebel and guerilla ranks became fodder for local newspapers. The *Liberty Tribune* in Clay County and the *Kansas City Daily Journal of Commerce* wrote articles about the increased size of Confederate and guerrilla troops in

---

[84] Under Quantrill the James & Younger boys would learn to fight. Goodrich, Thomas. "Black Flag: Guerrilla Warfare on the Western Border, 1861-1865." (Bloomington, In.: Indiana University Press, 1995), 252.

[85] O.R. 8:1:346; Hale, 40.

the region. Some reports even estimated that well over twelve hundred Confederates and bushwhackers had been seen forming in the woods of Jackson County. On Sunday, August 3, 1862, the *Daily Journal* warned that an attack by bushwhackers was imminent and that loyal men needed to organize immediately. Six days later the *Daily Journal* insisted that reliable sources had reported that more than eight hundred rebels under Quantrill and Colonel Upton B. Hays of Westport were threatening Independence and other communities in the area.[86]

Upton B. Hays

These newspaper reports soon proved to be reliable and true. In the early morning hours of August 11, the combined forces of Quantrill guerrillas and Confederates captured Independence, in the first Battle of Independence. Four days later Confederates defeated Union troops at the Battle of Lone Jack in southeastern Jackson County. Unlike the Battle of Independence, Quantrill and his raiders did not take an active role in the Battle of Lone Jack. They did not arrive on the bloodstained streets of the town until the end of the engagement. However, their late arrival convinced Union officers that standing to fight any further could prove disastrous, and they abandoned Lone Jack to the Confederate troops while the bushwhackers rode into the town.[87]

As the summer drifted towards fall, the freshly recruited Confederate troops retreated south towards Arkansas while Quantrill and his followers continued their retaliatory strikes into Kansas border towns. In September they attacked Olathe killing at least ten citizens and capturing over one hundred Kansas troops stationed in the town. Quantrill justified the raid because of the death of a member of the gang who had been executed at Fort Leavenworth. On October 17, Quantrill and his raiders put the torch to Shawneetown leaving in their wake at least seven civilians and fifteen soldiers dead.

---

[86] http://shs.unsystem.edu/newsaper/newspapercatalog/pdf/jackson/pdf. It is said Quantrill went by the name of Charlie Hart in the 1850's. Guerilla William H. Gregg during the Civil War said the bushwhackers called him "Charlie."

[87] For accounts of the First Battle of Independence and the Battle of Long Jack, see, Monaco, Ralph A. II. "Blood on the Streets, the Civil War Comes to Jackson County, August 1862," (Independence, Mo.: Monaco Publishing, LLC and Jackson County Historical Society, 2012); Matthews, Matt, and Kip Lindberg. "Shot All to Pieces, The Battle of Lone Jack, Missouri, August 16, 1862." *North & South* (January 2004): 7: 1; and Hale, 44-46.

R. H. Graham, Colonel of the Eighth Kansas Regiment and Provost-Marshal-General for Kansas, was mortified by the outrages along the border. He blamed the uncontrolled and rampant insurrection and violence upon the clandestine guerilla raids into Kansas and the depredations inflicted in Missouri by "the horde of jawhawers." Graham recognized that he had insufficient military support to curb the mayhem and murder being caused by both sides of the conflict.[88]

The only respite that chilled these persistent attacks and counterattacks was the weather. Toward the end of the 1862 fall season, Quantrill's band departed for warmer land in the south while the guerilla chieftain traveled to the Confederate Capitol in Richmond, Virginia. The purpose of his mission was to meet with the Confederate Secretary of War James A. Seddon to receive a commission in the Confederate Army.

The Confederate Congress in April 1862 had enacted the Partisan Ranger Act. Confederate President Jefferson Davis promptly signed it into law. This Act authorized the organization and promotion of unconventional warfare designed to annoy the enemy and to cut off, harass and disturb Federal troops, scouts, communications, telegraphs and supply lines. Guerilla activity was specifically sanctioned under the new Partisan law and Quantrill wanted to receive a colonel's commission in a regiment in the Partisan Ranger's.

Seddon had grave reservations about Quantrill. He had received too many rumors and various reports about Quantrill's ruthless methodology of waging war along the borders of Missouri and Kansas. Seddon feared that by bestowing an official Partisan Ranger commission to Quantrill the North would retaliate against the South. Quantrill emphatically justified his methods. He insisted upon the value of waging a campaign that covered the field in the blood of his enemies and making surrender impossible.[89] There is no official record or extent account of Quantrill ever receiving a Confederate commission; this did not deter Quantrill. From that time forward, he proclaimed he had been officially appointed as a colonel under the Partisan Ranger Act.

The year of 1863 would become the most sensational and haunting time-period of Quantrill's reign of terror. His paramilitary Partisan Rangers would grow in numbers, reportedly as high as five hundred men. Quantrill's designs and plans as he reentered the war on the border were to infest the thickly wooded areas and deep ravines of the western counties of Missouri. From there he would launch his guerrilla attacks relying upon the support of the many inhabitants of these areas who had previously "aided in every way" his war policy. He knew that Union forces had limited manpower and would be virtually stymied in their attempts to capture or drive his compatriots out of the region.[90] Quantrill was desirous of retaliating against the Federal forces, and principally Kansas Jayhawkers

---

[88] O.R. 22:1:335-336; and Castel, "Frontier State," 103-104.

[89] *New York Times Opinion:* OPINIONATOR | DISUNION: 'Another Name for Death', Nicole Etcheson, March 24, 2013, http://nyti.ms/Y6G3b.

[90] O.R., 22:1:15.

and Red Legs, for what he considered to have been their murderous, destructive and atrocious willful acts against innocent people and towns.

As the spring season opened, reports were rampant that like a cyclone "Quantrill is back." Loyal citizens and Union officials were greatly afraid that Quantrill was preparing to inflict the most revolting hostilities along the western border of Missouri and into Kansas. The year 1863 would produce a storm of tumultuous activity by Quantrill and his Missouri guerrillas in Jackson County and the District of the Border never seen in all the previous war years dating back to 1854 combined.

CHAPTER V

General Schofield and District of Frontier

In September 1862, Lincoln replaced General John McAllister Schofield and turned the unstable Department of Missouri to General Samuel R. Curtis. The District Curtis assumed encompassed Missouri, Kansas, western Arkansas and the Indian Territory (Oklahoma today). Curtis' reign would last only eight months. Throughout the duration of his tenure, numerous issues diminished and undermined his ability to perform his military duties effectively. He became embroiled in direct confrontations with Governor Hamilton Gamble on such issues as use of the state militia, assessment of taxes, treatment of southern sympathizers and emancipation of slaves.[91]

In the end, President Lincoln grew exhausted "beyond endurance" by both sides in the Missouri political controversy. There was nothing within Lincoln's powers to remove Gamble, but he could take measures to impact the unsettling discord. Having grown weary over these never-ending squabbles, Lincoln took matters into his own hands. On May 24, 1863, Curtis was removed and Schofield was re-instated as the commander of the District of the Frontier.[92]

John McAllister Schofield

Schofield's background had well prepared him for a career in the military. He was born on September 29, 1831, in Geary, New York and graduated from the United States Military Academy in 1853, ranked seventh in his class. He briefly served in Florida after his graduation before returning to West Point to teach philosophy. The year before the Civil War broke out Schofield had taken a leave of absence from the army to teach at Washington University in St. Louis, Missouri.

[91] O.R. 22:1:271 and 281; and http://mocivilwar150.com/history/figure/291.

[92] O. R., 22:1:4 and 12.

At the outset of the War, Schofield served as the mustering officer for the State of Missouri. He was then promoted to major of the 1st Missouri Infantry and served as chief-of-staff under General Nathaniel Lyon. After Lyon was killed during the Battle of Wilson's Creek (near Springfield, Missouri) on August 10, 1861, Schofield assumed command of the Union troops and promptly and appropriately retired his outnumbered men from the battlefield.[93] On November 21, 1861, and about three months following the Federal loss at Wilson's Creek, Schofield was promoted to Brigadier-General in charge of the Union militia in Missouri.

Lincoln on May 24, 1863, then elevated Schofield to the rank of Major-General and reappointed him as the commander of the Army of the Frontier. Schofield's headquarters was located in St. Louis and some two hundred and fifty miles east of the guerilla war along the Missouri and Kansas border.

At the time Schofield resumed command, Lincoln warned him of the dangerous political intrigue and competing interests he would discover in Missouri, and most particular along the western border. In Lincoln's letter to Schofield on May 27, he felt it incumbent upon him to explain the reasons he had removed Curtis and renamed Schofield. Curtis, Lincoln insisted, had not been removed because of any wrongful act or deed. The President made the decision because of the "pestilent factional" fights within Missouri. Lincoln also reminded Schofield to exercise his own judgment "and do right for the public interest." Schofield's principal duty Lincoln explained was "to repel the invader and keep the peace." Lincoln concluded his letter to Schofield with sage political advice and direction: "If both factions, or neither, shall abuse you, you will, probably, be about right. Beware of being assailed by one and praised by the other."[94]

Schofield was not only cognizant of Lincoln's warnings. He was fully aware of the serious difficulties General Curtis had experienced in the District. Schofield also knew his problems were not limited to potential political backstabbing and wrangling. He faced the daunting challenge of meeting the horrific problems on the borders of Missouri and Kansas with insufficient troops. On June 2, Schofield had been ordered to dispatch nearly twelve thousand troops from his department to reinforce General Grant's campaign to capture Vicksburg on the Mississippi River. This reduction in force greatly compromised Schofield's ability to properly garrison and to fully protect the border from guerilla insurrectionists on both sides of the state line.

Schofield knew the war along the western border had continued unabated, remained tremendously unstable, and was virtually out-of-control. With limited troop availability, he chose to realign the border district. He knew his reduced military force was both insufficient and too ill equipped to combat the persistent guerilla activity in the

---

[93] By his death, Lyon became the first Union general killed in battle during the Civil War. Later in his military career, Schofield received the "Medal of Honor" for "conspicuous gallantry" at the Battle of Wilson's Creek.

[94] O.R. 22:1:293.

region. To meet these military difficulties, Schofield needed a cavalry not foot soldiers. He ordered the Missouri and Kansas veteran infantry regiments to be mounted. The bushwhackers rode strong horses and Schofield's men had to be as equally equipped to meet their challenges.

On June 9, Schofield then realigned the western border of Missouri and the eastern regions of Kansas. He divided the former District of Kansas into two separate districts. The first district embraced the northern portion of Kansas and several border counties of Missouri south of the Missouri River and north of the Osage River; the second district consisted of the southern portion of Kansas, the Indian Territory and Western Arkansas. Two days later General Schofield placed Major-General James G. Blunt in command of the latter district, headquartered at Fort Gibson.

The new District of the Border contained two tiers of western Missouri counties. The Tier I counties included Jackson, Cass, Bates and Vernon, north of the Osage River; Tier II counties consisted of Lafayette, Johnson, and Henry. While these new districts were novel in their scope and purpose, nothing was as eye opening as to whom Schofield selected to assume command of the new Border District. To the shock, surprise and dismay of almost everyone, Schofield named Brigadier-General Thomas Ewing, Jr. as the military commander with headquarters in the Pacific House at Kansas City.[95]

To General Blunt his reassignment was a slap in the face. He loathed the thought of being isolated in the remote portions of southern Kansas, western Arkansas and the Indian Territory. He was further incensed that Ewing an upstart had been assigned command of the more important District of the Border. Military protocol supported Blunt's fury. Ewing had no military training; his army career had been less than nine months; his appointment to colonel was political; his promotion was only to Brigadier-General; Blunt outranked him as a Major-General. As Blunt stalked off to his new command, General Ewing took charge of the Border. Schofield could only hope that this new District and its fresh general would be able to restore some calm to this long-term, tormented section of the country. The backlash Schofield received from Blunt, Lane and others over Ewing's appointment would pale in comparison to the negative repercussions he would garner within less than seventy-five days.

[95] O.R. 22:1.15; and, Niepman, Ann Davis. "General Orders No. 11 and Border Warfare during the Civil War." (January 1972): 66: 2: 187.

CHAPTER VI

"A Little God"
Provost Marshal Abram Comingo

On June 2, 1863, Abram Comingo was appointed the 6th Congressional District Provost Marshal, with the rank of Captain. Based upon the 1860 Census, Missouri's population had increased to the highest mark in its history. These new totals enlarged the state to nine Congressional Districts and eleven electoral votes. Austin A. King represented the 6th District in Congress, which included Jackson County.

The 6th District's jurisdictional boundary was expansive. It consisted of the following counties south of the Missouri River: Jackson, Lafayette and Saline. The other counties were north of the river: Clay, Platte, Caldwell, Ray, Carroll and Chariton. Only two of these counties were located within Schofield's new District of the Border, Jackson and Lafayette.[96] The appointment of Comingo as Provost Marshal gave him the military and civil authority over all of Jackson County, including Independence—his adopted home since 1848. He would soon recognize that the power of his new post was controversial and imposed upon him the conflicting burden of balancing his official duties against his personal beliefs and relationships.

The rule, command, magnitude, reign and power of the provost marshal during the Civil War is best defined by historian Joanne Webb Chiles Eakin under the nomenclature: "The Little Gods."[97] By pure definition, the duties of the provost marshal appeared to be

---

[96] The other Missouri counties within the Border District (Cass, Bates, Henry and Vernon) were located in the 5th Congressional District, "Proceedings of the Missouri State Convention Held in Jefferson City." (St. Louis: George Knapp & Co., 1862), 15.

[97] The name, aptly put, as found in: Eakin, Joanne Chiles. "The Little Gods: Union Provost Marshalls in the Civil War." (Independence, Mo. Two Trails Publishing, 1996).

magisterial. In application, the position held omnipotent dominion. The provost marshal was empowered to administer and enforce civil and criminal law. He was vested with regulating public places, conducting searches and seizures, making warrant less arrests, issuing passes to citizens, recording and investigating complaints, taking custody of prisoners and deserters, maintaining records of available troops and handling all other issues of law and order brought to his attention. In practice these assigned duties became more than regulatory assignments but evolved into dictatorial and "godlike" authority.

> *No graver question was ever considered by this court; nor one which more nearly concerns the rights of the whole people; for it is a birthright of every American citizen charged with crime to be tried and punished according to law . . . If there was a law to justify the military trial it is not our province to interfere; if there was not, it is our duty to declare the nullity of the whole proceedings.*[98]

The position and enumerated power of the U.S. Provost Marshal evolved over the course of the first two full years of the Civil War. It began on July 18, 1861 when Union General Irvin McDowell issued General Order No. 18. As the Union military campaign moved southward, McDowell delineated the authority of the provost marshal in the Army of Northeast Virginia. In 1862, General George B. McClellan assumed command of the newly formed Division of the Potomac (which included the departments of Northeast Virginia and Washington). He soon established the position of provost marshal within a field army. From that time until the war's end, each division, brigade, and corps of the Union Army had a provost marshal. McClellan installed the provost marshal and his guard the chief responsibility of the preservation of order. Before McClellan's order, the position held "duties which had not before in our service been defined and grouped under the management of a special department." The provost became responsible for suppression of marauding, deterrence of depredations upon private property, prevention of straggling, suppression of gambling houses or other establishments prejudicial to good order and discipline, and supervision of hotels, saloons, resorts and amusement venues.

The next revision to the post took place in September 1862. At that time, the Adjutant General's office issued General Order No. 140. This new order created "Special Provost Marshals" for each state. These special marshals were given assignments to investigate charges or acts of treason, and arrest deserters, spies, and disloyal citizens.

The final reorganization of the position of provost marshal was made by the United States War Department in early 1863. Under this order the position of the Special

---

[98] *Ex Parte Milligan*, 71 U.S. (4 Wall.) 2 (1866). The decision of the United States Supreme Court in Milligan struck down certain military tribunals including the dominion of the provost marshal. However, the High Court's ruling failed to address the overwhelming authority of the provost during the Civil War. The Court did not weigh in on this thorny constitutional matter until the war had ended.

Provost Marshal was eliminated and was replaced by a new chain of command. The national office of Provost Marshal General "(PMG") was established with headquarters in Washington. Each state had an Assistant Provost Marshal General ("APMG") who held control over the state and reported directly the PMG. Within each state, a District Provost Marshal ("DPM") was appointed to enforce military order within each Congressional District in the state. The Congressionally appointed DPM reported directly to the APMG of that state. Finally, within each Congressional District a deputy provost marshal was named for each county located within the District. The deputy provost marshal of the county was then accountable to the DPM and then up the chain of command to the APMG and ultimately to the PMG in Washington.[99]

Under this new act, the duties of the APMG, DPM and the county provost marshal's had remained much the same. In addition, these officers were assigned the jurisdictional power to maintain troop discipline, assume custody of prisoners and deserters, administer punishment, and suppress any disturbances caused by troops or individual soldiers. The DPM and his county provost marshals was empowered to conduct searches and seizures; take custody of deserters; issue passes to citizens; hear complaints of citizens; arrest persons even on mere suspicion; and incarcerate without conducting a trial or granting a writ of habeas corpus.

Every state was assigned an APMG regardless of the level of active warfare within that state or congressional districts. Where there was little or no military or guerilla activity, it was not uncommon for the provost marshal law to be suspended with the ordinary judicial process remaining in effect. Districts where there was warfare or fighting, the provost marshal's duty was expanded to include limiting marauding against citizens, preventing stragglers on long marches, and generally suppressing gambling or other vices not conducive to good order and discipline.

In regions like Jackson County in which there were vigorous guerilla insurgencies, the DPM and his county deputy provost marshals acted independently and without any real supervision. They were authorized and empowered to dispense with the rules of civil or criminal procedure. In many cases they simply ignored Constitutional guarantees all together. Overall in such areas a provost marshal acted as chief of police, judge, jury and executioner. He held military power unencumbered by the regular checks and balances of ordinary civil or criminal jurisprudence that had been guaranteed and

[99] The new order and chain of command of the Provost Marshal was created by the United States War Department by an act of March 3, 1863 (12 Stat. 731). This same War Department measure created the positions of Enrollment and Disbursing Divisions, Adjutant General's Office (AGO); Surgeon General's Office. The United States War Department by an act of March 3, 1863 (12 Stat. 731) abolished on August 28, 1866, by General Order No. 66, War Department, August 20, 1866, implementing an act of July 28, 1866 (14 Stat. 337).

ensured by the United States Constitution.[100] Into this new dictatorial position, Abram Comingo was appointed DPM in late spring 1863.

The origin of his surname Comingo originated from the name Comingore—which was Pennsylvania Dutch. He was born in Harrodsburg, Mercer County, Kentucky on January 9, 1820. Harrodsburg was the oldest settlement in the Commonwealth of Kentucky. Comingo was principled and reared in the Presbyterian Faith at the old "Mud Church" in the town of his nativity. He attended Centre College in Boyle County, Kentucky and studied law and in 1847 he was issued his license to practice law.

William Chrisman

In 1848, Comingo married Lucy Jane Morton of Mercer County. Lucy was the first cousin of Lucy Lee and both women were descendants of the famed Lee Family of Virginia. Lucy Lee had been courting William Chrisman, a life-long friend of Comingo. The two couples were married in a double wedding ceremony held in Harrodsburg; Abram married Lucy Morton and William Chrisman married Lucy Lee. Each man served as the bondsman for the other. In the year of their marriages, the Comingo's and Chrisman's relocated to Independence and formed a law partnership with Sam Woodson. The firm became known as Woodson, Chrisman and Comingo.

The decision to include Woodson in the new business venture was sound. Woodson had been born in Kentucky; studied law at the University of Transyvalania in Lexington, Kentucky; and had settled in Jackson County around 1840. Woodson had experience practicing law in the area and was well respected in the community. By 1850, Woodson held the first federal postal contract between Independence, the county seat of Jackson County, and Salt Lake City.[101]

Sam Woodson

The law firm of Woodson, Chrisman and Comingo during the decade of the 1850's was considered one of the leading political firms in western Missouri. Their office was located in the northeast corner of the county courthouse on the square in Independence.[102] All three partners had been reared on the

[100] Fellman, Michael. "Inside War: The Guerilla Conflict in Missouri During the American Civil War." (New York: Oxford University Press, Inc., 1989), 42; http://www.sos.mo.gov/archives/provost/#history.

[101] The Sam Woodson house built in the 1850's still stands on Winner Road—it was known as Lexington Road during the 19th Century, in Independence.

political philosophy of the Whig Party, and were staunch adherents to internal improvements, gradual emancipation and a strong central banking system. By the Kansas-Nebraska Act, the Whig Party slowly and irrevocably divided and eventually dissolved. Comingo and his law partners pledged their political allegiance to the American Party, commonly known as the Know Nothings, before finally settling upon the Democratic Party by the end of the decade. Comingo served two terms as Mayor of Independence (1852 and 1857), and Woodson served two terms in the U.S. Congress (1856-1861).

Two children were born of the Comingo marriage. Sally was born in 1851, and her brother John about two years later, but he died in infancy. At the time Comingo was appointed provost marshal, the family resided at 6th and S. Main in Independence, in a house named the "Gudgell Place." Their home was conveniently located only four blocks south of Comingo's office in the county courthouse. In keeping with this Whig mindset, the Comingo's did not own slaves. Rather, they leased from Joe Hoecker two slaves, a man and a woman. Hoecker a slave owner in Jackson County leased his two slaves on an annual basis and Comingo paid Hoecker about one hundreds per year for each slave.[103]

Comingo's office was in the Jackson County Courthouse

By the time secessionism broke out across the South, a similar division occurred within Comingo's law firm. Both he and Chrisman joined the ranks of the Constitutional Unionist Party. Their law partner Congressman Sam Woodson, a staunch southern

---

[102] Jackson County Court Records, July Term, 1 July 1862. The county records for that date reads as follows: "Whereas Chrisman and Comingo attorneys at law, have heretofore, paid to the county the sum of One Hundred Dollars for the rent of the office now occupied by them in the N.E. Corner of the Courthouse and the said Chrisman and Comingo of having paid the rent for said room up until July the 1st 1861 the Court doth order that the Treasurer receive from said Chrisman and Comingo the sum of Fifty dollars for one years rent up to the 1st day of July 1862."

[103] Sallie was born in 1851. She later married Dr. Jefferson D. Griffith on January 29, 1880 in Independence, and one of their children was named after her father, A. Comingo Griffith, (who eventually like his father became a physician). John died in 1853 and he was buried on June 12, 1853 (later his body was re-interred at Elmwood Cemetery in Kansas City). According to the 1850 and 1860 U.S. Censuses the Comingo's had two slaves living in their household whom they leased from Joseph Hoecker. An image of the Comingo home as it appeared in 1877 is in the "Illustrated Historical Atlas of Jackson County, Missouri."

supporter, left the firm and relocated with his wife Margaret J. (Ashley) to Texas when his term in Congress ended in March 1861.

During the presidential election of 1860, Comingo and Chrisman (like Bingham) rallied behind Senator John Bell of Tennessee. It was the political conviction of the Constitutional Unionists that the principals of the Founding Fathers were paramount and secession was illegal. The preservation of the Union outweighed the benefits derived from slavery or secession. Comingo did not believe Lincoln's election signaled the automatic dissolution of the United States. He remained steadfast to the platform of the Constitutional Unionist Party and the party's firm opposition to secession sentiments.

Governor Claiborne Fox Jackson did not share the views of Comingo and other like-minded Missourians. The Governor was certain that Missouri would join her sister southern states and secede from the Union. He called for a constitutional convention to consider the relationship between the State of Missouri and the Federal government. Three delegates were elected from each of the state's twenty-eight senatorial districts along with eight from St. Louis, the twenty-ninth Senatorial District, for a total of eighty-nine representatives. Jackson County was part of the 14th Senatorial District. In February 1861, Comingo along with Judge James K. Shelley of the Jackson County Court of Common Pleas and Robert A. Brown, a large slave owner from Harrisonville, Cass County, were elected to represent the 14th District at the Convention.[104]

On February 28, 1861, the Convention opened in the Cole County Courthouse in Jefferson City. Upon motion of Sample Orr of Greene County, Hamilton Gamble of St. Louis was named the temporary chairman until a permanent chairman was elected. Eventually, Sterling Price of Chariton County was voted President of the Convention. The choice of Price seemed obvious. He had served as a general in the Missouri State Militia during the Mexican War, and as governor of the state from 1852-1856. Comingo and the other delegates from the 14th Senatorial District cast their support for Price.

After the session's second day, the members approved relocating the Convention from the State Capitol to St. Louis. This was based upon fears and apprehensions the delegates' had about Jefferson City—which they saw as too unstable, unsafe and volatile. For these reasons, the members voted to relocate to the Mercantile Library Hall inside the courthouse in St. Louis on Monday March 4.[105]

Over the course of the next few weeks, questions of how to avoid a civil war, whether states had the legal right to secede from the Union and if Missouri should join her sister southern states and depart from the Union consumed the debate. On March 15,

---

[104] "Journal and Proceedings of the Missouri State Convention Held at Jefferson City and St. Louis," (St. Louis: George Knapp & Co., Printers and Binders, March 1861), 10 (hereinafter cited as "Journal & Proceedings"); and, Union Historical Company. "The History of Jackson County, Missouri." Originally published by Birdsall, Williams & Co., 1881. Indexed edition by Felix Eugene Snider. (Cape Girardeau, Mo.: Ramfre Press, 1966), 467.

[105] Journal & Proceedings, 18.

the thirteenth day of the Convention, Comingo addressed the body. He asked and answered the salient question of the day: "But what shall Missouri do at this time? Shall she secede at this time? No."[106] Four days later an amendment was offered during the debate that called for a vote on the question of secession. Comingo and eighty-seven other members of the body, including future Confederate General and Convention Chairman Sterling Price voted in opposition to the amendment.[107] Three days after the vote on the secession amendment, the Convention adjourned on March 22.[108] Missouri was not to join her sister southern states in the rebellion.

The work of the State Convention did not conclude when it voted against secession on March 15. Two additional sessions were held in 1861. For eight days, from July 22 through July 30 delegates assembled in Jefferson City, and again for eight more days from October 11 through October 18, 1861 in St. Louis. Comingo did not attend either session.[109] However, when the Convention reconvened the following year in Jefferson City on June 2, 1862, Comingo took an active role in the debate over such issues as the loyalty oath, taxation and formalization of the redistricted Congressional District lines based upon the 1860 Census.

During the June 1862 session of the Convention, Comingo articulated practical and moderate approaches in confronting the war. On taxes, he proposed in "good faith" an ordinance releasing "the citizens of Cass, Jackson and Bates counties from the payment of their State taxes for the years 1861 and 1862." His motive was predicated upon the tragic realities these counties had experienced during that time period. He set forth several historical facts to support his position: the government had been unable to protect the citizens; they had become "greatly indebted" and "almost every man in those counties is more or less in debt;" their 'great indebtedness" precluded them from paying "the taxes which are due the State of Missouri;" the "people of these counties have suffered more from the violence of the times than any other counties" in the state; their property had been greatly "depreciated" and "diminished" in value—not even the once affluent was able to pay taxes; and "an immense amount of property" had been seized, stolen and removed.[110] Comingo then artfully described conditions in these three counties to justify the exemption from taxation.

---

106 Journal & Proceedings, 149.

107 Journal & Proceedings, 216; and Monaco, "Blood on the Streets," 6. Fifty-one year old George Y. Bast, a physician and farmer, born in Kentucky and moved with his parents to the Missouri River Valley in Montgomery County, cast the lone vote in support to the amendment of secession.

108 Journal & Proceedings, 264-269.

109 "Journal and Proceedings of the Missouri State Convention, Jefferson City, Second and Third Sessions." (St. Louis: George Knapp & Co., 1861).

110 "Proceedings of the Missouri State Convention Held at Jefferson City and St. Louis." (St. Louis: George Knapp & Co., Printers and Binders, June 1862), 24-27.

> [F]*rom time to time, immense cavalcades of mules, horses, wagons, and almost every imaginable kind of property, was removed from these counties out of the State and it will never be returned. For miles and miles the road would be lined with the property of the citizens of these counties which had been removed from them by force and violence, and which has never yet been returned and never will be...*
>
> [T]*hey have been of the very necessaries of life. Their houses have been entered, and the silver ware, jewelry, bed clothing, and the clothing with which they covered their nakedness, have been taken from them, and they have been made desolate. Their homes have been made desolate; the houses which sheltered them from the storms have been burned, they being first driven from them in midwinter, and left destitute of the very necessaries of life. The value of their real estate has thus been greatly diminished.*[111]

Comingo passionately begged and implored his colleagues to travel through these counties. He defied "any man to go twenty miles in any direction without seeing incontestable evidence of the violence." He concluded his argument with the assertion that at least "one thousand families" had lost everything. Others had been swept by the "bosom of destruction, so that, in thousands of instances, no crops" could be raised as fences had been destroyed; they lacked horses and "implements of agriculture" to even conduct business. These conditions Comingo warned had placed a "duty" on the delegates "to pass an ordinance exempting these people from the payment of taxes for 1861-62." Comingo's efforts at least forced the referral of his tax relief request to the Committee on Finance for further deliberation.[112]

Comingo next demonstrated his more judicious viewpoints when he addressed the issue of the loyalty oath then under reconsideration. The delegation within the radical faction of the Republican Party had proposed to greatly expand the standards of the loyalty pledge that had been adopted by the Convention and General Halleck the previous year. The new oath would require all Missourians to submit to a loyalty oath in order to vote regardless of their previous stance on secession or the Union. Comingo and other moderates were troubled by these expansive conditions on the franchise right. Comingo insisted that the proposed bill violated the *ex post facto* laws of the Constitution.[113]

[111] "Proceedings of the Missouri State Convention Held at Jefferson City." (St. Louis: George Knapp & Co., June 1862), 25.
[112] Ibid, 27.
[113] An *ex post facto* law renders an act penal or criminal which was not a crime when committed; makes the grade of an offense greater than when committed; attaches additional punishment to an offense already committed; or punishes an offense in a manner or to an extent in which it could not, by existing law, have been punished when committed. "Journal of the Missouri State Convention Held at Jefferson City." 27-39,

Comingo artfully and skillfully weaved his constitutional argument in opposition to the language proposed by radicals.

> *We, sir, are attempting to do what all the courts of the country, aided by all the Legislatures, could not do; we are attempting to punish before conviction, without trial, without any of the established forms of law, and to an extent and in a manner wholly inadmissible when the supposed guilt was incurred...*[114]

Comingo proposed a compromise to require only those holding public office or seeking such office to submit to a loyalty oath. He insisted that requiring all citizens to take the oath, as a condition precedent to the right of franchise was onerous and unconstitutional. The more moderate proposal offered by Comingo was overwhelmingly rejected by a vote of forty-four to fourteen.[115] The vote further revealed that the radical delegation had assumed dominion over the Convention and had demonstrated an overt goal to inflict extreme measures in combating the rebellion.

Abram Comingo

As the year 1863 opened, Comingo experienced first-hand the intensity, insanity and rabidity of the climate in Jackson County. He also witnessed the complete collapse of the judicial process in the county. He was equally aware that conditions had deteriorated to the point that no one confidently knew who was a friend or foe. It was during these troubled times that Comingo and Chrisman by mutual consent dissolved their law partnership.[116]

By the late spring of 1863, the new provost law had been implemented across the country. Comingo's political experiene, opposition to secession, knowledge of Independence and Jackson County and conciliatory efforts at the state convention were integral considerations leading to his appointment as Provost Marshal for the 6th Congressional District. The provost marshal system was fundamental to the successful

(St. Louis: George Knapp & Co., St. Louis, Missouri, June 1862), 27-39; and "Proceedings of the Missouri State Convention Held at Jefferson City." (St. Louis: George Knapp & Co., June 1862), 119.

[114] "Proceedings of the Missouri State Convention Held at Jefferson City." (St. Louis: George Knapp & Co., June 1862), 120.

[115] "Journal of the Missouri State Convention Held at Jefferson City." (St. Louis: George Knapp & Co., June 1862), 29.

[116] Union, 870.

operation of martial law, and Comingo had taken center stage within this new military structure in Jackson County.[117]

Thirteen days after his appointment Comingo, now Provost Marshal Captain Comingo, returned to Jefferson City to represent his senatorial district at the state convention.[118] Within three months, he would find himself embattled over Order No. 11 and the ominous responsibility of enforcing an order in which he would eventually openly oppose.

---

[117] Parrish, William E. "A History of Missouri." Volume III, 1860-1875. (Columbia, Mo.: University of Missouri Press, 1973), 65.

[118] "Journal and Proceedings of the Missouri State Convention Held at Jefferson City." (St. Louis: George Knapp & Co., June 1863).

CHAPTER VII

General Thomas Ewing and His Return to Kansas

On June 9, 1863, Brigadier-General Thomas Ewing, Jr. took command of the new District of the Border. The assignment would have been a daunting task for anyone, but for a man considered by western Missourians as a Kansan, it would become most disturbing. The citizenry Ewing encountered consisted of every combination and permutation of political philosophy and loyalty. Those who supported the Union were divided amongst themselves: Unconditional Unionists (supportive of all measures to successfully prosecute the war); Conditional Unionists (against secession but against war); pro-Union but anti-federal troops; pro-Union, pro-federal and pro-Missouri militia but anti-Kansas troops; or pro-Union and willing to allow the southern states to secede. Southern supporters were as equally divided: pro-southern but anti-secession; pro-southern and pro-secession; pro-southern but anti-bushwhackers; or "fire-eating" southerners who detested the Union. Between these diverse positions were those who simply wanted to remain neutral and uninvolved. Many others had become unwilling participants who directly or indirectly were compelled to give support or aid to rebel guerrillas or jayhawking raiders out of fright and fear of both sides.

Thomas Ewing, Jr.

A population of former war veterans who had returned to their homes and farms was as equally problematic to Ewing as was the political mixtures he faced. These men consisted of those who had been discharged, wounded or deserted. Some had been in the former Missouri State Guard, Confederate Army or in the Union Army. From both armies, many had joined various bushwhacking organizations (rebel or Union). There were former Union soldiers turned rebel guerrillas; former Confederate or guerrillas turned Home Guards (formed to defend local towns and cities against rebel intrusion—some were loyal to the Union and some known as "Paw Paws" were not).

The division between loyalties was as equally diverse as the infestation by ruffians, guerrillas, bushwhackers, Jayhawkers, Red Legs and criminals. No one was safe. Southern sympathizers feared the sword of Kansas and federal troops. Union loyalists trembled with fear of another guerrilla raid. Most daunting was the stark reality that civilians did not know whether mounted troops approaching their farmhouse were rebels or federals. The wrong answer could make a difference between life, death or destruction of property or not.

Ewing's predicament was not limited to a frightened citizenry of diverse perspectives. He faced political challenges as well. The Radical Republicans in Kansas openly opposed his appointment. Blunt's indignation was shared by Lane. Missouri politicians were equally critical of a Kansas man taking command of any portion of Missouri, let alone the area on the western edge—where the people had once participated in the slavery controversy during the days of Bleeding Kansas. They reasonably feared that Ewing, the Kansan, would impose his wrath and retaliatory measures upon Missouri.

The internal factional squabbling and political dialogue over his appointment was nothing more to Ewing than skirmishing issues. They were substantially subordinate to the real problem. Ewing had to corral a guerilla force that had been waging war in an unconventional style unequaled anywhere in the United States. Of these problems, the central and most compelling was how to impede, interfere, control and curb rebel guerilla activity along the border. General Schofield conceded that preventing the assemblage of Quantrill rebels during the "summer season" was impossible. Ewing was convinced that it would require at least five times his existing forces to have any semblance of an opportunity to restrain and control the guerrillas.[119] Such was the district Ewing had assumed command on June 9, 1863.

Ewing's rise in the military ranks was unquestionably facilitated by his father's strong connections in Washington. Ewing was born in Ohio on August 7, 1829, to a powerfully influential family steeped in the tradition of Henry Clay and the philosophy of the Whig Party and its commitment to internal improvements. He was one of seven children born of the marriage of Thomas Ewing, Sr. and Maria (Boyle) Ewing. His mother was a devout Catholic, while his father was Presbyterian. His father accepted his wife's Catholic Faith only because she had insisted upon it. Three of their children became devout Catholics, while Tom, Jr. developed a more deistic Christian theology similar to that of his father's.

---

[119] O.R. 22:1:491.

In the year Tom, Jr. was born, his parents took into their family a nine-year old foster child: Tecumseh Sherman. The young boy's father Charles had died suddenly and his mother Mary was unable to care for the child. His father's death compelled young Sherman's mother to divide her eleven children with neighbors and other family members. Mrs. Ewing approved of young Tecumseh moving into the Ewing household only if the lad was given a proper Christian name. For this reason, redheaded Tecumseh was given the name William Tecumseh Sherman.

William Tecumseh Sherman

Whig politics and the practice of law had been chosen by the elder Ewing as his career paths. The Ohio General Assembly elected him to the United States Senate in 1831, and before his term ended, he ensured that his foster son Sherman received an appointment to West Point. In 1841 President William Henry Harrison appointed Tom Sr. as Secretary of State. His post was temporary as he and other members of Harrison's cabinet resigned in masse when the President died.

The Whig fortunes of the Ewing family were re-sparked by the presidential election of General Zachary Taylor in 1848. The family patriarch returned to the cabinet, becoming the first Secretary of the Interior. He soon moved the family to Washington City.[120] From that point forward, Tom, Jr., then only twenty years old, became fully entrenched into his father's legal and political career path. Young Tom soon became the private secretary to the new president.

Within a year, the younger Ewing was given a position in the U.S. claims office, a post within the Department of Interior. This assignment afforded him the opportunity to witness first-hand the great Senatorial debates of Henry Clay, Daniel Webster and John Calhoun over the Compromise of 1850. During this time-period, Ewing's father won a large lawsuit earning him a fee of $100,000. The family patriarch invested his earnings in property located in St. Louis and titled the land in the names of his sons and daughters. The future of Ewing was well under way and the dynasty of his family was fully entrenched at the nation's Capitol and in the Midwest.[121]

By the time the Kansas-Nebraska Act became the law of the land, twenty-five year old Ewing had developed an intense antislavery political philosophy, but he was no

[120] The United States Capitol was called Washington City. Washington did not receive the designation as the District of Columbia, or Washington D.C. until the early 1890s.
[121] Smith, 17-18.

abolitionist. Ewing had attended Brown College and law school in Cincinnati, but he did not finish the curriculums in either school. He was more interested in the fury-taking place in Kansas. He desired to see the new territory develop into a free state.

Ewing married Ellen Cox on January 8, 1856. With his new bride and a strong ambition to enter the political turmoil in Kansas, he and his brother, Hugh (also an attorney) decided to seek their fortunes in the new territory. They settled in Leavenworth, Kansas Territory, a river town on the west banks of the Missouri River directly across from the slave town of Weston, Missouri.

There were only a few lawyers in the Kansas Territory when the Ewing brothers arrived. They knew their father's political clout had a reputation that preceded them. The Ewing's speculated in land development and practiced law—later they were joined in their practice by their foster brother and now brother-in-law, William Tecumseh Sherman and later by Daniel McCook, Jr.[122] Despite having had monetary successes in Leavenworth, Ewing chose to actively participate politically in the Kansas free-state movement. Because of his name, other supporters of the cause turned to Ewing for advice and direction. He encouraged them to oppose the pro-slavery agenda, but he protested against warfare to settle the dispute, professing that their mission for a free state should be won at the ballot box.[123]

During the days of "Bleeding Kansas," Ewing grew to believe that abolitionists led by such men as John Brown and James Montgomery were as dangerous to Kansas as were pro-slavery advocates. He held fast to the Whig mantra of gradual rather than immediate emancipation.[124] Despite his political differences with Lane, Ewing drew upon his successful defense of Lane in the Jenkins murder case in 1856 to temper Lane's political differences. Ewing also used his prior relationship with Lane to garner the Grim Chieftain's support for an election bid as Chief Justice of the Kansas Supreme Court. Ewing's election bid was successful—a position in which he assumed when Kansas became a state on January 29, 1861.

Within months of the commencement of the Civil War, Ewing found himself as the only member of the old law firm not engaged in the rebellion. His brother Hugh had become a colonel in the Thirtieth Ohio Volunteers; his foster-brother/brother-in-law William Tecumseh Sherman was serving with General Ulysses S. Grant in Tennessee; and Daniel McCook, Jr. had assumed a staff position with his brother Alexander McDowell McCook in the Army of Ohio. Ewing had eyed the senate seat but that had gone to Jim Lane. As the war's first year concluded, Ewing had grown anxious to join his

---

[122] The Leavenworth law firm would eventually be known as Sherman, Ewing & McCook (William T. Sherman, Thomas Ewing, Jr. and Daniel McCook, Jr. were the eventual partners in the firm).

[123] Smith, 20-66. Ewing's political stance differed drastically from the vulgar war mongering voice and tempestuous spirit of Jim Lane.

[124] Smith, 68.

former law partners in the war effort. He had grown weary and impatient with the tedious work on the Kansas high court.

When President Lincoln called for another three hundred thousand volunteers in the summer of 1862, Ewing let it be known he was interested in vacating his Supreme Court seat to become part of forming a new Kansas regiment. He leaked word to Lane of his interest in leaving the bench for an officer's commission. Lane supported the transition; his motives in doing so were pure Lane. He could control Ewing better in the military than he could while Ewing performed judicial duties.[125] Lane also knew Ewing was a political creature and potential senatorial opponent. An officer's commission would prevent Ewing from seeking Lane's senate seat—it was too fresh in Lane's mind to have forgotten his own practical problems the previous year when he had attempted to hold a general's commission and serve in the senate at the same time.

Ewing knew the Eleventh Kansas Volunteers would soon be established, and he wanted to be an integral part of the enlistment process. Before doing so, Ewing needed the consent of his powerful father. The elder Ewing initially opposed his son's desire to become an officer to form his own regiment. Ewing's father reminded him that unlike his older brother Hugh and older foster-brother Sherman, Ewing had no military service or training. In a letter to his father, Ewing explained his rationale to serve in the Army. He told his father that the paramount cause of the war was to end slavery. His failure to take an active part in its eradication would severally compromise his political future once the rebellion ended. Ewing's logic and reasoning was convincing to his father, who soon acquiesced.[126]

When the Eleventh Kansas was commissioned into federal service, Ewing despite no military experience was named its first colonel. On September 29, 1862, Ewing marched his Eleventh Kansas regiment to Fort Scott, Kansas for formal training and drilling.[127] Within thirty days, he formally resigned from the Kansas Supreme Court. His political motivation in becoming actively engaged in the Civil War had come to fruition. Ewing's regiment became part of the Army of the Frontier under General Samuel Curtis. Curtis soon directed Ewing's regiment to join forces with General James G. Blunt in Arkansas. Ewing had previously worked with Blunt earlier that fall in establishing the Red Legs and assigning them the defensive duty of protecting the rear of Blunt's forces along the Kansas-Missouri border.

Once Ewing's regiment joined Blunt in Arkansas, it would not be long before he and his men would "see the elephant."[128] On November 28, they missed the opportunity

---

[125] Smith, 163-164.

[126] Goodrich, 235; and Smith, 166.

[127] Smith, 176.

[128] The term "Seeing the Elephant" was a reference used by Union and Confederate soldiers during the Civil War marking the first time a soldier experienced battle. There are some historical accounts that William Quantrill had participated in the Battle of Prairie Grove as well.

at the Battle of Cane Hill some twenty miles southwest of Fayetteville in Washington County, Arkansas. Ewing's regiment had been held back in reserve and saw no action. This would all change at the Battle of Prairie Grove on December 7. During this battle, Ewing and his regiment played a significant role despite the greenness of both commander and his men. The action of the Eleventh Kansas was so impressive under fire that Blunt saluted Ewing and his regiment. Some even described Ewing's Eleventh as being one of the best Kansas regiments.[129]

Ewing's early success in the fall of 1862 and his political relationships spurred his meteoric military rise. By April 1863, he was promoted from colonel to brigadier-general. There is no documented evidence of how or exactly when Ewing became a general after only seven months of military service. The first record of Ewing being referenced as a general was in a letter from Major General Sam R. Curtis to Blunt in Arkansas dated April 23, 1863. From that day forward Ewing was called general. Curtis named Ewing as the General of Division I of the Army of the Frontier.[130]

Ewing's continued movement up the military chain of command was further heightened after Lincoln removed Curtis as commanding General of the Army of the Frontier and replaced him with Schofield. Following Schofield's realignment of the Districts of Kansas, he promoted Ewing to the District of the Border. As the controversy over Ewing's appointment circled the western front (and consumed Lincoln's desk with protest letters and wires from Blunt and Lane), Ewing on June 16 established his headquarters at the Pacific House on Delaware Street near the river landing on the southern banks of the Missouri River in Kansas City.[131]

Upon taking command of the District of the Border, Ewing faced a Constitutional predicament. He had to balance the civil rights of its citizens while protecting them from terrorism. Residents along both sides of the state line experienced victimization by rebel guerrillas and bushwhackers and their counterparts, the Kansas Jayhawkers and Red Legs. These paramilitary organizations were mutually guilty of preying and warring upon the innocent.

For nearly two years each side had crossed the state boundary-line to attack the "enemy." There was no front line, only an imaginary state line. Everyone's farm or boundary line had become a war front. To the Kansas troops anything or anyone in Missouri was deserving of punishment. They insisted Missouri slaveholders had caused the War and were guilty of treason.

---

[129] Castel, "Frontier State," 89; and Smith, 183-184.

[130] O.R. 22:2:245 and 252.

[131] O.R. 22:2:319. Eakin, Joanne Chiles and Donald R. Hale. "Branded as Rebels." Volume 1. Second Printing with corrections. (Independence, Mo.: Print America, 1995), 378, quoting *Kansas City Daily Journal*, 30 Aug.1863, "Mr. E. B. Chadwick, late proprietor of the Pacific House, has opened a saloon in the basement of that house, and will be pleased to entertain all who may give him a call."

Pacific House on Delaware Street, Kansas City, Missouri

The guerilla, rebel or southern sympathizer held the same viewpoints towards the people of Kansas. They were considered to be free-soil men or even worse "abolitionists" who had unmercifully warred against women and children by crossing into Missouri to inflict every atrocity known to mankind. The horror and terror on the Border was without equal in the country, and it had only grown worse by the time Ewing ascended to power. It was like a house fire that had spread into a conflagration throughout the countryside. Such was the emotions and realities of civil war, and the question was whether Ewing would be up-to the challenge.

Ewing was afforded no opportunity to quietly settle into his new position. On the day he assumed his duties, rumors were rampant that Quantrill and other rebel forces planned to converge upon Independence and Lexington. The validity of these reports Ewing was certain. He notified Schofield of these guerrilla threats and activities, and Schofield ordered General Egbert E. Brown at the State Capitol in Jefferson City to dispatch reinforcements to Ewing.[132] This would mark the first but certainly not the last time in which Ewing's new District would feel the wrath and power of guerilla intrigue. To combat "the ubiquitous rebel guerrillas on the border," he was determined to keep the

[132] O.R. 22:2:320.

military forces at his command well engaged. His main goal and intention was to afford sufficient protection to the citizens.[133]

Rebels in Jackson County would remain the main focal point of Ewing's attention throughout his days in the District. Shortly after he took command some eighty guerrillas ambushed the Ninth Kansas Cavalry near Westport in Jackson County. Ewing was hog-tide to prevent such attacks. He had a limited number of available troops. He could not find a sufficient manner in which to defend against three different companies of "bushwhackers" numbering at least eighty members each. To Ewing, they seemed to hold omnipotent power to patrol the county with impunity.

The widow of Colonel Upton Hays wrote a letter to her mother shortly after the engagement near Westport on June 17, 1863. She provided an account of the conflict and horror on the border—the same one that Ewing had been assigned to control (the grammar, spelling and punctuation of the original letter have been retained to provide an authentic presentation of Mrs. Hay's own words):

> *Dear Mother*
>
> *It has been about two weeks since I received your kind letter, we are all well at this time. I have had a very bad cough for about two months but are better now. We have had a very wet Season this summer, some of the coldest weather i every knew for this time of year. Crops of all kinds look well, very fine wheet crop. As for our country troubbles not any better and no prospect of it being any better soon. Their has been a great many of my friends Banished and a great many killed. Some two or three weeks ago we had a fight within a mile of our house.*
>
> *The Bushwackers [sic.] passed by Linvels [Linville Hays] house, they had started to go to Westport but got sight of some Feds going to town. When they saw them they gave a yell and went in as fast a speed as their houses could go to head them, they headed them about a mile from Westport and drove them back to Wornels lane, Strewing the road with dead men.*
>
> *When they got to the lane they were fighting hand to hand, the Feds so excited being attacted so near town that they could not fight. They killed one man then drove them on the perraria near a Barn, they then fired on the Bushwachers, killed two of their men but the Bushwackers*

---

[133] "The War in the Far West," New York Times, July 8, 1863 referencing "Leavenworth, Kansas, Thursday, June 25, 1863"; Jackson County (Mo.) Historical Society Archives, Document ID Vertical File Newspaper Subject Files: Military/Wars: Civil War: 1863-"Order No. 11".

> *made it two warm for them to Stand it long so they retreated as fast as they could get away.*
>
> *They had thirty three buried in Westport out of Seventy. Bushwackers has sixty men, lost three men, one of them was killed accidentally by one of their own men.*[134]

While these casualties alarmed Ewing (as they had alarmed Mrs. Hays), he was more outraged by the fact that there were numerous "haunts" in which "bushwhackers in Jackson County" could seek refuge and protection. These sanctuaries had given these counterinsurgents venues in which to launch their clandestine attacks and thereafter return for succor and support. Ewing voiced frustration that his own forces were unable to secure the safety of the innocent citizens or the mail routes leadings westward out of Westport. He was firmly convinced that he would be unable to protect the border without at least six cavalry units; however, there were simply not enough troops available for Schofield to assign to Ewing.[135] These frustrating facts were the genesis of Ewing's plans to impose drastic measures. No-where in the country had the rebellion been "fought with more unremitting iron-fisted savagery than on the Kansas-Missouri border."[136]

Ewing was not only confronted with the insurrections and insertions by the guerrillas, he faced political challenges and intrigue as well. General Blunt was not the only person to voice complaints about being superseded by a subordinate. Senator Lane, General Curtis, the district's former commander, and other Kansas politicians, including the Mayor of Leavenworth, Kansas, Dan Anthony, took aggressive measures to undermine Schofield's decision to reorganize the District and place it under Ewing's command. They soon conspired and stringently presented their overt objections to Lincoln.[137] These antagonists were embittered that Schofield had named Ewing out of political friendship. Newspaper stories reported that a considerable dissatisfaction existed over Blunt's removal and his being replaced by Ewing. They wanted the District of

---

[134] The Colonel Upton Hays and Margaret Jane (Watts) Hays Overstreet Family Collection, 1849-1865 [early 1900s], Jackson County (Mo.) Historical Society Archives, with gratitude to Marian Franklin and family, who donated the originals for permanent conservation, for annotating transcriptions, and for sharing them online at wattshaysletters.com. Also, Ashby's casualty report to Ewing dated June 18, O.R. 22:1: 373.

[135] O.R. 22:2:328 and 365.

[136] Smith, 189.

[137] O.R. 22:2:388-389. Dan Anthony was no stranger to controversy. As the younger brother of suffragette, Susan B. Anthony, he had been actively involved in the Kansas free-state movement during the 1850s and had been a participant in military activities during those border war years. Anthony had also led Kansas' troops into Cass County, Missouri on January 1, 1862, and burned down the Town of Dayton. Anthony launched not only personal attacks against Ewing he openly interfered with Ewing's military orders. Anthony's open defiance resulted in Ewing on July 18 declaring martial law in Leavenworth and having Anthony arrested.

Kansas restored as it had been prior to Schofield's reorganization and they wanted Ewing removed.[138]

Lincoln had grown emotionally drained by these Kansas politicians. He also realized he could not politically ignore them either. Although Lane' persistent needling exhausted the President, Lincoln held a certain trust for Lane's political dependability and reliability. Instead of ratifying Lane's protests, the President delegated the situation to his General-in-Chief in Washington, Major-General H. W. Halleck. "Old Brains" then presented these complaints to Schofield. In a letter to Halleck dated July 3, Schofield not only defended his decisions but insisted that Ewing "was a most excellent officer, who had the confidence of the people of Kansas." He was adamant that peace could never be restored within the District of the Border unless the "country on both sides of the line be under the same command."[139] Schofield may have been correct that Ewing had the confidence of the people of Kansas, but his letter ignored the viewpoints of the citizens of western Missouri. In the end, Ewing retained his command.

Ewing grew increasingly outraged and angered over what he perceived to be the total dominion held by rebel guerrillas over Jackson County and the other border tiered counties in Missouri. He was firmly convinced that about "one-half of the farmers" had entered the rebel service to fight against the Union. Many of them Ewing insisted had either left or deserted the Confederate Army and returned to Jackson County or the other counties in the District and had "gone to bushwhacking." Ewing maintained that the bushwhackers had driven "almost all avowed Unionists out of the country or to the military stations." He estimated that "squads of a dozen and sometimes in bands of several hundred" scoured "the country, robbing and killing those they think unfriendly to them, and threatening the settlements of the Kansas border and the towns and stations in Missouri."

Ewing had become overwhelmed by the strength of the guerrillas and their ability to secretly move about the "rugged and heavily timbered" places to conceal their location. The guerrillas he complained were too well mounted, unable to be surprised and never engaged in battle unless all the odds were on their side. Ewing was convinced he needed at least three times his military strength to properly protect the District. This left Ewing with no other option but to utilize a large portion of his troops in guarding the Kansas border towns and stations in Missouri.[140]

Ewing further blamed large numbers of people in the Missouri counties for the ongoing success of the bushwhackers. He questioned the character of these citizens. To Ewing, they were nothing more that vigilant supporters "of the enemy" two-thirds he estimated were "kin to guerrillas" who had "actively and heartily engaged in feeding,

---

[138] Jackson County (Mo.) Historical Society Archives, Document ID Vertical File Newspaper Subject Files: Military/Wars: Civil War: 1863-"Order No. 11".

[139] O.R. 22:2:351.

[140] O.R. 22:2:428.

clothing, and sustaining them." The presence of these families had given guerrillas the ability to wage war and ambuscade, and Ewing saw "no prospect of an early and complete end to the war on the border" while these families remained throughout the countryside. Ewing declared these enemies of the state had run all loyal citizens from their homes. Unless he took drastic measures against them, Ewing was convinced guerrilla warfare would continue unmolested. To curtail the activities of these southern aiders and abettors, Ewing began having them arrested and imprisoned.

By the middle of the summer of 1863, traitors, raiders, abettors, criminals and other disloyal people had been banished from Jackson County or imprisoned. The Jackson Jail in Independence and other prison facilities, including the Union Hotel had become overflowing with male and female inmates. There were reports that some of these facilities were infested with rats and vermin as to render them unfit. Union guards even appealed to General Ewing about the horrific stench and parasites in these structures, many of which were located in the levee area of Kansas City along the Missouri River. These repulsive conditions resulted in Ewing commandeering, without consent, the vacant Bingham property and converting it into a woman's prison.[141] Many of the women inmates at the Union Hotel prison were then transferred to the Bingham house on Grand and placed under the supervision of the 11th Kansas Regiment.[142]

Ewing's next decision was to inflict pain and wage war upon the kinfolk of known or suspected rebels living in the Missouri border counties. Around August 3, Ewing dispatched a letter to Schofield in St. Louis requesting authority to remove all families of guerrillas living in the border counties of Missouri.[143] He also wanted authorization to emancipate slaves of guerilla supporters who desired freedom. It was Ewing's plan to use military forces to escort the slaves off the property of secessionists. In other words, slaves owned by those engaged in rebellion would be granted their freedom. Ewing considered this action as being within the spirit of the Emancipation Proclamation issued by Lincoln on the first day of the year. Ewing was convinced that this measure would further compromise the guerrilla activity in the western Missouri border counties and especially Jackson County—the county from which the majority of recruits had been enlisted by the Confederates or guerrillas the previous summer.

While Ewing awaited Schofield's reply, Kansas City experienced and suffered a singular tragedy that would send shockwaves throughout the region and would thrust daggers into the hearts of Quantrill and his raiders.

---

[141] Jackson, "LOCK DOWN," 48; Rash, 189; *Kansas City Star Wednesday Magazine*, November 6, 1963.
[142] Brownlee, "Gray Ghosts," 119; and Petersen, Paul R., and David W. Jackson. "Lost Souls of the Lost Township: Untold Life Stories of the People Buried in the Davis-Smith Cemetery, Kansas City, Missouri." (Kansas City, Mo.: The Orderly Pack Rat, 2011), 43.
[143] O.R. 22:2:428 and 450.

CHAPTER VIII

Death and Calamity in Kansas City

After Bingham's second wife inherited her late father's 14th and Grand property in Kansas City, the Bingham's converted the Thomas Property (the name in which the building was then known) into their personal residence. Within a year, Bingham added a third floor, which he used as his art gallery and studio. The Bingham's resided there until January 2, 1862 when they moved to Jefferson City following Bingham's appointment as the State Provisional Treasurer of Missouri. The Thomas Property would remain unoccupied for the next eighteen months.

Ewing ordered the seizure of the vacant Thomas Property for conversion into a prison for "disloyal" women. A Jewish merchant utilized the first floor as a general store and liquor shop. The second level housed the female prisoners and young children while the third floor remained unoccupied.[144] The women incarcerated at the new facility were either directly or indirectly related to known or suspected rebel guerrillas or had provided to them aid, succor or support. Some of the lady captives had been arrested while allegedly carrying food, supplies and information to bushwhackers.

Many of the women detainees were closely related to rebel guerrillas. Mrs. Charity McCorkle Kerr, Mrs. Nancy Elizabeth Ellen "Nannie" (Harris) McCorkle and the sisters of Bill and Jim Anderson were among the female relatives. Charity's husband, Nathan Kerr, and her brothers, John and Jabez, rode with Quantrill. Nannie had been married to Jabez, but she had become a young widow earlier in the year when her husband had accidentally and tragically shot himself to death. John McCorkle provided the following story of the arrests of his sister and sister-in-law:

> [They] *went to Kansas City in a wagon, driving a yoke of oxen, with a load of wheat to exchange it for flour, the women then having all the buying to do. When they had procured their flour and were ready to start home, Anderson Cowgill, a neighbor, who had known these girls all their lives, and the same man who refused to speak to me when we paroled him at Independence, saw these two girls and reported to the authorities that these two women were rebels and were buying flour to feed the*

---

[144] Petersen, Paul R. "Quantrill of Missouri: The Making of a Guerilla Warrior." (Nashville, Tn.: Cumberland House, 2003), 251.

bushwhackers. They were immediately arrested and placed in jail with some other girls, who had been arrested and sentenced to be banished.[145]

By Thursday August 13, the number of prisoners in the jail ranged from as few as nine and as many as twenty-seven women and one boy.[146] The building had not been designed for prison use and rumors had been swirling for at least two weeks that the structure was showing signs of collapse and dilapidation.[147]

The makeshift Federal Women's Prison (aka. The "Thomas Property") was about the middle of the Metropolitan Block on Grand Avenue between 13th and 14th Streets, Kansas City, Missouri

---

145 Baron, O.S. "Three Years with Quantrill: A True Story Told by His Scout John McCorkle." Republished. (University of Oklahoma Press, 1992), 120. Nannie Harris McCorkle would eventually remarry James Lilley who had also been a Quantrill man.

146 The actual date of the jail's collapse is certainly a topic of confusion and debate. The *Kansas City Journal of Commerce*, stated that it fell on a Thursday and it also reported that if fell on August 13; this date has been identified by other primary and secondary sources, including Leslie, 194 and Nagel, 111. The official historic landmark located near the site of where the building once stood, and just south of the Sprint Center on Grand, and other sources, including Richard Brownlee in his seminal work, Brownlee, "Gray Ghosts," 118, Yeatman, Ted P. "Frank and Jesse James: The Story Behind the Legend." (Nashville, Tn.: Cumberland House, 2000), 43; and Hale, 50 give the date as being August 14. To further demonstrate the confusion as to when the prison fell, Petersen in "Quantrill of Missouri," (p. 252) states it occurred on August 14, and August 13 (p. 262). If the jail fell on a Thursday, then the date would have been August 13. For purposes of this book, we will use the date August 13. Perhaps someday the debate over the correct date, the number of victims and the cause for the building's collapse will be resolved, but this is extremely doubtful.

147 Goodrich, "Black Flag," 73.

Earlier that day, a member of the provost guard informed the acting Captain John Singer of the renewed concerns about the structural integrity of the building. Singer conducted another inspection. He discerned cracks in the walls and ceilings with mortar dust on the floor. Singer dispatched his findings to Ewing who immediately sent Major Harrison Hannahs, adjutant general of his staff, to further evaluate the situation. Hannahs' findings reportedly failed to uncover any serious deficiencies with the prison; no further action was taken.

Hannahs' summary conclusion did not dispel rumors of the building's impending doom. These mounting apprehensions were further bolstered when the store merchant on the first floor was seen removing his goods. He, too, had apparently noticed the same cracks. Union troops garrison on or near the building had also purportedly been seen abandoning the dwelling that same day.

Nannie Harris and Charity Kerr

It was a hot, humid afternoon around two o'clock when the prisoners were served dinner. One inmate inside the jail, Charity McCorkle Kerr, was very ill suffering with a fever allegedly caused by exposure and poor diet. Reportedly, Charity had asked her sister-in-law Nannie "to get the guard to bring her a drink. She waited at the second story window." At about the same time, another military inspection of the building was taking place. As the sentry returned with Charity's glass of water, he and the inspectors sensed imminent danger and reacted promptly. They heard disturbing sounds and rushed out the building or jumped through the windows. The soldier delivering the water began shouting to the girls to jump. Moments later, the three-story brick building collapsed, burying beneath its walls a number of inmates, four of whom were dreadfully mangled and crushed to death. The magnitude of the building's destructive force caused the adjoining building to its south and owned by Mrs. Elizabeth Cockerel to likewise fall.

The disaster sent shock waves across the city. Many women prisoners were buried under the brick and mortar. Soldiers and citizens rushed to the scene, but dust clouds ascended into the air precluding any visual acuity into the debris field. Soon like a phoenix several women began to crawl out of the ruins. When the smoke began to dissipate, there was a charge to attempt to free the women beneath the rubble. Women caught under the bricks and walls could be heard crying, yelling, begging, imploring and praying to be freed. Some were, but some had been buried alive—the weight of the structure having proved fatal. The crowd that gathered in the haze of smoke became intensely and openly hostile and outraged. Ewing feared a riot would erupt and sent in troops to garrison the vicinity. No violence broke out, but the echoing cries from the injured, maimed and killed fostered intense revengeful hearts throughout Jackson County.

Several newspaper stories and first person accounts regarding the catastrophe have been written and told. They vary from the actual number of those incarcerated to the cause of the building's downfall. The *Western Journal of Commerce* on August 14 reported there were nine women prisoners, two children and one man. Another newspaper said there were at least eighteen women and children inside the structure when it collapsed. Regardless of the actual number of prisoners, four women were killed, one was mortally injured and many others badly hurt.[148] The lead storyline in the *Kansas City Journal of Commerce* read as follows:

> *The Female Prison at this place fell in on Thursday morning, burying in the ruins eleven women. Four were killed, one mortally wounded, and six slightly injured.*[149]

The names of the women killed were subsequently identified as fourteen year-old Miss Josephine Anderson, Mrs. Armenia Crawford Selvey, Mrs. Susan Crawford Vandevere and Mrs. Charity McCorkle Kerr. The fifth fatality was Mrs. Wilson who had been mortally wounded and died shortly thereafter.

---

[148] *Kansas City Daily Journal of Commerce*, "Untitled," August 15, 1863; Goodrich, Thomas. "Bloody Dawn, The Story of the Lawrence Massacre." (Kent State University Press, 1991), 8.

[149] *Kansas City Journal of Commerce*, August 14, 1863; Jackson County (Mo.) Historical Society Archives, Document ID Vertical File Newspaper Subject Files: Military/Wars: Civil War: 1863 "Order No. 11". Much has been written about this seminal event, including, "Lost Souls of the Lost Township," by Paul Petersen and David W. Jackson (Kansas City, Mo.; The Orderly Pack Rat, 2011); the "Collapse of a Guardhouse," *Kansas City Star*, 19 Nov. 1911; "Catalyst for Terror: The Collapse of the Women's Prison in Kansas City," by Charles F. Harris, Missouri Historical Review, Vol. 89, 299-306. (The immediate surnames involved were: Harris; Kerr; Anderson; Vandevere and Selvey): "Guardhouse," Kansas City Journal of Commerce, 14 Aug 1863; "Untitled," Kansas City Daily Journal of Commerce, August 15, 1863; "Untitled," Kansas City Daily Journal of Commerce," August 18, 1863; and Daily Times [Leavenworth, Ks], August 19, 1863, 3, c. 2.

Several women required medical care or hospitalization. Nannie jumped out the second story window and though injured survived. By her leap, she avoided the fate of her sister-in-law Charity who was "caught in the crumbling walls."[150] Twenty-three year-old Molly Granstaff suffered several broken bones and had to be rescued from the debris as had Alice Fay Vanness. Within a few days, the injured were said to be slowly recovering.[151]

Alice Fay Vanness, formerly of St. Joseph, Missouri also sustained what was initially considered to be mortal injuries. The *Daily Times* in Leavenworth, Kansas on August 22 provided an unflattering account of Alice. Demonstrating the newspapers Union bias, the *Daily Times* described Alice as having once been as "pure as the beautiful snow" but had become soiled by the southern cause, which had resulted in her capture. The *Daily Times* reported that after Alice's father died in St. Joseph in 1859, her widowed mother and fatherless children were sadly "reduced" in "circumstances." As hostilities erupted during the Civil War, "Alice took an irrevocable step downward" causing her mother to "hide in shame" and relocate to Leavenworth. "The mother, and we believe two of the children, have since died of small pox, and if Alice has now gone, the family is extinct." Although Alice did not die in the disaster, she certainly was the sole survivor of her family.

While the actual number of inmates will likely remain undetermined, there certainly were a large number of prominent Confederate soldiers or guerrillas whose woman folk were injured or killed on that scorching August afternoon. In addition to Charity and Nannie, one of the three sisters of the infamous bushwhacker, Bill Anderson (later to be known as "Bloody Bill"), fourteen year-old Josephine, was crushed to death. For a while she could be heard yelling and begging to be rescued from the heavy bricks that were on top of her head. Bloody Bill's two other sisters were seriously injured. Eighteen-year-old Miss Mary was left crippled for life. The youngest sister, 10-year-old Margaret had one leg shackled to a twelve pound ball as punishment for some alleged misconduct; she had gone down with the building fracturing both legs, injuring her back and suffering severe facial lacerations. By their injuries and death, Bloody Bill became maniacal and spent the next fourteen months of his remaining life wantonly and maliciously avenging what happened that day in Kansas City.

---

[150] Wilcox, Pearl. "Jackson County Pioneers." Reprint. (Independence, Mo.: Jackson County Historical Society, 1990), 351; and Petersen, "Lost Souls," 46.

[151] "Falling of a Female Prison," *New York Times*, August 16, 1863: citing *Kansas City Daily Journal of Commerce*, 15 Aug. 1863; and *Daily Times*, August 22, 1863. Despite newspapers on both sides of the state line covering the tragedy, the *Liberty Tribune*, a weekly publication in Clay County, Missouri, provided absolutely no coverage of the jail collapse in its August 14, August 21, August 28, September 4 or September 11 editions of their paper. Alice Fay Vanness would later marry and her granddaughter would become the famous actress Fay Templeton.

Other relatives of rebels and guerrillas held in the prison included Miss Molly Granstaff, Mrs. Armenia Selvey and the orphaned sisters Susan Munday, Martha "Mattie" Munday and Mrs. Lou Munday Gray. The brother of the three Munday sisters was serving in the Confederate Army under General Sterling Price, and Mrs. Lou Gray's husband was bushwhacking in the area. Twin sisters, Susan (Crawford) Vandevere and Armenia "Mina" (Crawford) Selvey had both been killed. Their brother Riley Crawford was one of the youngest recruits in Quantrill's band. The two husbands of the dead twin sisters were far removed from the district as they were serving in the Confederate Cavalry under Jo Shelby (they were also cousins to the Younger family). According to Cole Younger, three of his sisters Josie, Caroline and Sara "Sally" were prisoners and they along with Miss Hall were able to escape the tragedy. Cole further claimed that the injuries his sister Caroline "Duck" Younger suffered compromised her health so badly that it caused her death two years later at age 23.[152]

The bodies of Susan Crawford Vandevere, Armenia Crawford Selvey and Charity McCorkle Kerr were placed in crude wooden coffins. The following day they were transported by ox cart and buried in the Davis-Smith Cemetery on the outskirts of Raytown, alongside other victims of the War.[153] The other two women killed were reportedly buried in the soldiers' burial ground at Union Cemetery.[154]

The story of the deaths did not end with contemporary newspaper accounts. Stories continued down through the years. The *Kansas City (Mo.) Post* on May 2, 1912, provided an encapsulation of a chilling and macabre address given by Mrs. Flora Stevens as she stood at the grave of Josephine Anderson at Union Cemetery in Kansas City in which she recalled the Union prison collapse:

> *There were nine of the girls in the prison at 1409 Grand Avenue, when it fell. One of these was Josephine Anderson, her two sisters, Mollie, age 16, and Janie, 10-years-old, were prisoners with her, and it was these three especially that the Union soldiers wanted to kill because they were sisters of Bill Anderson, the guerilla. . . These girls, none of whom was more than 20-years-old, had been arrested and brought in by Union soldiers because they were southern sympathizers and were being held in prison while waiting to be banished. When the soldiers heard that Bill Anderson's*

---

[152] For sources on the tragedy also see: Grey Ghosts, 117-120, Goodrich, "Black Flag," 73-75; Baron, 120-123; History of Missouri Vol. III, 100; Leslie, 194-198; and Wilcox, 351-352. Younger, Cole. "Cole Younger, By Himself." Originally published 1903. (Springfield, Mo.: Oak Hills Publisher, 1996), 3. Cole references the incarceration of his three sisters, the death of his cousin, and opines that the death of Duck was connected to the injuries she suffered that afternoon. At the time of Duck's death she was married to George M. Clayton, a Quantrill guerilla, and they had two children, Leola Clayton and Gustavus Clayton. Cole also claims there were twenty-five women prisoners in the jail when it fell.

[153] Petersen, "Lost Souls," 49.

[154] *Kansas City Daily Journal of Commerce*, "Untitled," August 18, 1863.

sisters were in their power, they determined to kill them. The first inkling of the plot was when Mrs. B. F. Duke, who now (1912) lives at 1717 Wabash Avenue, but who then had a boarding house at Independence Avenue and Oak Street, heard some of the soldiers staying at her house speak of the progress they were making in tearing down a wall. Mrs. Duke was a cousin of Bill Anderson, but the soldiers did not know it and told her of their scheme, and how they had removed a large section of the wall of the women's prison. General Bingham, the owner of the building, had protested they said, but it did not stop them. The building did not fall the first day, so more of the wall was removed and it was at this time that Mrs. Duke learned of it. She was beside herself with rage and ordered all the soldiers from her house. With a number of friends she hurried to the military headquarters and begged that the girls be taken from the building before they were killed. Their pleadings were in vain and an hour later the building fell. The girls had been uneasy by hearing the people on the floor below moving out their stock of groceries and whisky which they took to a safe place. The plaster had been falling all day and the girls were in a panic. Nan Harris and Mollie Anderson had just gone out into the hall to get a bucket of water when they heard cries from the other girls that the roof was falling. The guards, evidently repenting at the last moment, carried the two girls to safety. Janie Anderson, who was the youngest, tried to escape through a window, but a twelve-pound ball had been chained to her ankle which held her back and both her legs were broken. The other girls went down with the ruins. There were groans and screams for a long time, and Josephine Anderson could be heard calling for someone to take the bricks off her head. Finally the cries ceased.[155]

Another description of the horrific event was also provided years later in a newspaper article dated November 11, 1911. Eliza Harris, a sister of Nannie, who was in the building when it fell, provided the reporter the following haunting version of what transpired after the tragedy:

I was a girl of eleven at the time as I remember that the Union men sent three caskets containing my cousins to Little Blue. With the caskets was the

[155] Harris, Charles F. "Catalyst for Terror: The Collapse of the Women's Prison in Kansas City," *Missouri Historical Review* (April 1995): 89: 3: 290-306; Jackson County (Mo.) Historical Society Archives, Document ID Vertical File Newspaper Subject Files: Military/Wars: Civil War: 1863-"Order No. 11"; *Kansas City Post*, May 2, 1912; *Kansas City Journal of Commerce*, 14 Aug 1863; Lost Souls; and Baron, 120-122.

> *satchel of trinkets and dry goods that my sister and Charity had gone to town to buy. The dead were buried in the old Smith Cemetery near Raytown. We didn't have much time for funerals in those days and the three were buried in one grave. A little after the guard house fell, Order No. 11 was issued and our house at Little Blue was burned. We went down near Glascow and my sister Nannie walked thereafter she got away from the Union guard house-almost 100 miles. After the guard house fell, the remaining women were taken to some hotel and kept for a period of time.*

Nannie's personal tragedies were not limited to the horror of the jail. She also felt the destructive power and terrorist conduct of the Kansas Red Legs after the issuance of Order No. 11. She saw her family home burned and their gristmill located over the Little Blue River in eastern Jackson County destroyed.

Martha Ann "Mattie" Lykins (who would later become the third wife of George Caleb Bingham) prior to her death in 1890 provided the backdrop surrounding the seizure of Bingham's building and its compromised condition.

> *Thomas Ewing . . . did not notify General Bingham that his building was needed for a prison, but simply took possession of it and ordered the women prisoners transferred to it and confined in the third story which General Bingham had used for his studio.*
>
> *The . . . block was so constructed that the sleepers of the floors served the purpose of braces to the dividing walls of each building. The sleepers of the first floor rested on brick pillars in the cellar . . . I was told that on investigation it was learned . . . that the building had been weakened by the removal of the brick pillars which support the first floor and further that some of the sleepers of the adjoining buildings on both sides had also been removed, thus weakening the dividing walls beyond safety. By whom this act was done was not known, for what purpose was left to conjecture.*

Although Mattie was not an eyewitness to the jail collapse, she soon heard reports of the catastrophe. Within one hour, Mattie had arrived at the scene and spoke to some of the victims. They told her how they had repeatedly informed the guards that the house was giving way and had heard sounds as if it was cracking. Mattie watched the removal of the living and dead from the debris. She also overheard a conversation someone had with Dr. Joshua Thorne, the acting U.S. Chief Surgeon of the hospital in Kansas City, while Dr. Thorne was at the scene tending to the wounded. Mattie related what she had overheard Dr. Thorne state:

While standing beside him [Dr. Thorne], *near the building, watching the removal of the living and the dead from the debris, someone remarked to him that they supposed some of the soldiers on guard would be found buried beneath the ruin. "No." replied Doctor Thorne. "Not a bluecoat will be found; every man who has been detailed to stand guard at this prison for the last few weeks, knew the house to be unsafe and have kept themselves at a safe distance from the trembling walls. I knew the building to be unsafe," he continued, "and notified the military authorities of the fact, and suggested the removal of the women prisoners, but my suggestion was not heeded and before you is the result."*[156]

Mattie Lykins (later Bingham)

The actual cause and responsibility for the destruction of the women's prison may never be known. Finger pointing from both sides began to permeate the community—and has continued throughout the ages. Southern sympathizers voiced utter shock, consternation and alarm. They strenuously asserted that the cataclysm was a direct result of Ewing having ordered his soldiers to remove the support beams under the building. They considered Ewing criminally culpable of murder. Elijah M. McGee and his son-in-law Charles H. Vincent both gave sworn statements corroborating this version of the story. They each claimed to have visually inspected the building days before the disaster and saw that the posts had been removed.

Those supportive of Ewing or oppositional to guerrillas leveled the blame upon the captives. Arguments ranged from accusing the women of culpability to infidelity. Those blaming the prisoner firmly held to the proposition that the prisoners had

---

[156] Goodrich, "Black Flag," 73-74; Petersen, "Lost Souls," 45-46; Findlen, Rose Ann. "Missouri Star: The Life and Times of Martha Ann "Mattie" Lykins Bingham." (Independence, Mo.: Jackson County Historical Society, 2011), 62; and, Bingham, Martha A. "Mattie" (Livingston) Lykins, "Recollection of Old Times in Kansas City," 67-page handwritten manuscript, ca 1883-1900. Gift of Robert Dewitt Owen, in memory of his grandmother, Mrs. Dwit Livingston (Ada Armintha Campbell) Owen, Jackson County (Mo.) Historical Society Archives, Independence, Missouri, Document ID 110F7.

compromised the walls by digging tunnels under the foundation during escape attempts. William E. Connelley, a Kansas historian in 1910, and pronounced antagonist to Quantrill, perpetuated the tunneling story as the causal connection for the building's demise.

Another explanation is that the women suffered from venereal diseases and had been confined in the cellar of the jail. Soldiers stationed in the adjoining building are said to have accidentally tunneled into the basement undermining the support beams and thereby caused the collapse.[157]

The most sensational claim was told by Dr. Thorne years after the tragedy. In an unflattering, offensive and prejudicially biased rationalization, he adamantly insisted that the openings under the cellar had been dug to provide the prison guard access to the women inmates in exchange for their sexual gratification. Dr. Thorne considered the house to be a den of prostitution where soldiers frequently intermixed with these women "of bad character," "diseased" and "beastly" intoxicated."[158] In other words, the female prisoners were prostitutes, or in the 19th Century contemporary vernacular, "soiled doves." No serious student of local history, then or now, truly considers Dr. Thorne's assessment to have been true or accurate.

When word reached Bingham of the event, his blood boiled. He promptly commenced his own investigation as to the cause. Bingham obtained sworn affidavits from "good intelligent and Loyal citizens" that established the causal connection for the loss of the building and those five lives. These statements placed the blame squarely upon the removal of "columns" beneath the building by Ewing's "soldiers." Bingham demanded full restitution for the casualty loss. His efforts were rebuked by Ewing who only certified "that the building was occupied as a military prison by his order, and while thus occupied fell." Bingham pressed his claim against the federal government and protested to Ewing's response.

> *Had I been a Kansas horse thief he would have certified to all the facts, proof of which I laid before him, or he would have ordered his Quartermaster to pay me, but he was not capable of doing justice to an honest Missourian. He certainly excels in meanness all of his Kansas predecessors.*[159]

It may be recalled that as early as August 3, Ewing had contemplated imposing more stringent military measures over the civilian population in western Missouri.[160] On

---

[157] Yeatman, 43.
[158] Leslie, 195-198; and Brownlee, "Gray Ghosts," 119.
[159] Bingham's letter to James S. Rollins dated December 21, 1863, "Bingham Letters," 283. There is no historical documentation that shows evidence that Bingham ever received compensation on his claim.
[160] History of Missouri Vol. III, 100.

August 14, Schofield authorized Ewing's implementation of his depopulation and emancipation plans. At the time of Schofield's Order, he was unaware of the disaster in Kansas City or the flames it had fuelled in the hearts of rebels. However, Schofield did warn Ewing not to permit military abuse and to avoid summary decisions in any "doubtful case" of a person's loyalty. Schofield wanted Ewing to remain within the spirit of the Emancipation Proclamation and not to make snap decisions. He did not want Ewing liberating slaves of those persons who were in support of the Federal government.

Despite emotional resentment surrounding the building's cave-in and the allegations of the general's culpability, Ewing did not allow time to pacify these sentiments before inflicting his intended retaliatory military measures against rebels and their sympathizers. On August 18, Ewing having received Schofield's authorization issued Order No. 9 and Order No. 10.[161] Order No. 9 applied to the emancipation of certain slaves while the scope of Order No. 10 was much more far-reaching.

Within Order No. 9 Ewing established the procedures for the implementation of freeing slaves owned by rebels and their supporters. Ewing's troops were to: (1) carefully identify only disloyal citizens; (2) prudently identify slaves owned by such untrue citizens; (3) circumspectly ascertain which of those slaves desired freedom; (4) legally provide public and private notice of those so identified; (5) cautiously escort those slaves out of Missouri or to Independence, Kansas City, Westport in Jackson County; and (6) carefully enlist freed slaves for military duty if they were willing to serve. The Fourth Regiment of the Missouri State Militia was to escort the slaves and offer food and shelter to them if possible. Ewing strictly enjoined and prohibited the involvement of Kansas' troops to enforce the Order.

The decision to free slaves and proffer military service to them had been part of Lincoln's military plans since he issued the Emancipation Proclamation. The Proclamation had authorized the enlistment of blacks "of suitable condition." They were to "be received into the armed service of the United States to garrison forts, positions, stations, and other places, and to man vessels of all sorts." Throughout spring and summer of 1863, the Proclamation's edict resulted in the formation of black regiments, including the 54th Massachusetts under Colonel Robert Shaw. These troops proved their mantle by demonstrating dependability and reliability in action.

Lincoln believed the program of enlisting black soldiers would also have a pronounced psychological impact upon the south. He was convinced the rebellion would not survive if a black military force could "take shape, and grow, and thrive." Lincoln envisioned one hundred thousand black troops being recruited and their success would

161 O.R. 22:2:450-451. The complete text of Order No. 9 is set forth in Appendix C and the complete text of Order No. 10 is provided in Appendix D.

"soon end the contest."[162] Order No. 9 was simply an extension of Lincoln's policy to include black soldiers in the overall strategy of defeating the Confederacy.

After waging war on the "chattel property" of rebels and their family members under Order No. 9, Ewing imposed even tougher measures against those traitors under Order No. 10. There were primarily two purposes of Order No. 10. The first was to depopulate from the Missouri counties in the District those citizens who had been disloyal to the Union since July 17, 1862. Officers enforcing the Order were specifically enjoined to expel only disloyal citizens. The soldiers had to be careful not to compel removal of loyal Missourians or those who by threats or fears had been coerced to aid rebels.

The wives and children of known guerrillas, and women, who were heads of families and willfully engaged in aiding guerrillas, were to be notified by such officers to remove out of the District of the Border and out of the State of Missouri. They were only permitted to take, unmolested, their stock, provisions, and household goods, and if they failed "to remove promptly" or refused to comply with the evacuation order they would be sent under military escort "to Kansas City for shipment south with their clothes and such necessary household furniture and provision as may be worth removing."

The second purpose of Order No. 10 was to provide a military escort for all loyal persons desiring to remove to Kansas or to a permanent military station. These faithful stewards were then to be taken to an assistant provost marshal or to the District Provost Marshal for further orders. Once again, only Missouri troops were authorized to provide escorts to these steadfast people, and the soldiers were enjoined from destroying property. As under Order No. 9 no one living in Kansas was permitted to enter into Missouri without prior authorization, and any and all Kansas' militia or other military organization was expressly barred from enforcing Order No. 10.[163]

Total war had become Ewing's battle plan in confronting the guerrillas and their family members. The overall aim and goal of Order No. 9 was to emancipate slaves owned by citizens who had supported the rebellion since July 17, 1862. Order No. 10 was had two principal purposes. The first was the outright depopulation of the counties of Jackson, Cass, Bates and Northern Vernon of any and all persons who were sympathetic to the Confederate cause or who had provided aid, succor or support to known guerrillas since July 17, 1862. The second basis for the Order was to provide safe transportation to lawful citizens desiring to escape the region from the terror and ravages of the war. Ewing had utilized his entire military authority to wage war against the civilian

---

[162] Emancipation Proclamation (Appendix A below); and Civil War History, A Journal of the Middle Period, Hattaway, Herman, and Archer Jones, "*Lincoln as Military Strategist*," Kent State University Vol. 26, Number 4, December 1980, 303.

[163] O.R. 22:2:428-429.

population—a population in Jackson County he considered to be at least two-thirds kin "to bushwhackers actively engaged in feeding, clothing and sustaining them."[164]

Orders No. 9 and No. 10 outraged many Missourians along the border regardless of their loyalties. Some loyal citizens disapproved of the emancipation of slaves as being in direct derogation to Lincoln's Emancipation Proclamation. Missouri had not seceded from the Union and Federal forces occupied the entire District of the Border. There was simply no rebellion in this part of the country to justify emancipation. To these antagonists, the Orders were simply illegal and violated the Constitution. If freedom to slaves was to become the law, they maintained it could only be accomplished through a graduated process, not by military order.

Not surprising, supporters of the rebellion were shocked and dismayed. Many had previously signed the detested loyalty oath and had remained uninvolved in the conflict. They had gravely suffered by the terrorist actions of Kansas Red Legs and Jayhawkers and were now being victimized by a government in which they had agreed to remain loyal or at least neutral.

Rumors of both Orders soon circulated throughout Jackson County. This prompted public concerns and a "committee was appointed, resolutions were drawn up, adopted, and signed by nearly all present" in which they proclaimed their grave concern over the potential "great hardship and ruin" the Orders would impose upon citizens, regardless of their station in society or their loyalties. The committee determined that their resolutions needed to be personally delivered to Ewing. Martin Rice, a known loyal sympathizer from Lone Jack, was assigned this mission. Rice arrived at Ewing's headquarters in the Pacific House on about August 19. He was informed Ewing had gone to Leavenworth for the purpose of evaluating the effectiveness of his martial law order that he had previously imposed upon that city. Rice presented the committee's report to Major Plumb, Ewing's adjutant. The resolution fell on deaf ears. Plumb instantly criticized and resented the resolution. The rebuke did not dissuade Rice who remained at the Pacific House for over an hour until Plumb assured Rice that he would present the resolution to Ewing once he had returned from Leavenworth.[165]

There is no extant record as to whether the committee's request was ever presented to Ewing or read by him. Even if he had read or seen it, intervening events would have made the report pointless and irrelevant to him. For Quantrillians, they would have cared nothing about what was contained within the resolutions. The Orders and the prison shock had emboldened and intensified their soul-filled passion and desire for revenge.

---

[164] O.R. 22:2:450-451.

[165] Martin Rice, *Rural Rhymes, and Poems from the Farm, Millett & Hudson* Kansas City, Missouri 1877; Martin Rice, *Rural Rhymes, and Tales of Olden Times Ramsey, Millett & Hudson, Kansas City, Missouri 1882, Martin Rice Remembers: What I saw of Order Number Eleven*, Missouri Historical Review, Volume 106, No. 3, April 2012; Union, 940-941; and http://www.historiclonejack.org/whatisaw.html.

Neither the two Orders nor the demise of the jail directly caused the subsequent attack on Lawrence. Over the last one hundred and fifty years, there have been historical innuendos, allegations, assertions and bald statements to the contrary. Among those who helped perpetuate this tale was Bingham himself (and later by his third wife Mattie) and bushwhacker John McCorkle. Bingham blamed Ewing for the loss of his property and McCorkle held Ewing culpable for the death and injuries to his family. Bingham demanded $5,000 from the federal government to compensate him for the military taking of his house and the subsequent property damage.[166] McCorkle proclaimed that "the famous raid on Lawrence, Kansas" was in direct response to the catastrophe in Kansas City. There were others who shared the arguments of Bingham and McCorkle.[167] Their conclusions overlook certain historical and pragmatic facts, which in the final analysis amply dispel the arguments that the Orders of August 18 or the heartbreak of August 13 directly caused the subsequent raid upon in Lawrence.

As a practical matter neither event could have caused the planning of the attack against Lawrence. Quantrill and other guerilla bands did not fight in one tightly fit organization. Instead their battle strategies included only limited number of men who utilized secretive hit and strike assaults against unsuspecting Federal troops or Kansas' towns. They secreted themselves in small packs in the woods, valleys and caves throughout Jackson County. They did not benefit by telegraph wires but communicated by fast riding horses and word of mouth. To have coalesced as many as four hundred and fifty men in less than a one week time period would have been virtually impossible.

The journey westward towards Lawrence began a few weeks earlier. It is unquestioned that Quantrill had been eyeing Lawrence for some time and long before August 13, and maybe as early as his springtime return to Jackson County earlier in the year.[168] Planning a raid of such magnitude required long-term preparations and analysis. Quantrill had initially sent Fletcher Taylor to scout and reconnoiter the town. Taylor disguised as a cattle trader stayed at the Eldridge House and spent a week in Lawrence.[169] After completing his surveillance work, Taylor around August 10 returned to Quantrill and presented his findings to a council of war held near the Garrol farm south of Oak Grove in eastern Jackson County.

---

[166] Letter from Bingham to James S. Rollins dated December 21, 1863, "Bingham Letters," 283; and Nash, 190.

[167] Baron, 122. Parrish in History of Missouri, Volume III (p. 100) suggests that the jail tragedy resulted in the raid on Lawrence raid. The opinion of Thomas Goodrich in, "Black Flag," (p. 74-75) seems to share McCorkle's viewpoint or at a minimum it became the justification for the attack. Houts (p. 57), combines both arguments stating that the tragedy in Kansas City and the revenge factor against Lawrence were the catalysts for the invasion.

[168] Castel, "Frontier State," 126.

[169] Younger, 46; Burch, 142-143; and Cummins, Jim. "Jim Cummins' Book, Written By Himself." Originally published in 1903. (Independence, Mo.: Two Trails Publishing, 1996), 20; and Eakin, "Recollections of Quantrill's Guerrillas," 56-57.

The conference reportedly lasted nearly twenty-four hours. During their deliberation, Quantrill and the other guerilla chieftains considered Taylor's full report. They evaluated the pros and cons of such a wide scale attack.[170] Taylor detailed the overall military defense of Lawrence and the town's citizenry. The detested Senator Lane had returned to his home in Lawrence; his death alone would be worth the risk of a clandestine attack. Quantrill then laid before his officers his plan, and after much debate the band agreed to march on Lawrence, the "Citadel of Stolen Property," home of "Kansas Abolitionism" and the root of all their problems since the enactment of the Kansas Nebraska Act nine years earlier. They agreed to rendezvous with their men within the coming days in western Missouri to start their westward trek.[171]

The most compelling evidence that neither the Orders nor the disaster caused the raid is the absence of evidence from at least four writers of the era—three guerrillas and one rebel apologist. Three bushwhackers who rode with Quantrill into Lawrence and subsequently wrote of their experience help illustrate this point: Captain Harrison Trow, Cole Younger and Jim Cummins. Trow was a Quantrill captain involved in the decision making process of marching into Lawrence; Younger a notable dependable guerrilla had cousins and according to him three sisters imprisoned in Kansas City; and Cummins a more recent recruit into the ranks of Quantrill had the reputation as a fierce fighter. Not one of these three notables made any representation, claim or statement that the annihilation of Lawrence was predicated upon the Orders or the prison debacle. In fact, the only one to mention the tragedy was Cole Younger and though he blamed Ewing for it, he did not attribute it as the cause of the Kansas raid.[172]

Even more convincing is the account (or lack thereof) by John Edwards in his work, "Noted Guerrillas." Edwards, unlike Trow, Younger or Cummins, was not among the raiders. Nevertheless if Edwards could justify the conduct and actions of Quantrill and his boys, he was never reluctant or bashful to do so. Edwards admired Quantrill and his "cause." He commonly depicted Quantrill and his followers as knights of the round on a mission to preserve the southern way of life. His book is written in an almost flowery prose, and he never failed to embellish any tale involving these "knights." Edwards wrote after the War but at no time did he even suggest or attribute the attack on Lawrence to the Orders or the deaths in Kansas City. It would have only seemed fitting for Edwards to have included such events as the justification of the raid—even if it had

---

[170] Webb, 274; Webb at 275 dates the meeting, August 15.

[171] It is unquestioned that Taylor's scouting expedition occurred well in advance of the prison disaster. Burch, 142; Noted Guerrillas, 188-190, Castel, "Frontier State," 132 and 124 for reference to "Citadel"; Quantrill's Guerrillas, 22; Leslie, 193; and Castel, "Frontier State," 126 both date the meeting on August 10; W. L. Webb in his work Battles and Biographies of Missourians 1900, 275 states that the council of war was held on August 15; and Goodrich, 235 where the author affirmatively states that "[e]ven before the prison collapse, Quantrill had decided to raid Lawrence."

[172] Burch, 141-161; Younger, 3; and Cummins, 103-104.

only a shred of truth in it. He certainly would have included such an epistle if it had been so—or at least if it would have furthered his defense of these knighted and sainted men embarked on a noble mission during the war.[173]

Frequently, it is said that what factually occurs in history is less significant than the perception of what occurred, which develops into "reality." Cummins understood this distinction when he wrote in 1903 of his fear that future students of the Civil War would never understand what prompted the attack on Lawrence. He did not what the progenies of Lawrence residents to call the raid as nothing more than a villainous act without justification or cause. Cummins felt compelled to explain why they had assailed the city from the perspective of southerners, Confederate veterans, rebels, guerrillas, bushwhackers and their abettors. To Cummins and his compatriots the cause and blame was simple: it rested squarely upon Lawrence itself. Either directly or indirectly all of the agony, suffering, grievance or sorrow suffered in Missouri had evolved out of Lawrence and had been brewing for years. Their victims had been grandparents, fathers, mothers, wives, children, relatives and friends and all of whom had been assaulted, insulted, imprisoned, injured, banished, robbed, pillaged, murdered, widowed or orphaned. Cummins insisted that they had suffered such other privations and deprivations by these bands from Kansas who had been like hordes of locusts in the Old Testament of the Bible. Western Missouri had been victimized by "murder after murder, robbery after robbery, fire after fire." Each of these indignities had been heaped upon an inoffensive people by radicals and their affiliate gangs inside Lawrence.[174] With such an explanation, Cummins did not spread the fallacy that the raid had been caused by the jail collapse.

The time for retribution had come—another conflagration would erupt. This time the fire would not be in Missouri; Lawrence was soon to experience the terrible swift sword and the rebel cry.

---

[173] Burch, 142; Noted Guerrillas, 188-190; and Leslie, 193.
[174] Cummins, 103-104.

CHAPTER IX

Black Friday

By the third week of August, Ewing was well into his third month in command of the District of the Border. Ewing's military sanctions appeared to be working and gaining public support. The headline of the *Liberty Tribune* on August 21 read: "The Peace of Missouri." The editor proclaimed that the "rebellion, in Missouri, is at an end, and the people now desire peace, without which no real happiness can be enjoyed." The paper was convinced that Ewing's effort of stationing large military forces in almost every county of western Missouri had been a major reason for this success. The editor was either speaking hopefully or naïvely as days earlier on August 15 the mail coach and passengers was robbed "by guerrillas near the Big Blue River, between Kansas City and Independence."[175]

The *Weston Sentinel* in Platte County did not share the same confidence as the editor of the *Liberty Tribune*. Its readers were warned that both "Jayhawker and the Robber, who are despoiling our people of their property, and threatening their lives" had to be exterminated. Although Ewing's jurisdiction did not extend into either Clay or Platte County, he was certainly aware that the diverse opinions of both newspapers were shared within his military domain.

A large percentage of the people in Kansas shared the view of the *Liberty Tribune*. They were convinced Ewing's action had weakened guerrilla activity. Ewing was applauded for having placed thousands of cavalry troops in the saddle and fully garrisoning each and every military post along the Border of Kansas. The *Leavenworth Daily Times* endorsed Ewing. "No general in command of the district has succeeded better" and had reduced the insurgency by both bushwhackers and Red Legs.

Inside Lawrence the vast majority of the population shared the *Liberty Tribune's* opinion that the war was winding down. Methodist minister Reverend George Paddock addressed his congregation pointing out how Confederates under General Robert E. Lee had been foiled in their quest to seize the north and had been resoundingly defeated at the Battle of Gettysburg; the Mississippi River flowed uninterrupted to the sea because of

---

[175] *Liberty Tribune*, August 21, 1863; Jackson County (Mo.) Historical Society Archives, Document ID Vertical File Newspaper Subject Files: Military/Wars: Civil War: 1863-"Order No. 11". "Robbery of a Mail Coach by Guerrillas," New York Times, 16 Aug 1863: "Kansas City, Saturday, Aug. 15. Jackson County (Mo.) Historical Society Archives, Document ID Vertical File Newspaper Subject Files: Military/Wars: Civil War: 1863-"Order No. 11".

General Grant's successful conquest of Vicksburg. These Union victories had only taken place one month earlier and were signs of the Good Lord's design, and thus "the beginning of the end unmistakably appears . . . Hope begins to smile over the land." Rev. Paddock insisted that no general had done more for the area, instilled better peaceful feelings and delivered greater security to the District than General Ewing. As the people of Lawrence retired on the evening of August 20, they were confident that there was no "special danger."[176]

William C. Quantrill

By the time anyone in Lawrence would be in a position to read the editor's opinion in the *Liberty Tribune* or re-read Parson Paddock's sermon, the citizens of the city probably would have considered tarring and feathering both of them. Their overall feeling of safety, victory and peace was rocked to the very core by an early morning raid on Friday, August 21: "Black Friday."

With a battle strategy finalized during their council of war, Quantrill began reassembling his forces. Between August 17 and 18 the largest band of bushwhackers united and gathered at their secreted rebel encampment on the farm of Captain Pardee located near the banks of Blackwater Creek in Johnson County, Missouri. The list of guerrillas included Bloody Bill Anderson, Jim Anderson, Arch Clements, William H. Gregg, Sam Hays, Joseph and Frank Lea, Fletcher Taylor, Captain Harrison Trow, Jim Cummins, George Todd, Dick Yeager, John McCorkle, David Poole, George Shepherd, John Ross, John and Tom Maupin, Cole Younger and Frank James.[177] These men Quantrill knew he could trust in a desperate fight.

---

[176] *Leavenworth Daily Times*, August 18, 1863; Smith, 193; Goodrich, "Black Flag," 76; Castel, "Frontier State," 123 and 126; and Goodrich, "Bloody Dawn," 27-28, 41 and 44.

[177] O.R. 22:2:486; Devil Knows How to Ride, 199; Edwards, 178; but see, Burch, 146, where Captain Trow claims the date was August 16 and not the 17th or 18th; Brownlee, "Gray Ghosts," 121; Eakin, "Recollections of Quantrill's Guerrillas," 57. Many of the Lawrence raiders would later become noted guerilla members of the James-Younger Gang. Edwards in *Noted Guerrillas* (202) provides an even larger list of guerrillas who were at Lawrence. Of those identified herein, Sam Hays was the brother of Colonel Upton Hays, who had been killed by Union troops on the western border of Missouri on September 15, 1862; Frank and Joseph Calloway Lea were sons of Dr. Pleasant John Graves Lea, a physician who lived in

By August 19, the guerrillas had broken camp, passed through Jackson County and rested along the headwaters of the Grand River in Cass County, Missouri. Their respite was short-lived and they were soon in their saddles continuing their westward trek. As they traveled, they encountered Confederate Colonel John D. Holt and his approximate 100 recent recruits from north of the Missouri River. Holt and his men promptly united with the guerilla chief. With these combined forces, Quantrill had over four hundred and fifty men under his command as they crossed into Kansas on the night of August 20. For the first time since blood splattered the streets of Independence and Lone Jack the previous August, Quantrill's men had become a large, united cavalry regiment.

By the predawn hours of August 21, the rebel forces were on the southern outskirts of the sleeping town of Lawrence. Quantrill perched upon his brown gilded horse gazed north towards the town and addressed his troops. "Boys, this is the home of Jim Lane and Jennsion." They had hunted "us" down and gave "no quarter." Quantrill had no need to remind his "boys" of this reality. They each had been given names of marked men to kill—number one on the hit list was Jim Lane. The raiders had their own personal reasons for the attack; from deep seeded hatred for these radical Kansans to revenge from the wanton inflicted injuries they and their loved ones had suffered at the hands of the Red Legs or Jayhawkers.

---

the area of today's Lee's Summit who had been murdered by Kansas Jayhawkers on September 12, 1862; and Cole Younger's father Henry Washington Younger, a loyal Union man, had been ambushed, robbed and killed by Kansas troops on June 20, 1862 (See, Amite County, Mississippi 1699-1890, Vol. No. 3: The Environs, Albert Eugene Casey, 1957; and Monaco, "Blood on the Streets," 23).

Dramatic images of Quantrill's raid, as depicted in publications from the 1870s and 1880s.

***Quantrill's Retreat*, a mural by Ernst Ulmer, as painted in the dining room of Bud and Dee Morris, Prairie Village, Kansas.**

In the close-up above, Mrs. Morris' ancestors, Mr. and Mrs. George N. Simpson, hide in a cornfield outside their home.

Lawrence burns in the background as cavalry retreats.

Quantrill's final order before the siege was to dispatch Lieutenant Gregg and other scouts to reconnoiter the outskirts of the town. The spies returned and gave the report Quantrill and his rebels had longed to hear. The town was asleep; no attack was suspected; the time had come to seize the initiative.[178] What they did not know would prove even more revealing. The previous night the Lawrence militia had been directed to stack their arms in the arsenal and get some rest. The town had "a hard time of it" and they had "stood their guard for months" and they were getting "used up for want of rest from drill and guard duties."[179]

Around 5:00 a.m. all the planning and preparation for the raid was in order. The cavalry was formed into columns of four and the sign to advance spread down the cavalry lines. As word reached the soldiers, they knew the hour had come to strike the terrible swift sword.

The guerrillas rapidly galloped and charged unmolested from south to north straight down Massachusetts Street (the Main Street of Lawrence) and parallel streets. Their rebel cry and chants of "*Remember Osceola*" or, "*Remember the Women in Kansas City*" bellowed from the Eldridge House to Mt. Oread.[180]

***Harper's Weekly*, 5 Sept. 1863**

The two to three thousand citizens of Lawrence experienced the most horrible wrath and destructive force ever waged or unloosed upon an American civilian population during the Civil War. Lawrence was robbed, looted, burned, pillaged and destroyed. Its residents were subjected to wanton wrath, murder and mayhem. The town offered neither defense nor resistance. Witnesses

---

[178] O.R. 22:2:15-16; Baron, 125; Goodrich, "Bloody Dawn," 82; Leslie, 201; Goodrich, "Black Flag," 76; and Yeatman, 44.

[179] Findlen, 65-66.

[180] Shalor Eldridge owned the Eldridge House. He had been a Free-State advocate dating to the days of Bleeding Kansas. As discussed earlier in this book, this was not the first time the hotel had been under siege. In May 1856, pro-slavery supporters had entered Lawrence in the first sacking of the town and had fired a cannon shot into the hotel before setting the torch to it. Mt. Oread was a hilltop in the western section of Lawrence that overlooked the town and named after property owned by Eli Thayer in Massachusetts. As previously covered in this work, Thayer like Amos Lawrence had been a leading force behind the Massachusetts Emigration Company and its settlement of Lawrence.

saw men slain in a "most fiendish manner." These included accounts of women on bent knees begging rebels to spare the life of a son or husband. Some did, most didn't.

Of all the desperados who took part in the Lawrence tragedy, the enraged and embittered Bill Anderson, Bloody Bill, imposed the greatest retribution for what had befallen his sisters the previous week. He reportedly spoke not a word until he had slaughtered fourteen men, many of whom he shot at point blank range in front of pleading, begging and praying mothers, wives and sisters; their pleas had been to no avail. Before exiting the town, Anderson reminded a female victim, "I'm here for revenge and I have got it."[181]

The rampage through the streets of Lawrence lasted about four hours. To the victims and their families it had been an eternal, relentless nightmare. Around 10:00 a.m., the annihilation and obliteration had ended. In their wake, Quantrill and his followers abandoned Lawrence in the grips of death, fire, smoke and ashes. Millions of dollars in property damage had befallen the ravaged town. Most haunting was the tragic reality that more than two hundred people had been injured, killed or mortally injured. In the end, the rebel assault became the most "barbarous act" of the entire Civil War.[182]

When dust clouds settled and the smoky haze dissipated, death and tragedy enveloped Lawrence. The bushwhackers left behind a town on fire, blood on the streets and about two hundred sent to their graves. It appeared to have been a ruthless victory of total war for the rebels. For Quantrill and his men their high risk invasion had not delivered complete justice to Missouri or to their cause: Senator Lane wearing only his night-shirt had escaped his house through a ravine and into a near-by cornfield where he remained concealed until the assault ended; Doc Jennison had the good fortune of not having been in town; and former Governor Robinson had left town for his early morning ride around the time rebel guns first opened on Lawrence. For Quantrill and his men their failure to bring an end to the reign of Lane and his brutal accomplices meant they still had unfinished work. Only one of the invaders died; he had become inebriated and separated from the main force. By the following day, Quantrill and his men had successfully retreated and eventually secretly scattered throughout western Missouri.

Those killed in Lawrence during the raid included soldiers and civilians alike.[183] Seventeen of the victims were black recruits shot during the opening stage of the attack, including one who had been shot while running across the street carrying a baby in his arms. As to the civilian population, 145 men were identified as dead or mortally injured. Included in this list was an unnamed baby identified as 'Baby Jones.' Cole Younger

---

[181] Castel, Albert, and Tom Goodrich. "Bloody Bill Anderson: The Short, Savage Life of a Civil War Guerilla." (Lawrence, Ks.: University Press of Kansas, 1998), 29.
[182] Leslie, 199-200; Goodrich, 236; Brownlee, "Gray Ghosts," 121; Castel, "Frontier State," 126-127; and Castel, "Bloody Bill," 29.
[183] http://history.lawrence.com/profject/commuinity/quantrill/victis.html. See also, Appendix B below for the complete list of the known and identified killed.

claimed one black woman had been shot and killed under the belief she was a man. Of all the victims only 165 victims were ever identified.

The actual number of slain was closer to 200. The higher death count is based upon the fact that of the 22 black soldiers found dead, only three were ever named or identified. The others were buried without identification. The unknown and unlisted included German immigrants and other visitors in Lawrence that day. Thirteen men were buried in a mass grave that included seven laborers and a foreman who had been in the town building a new windmill and bridge. Furthermore, there were a significant number of other men killed who were never identified because they were burned beyond recognition.[184] Contemporary accounts further support the higher death toll. Governor Thomas Carney's own official report to General Schofield estimated that "nearly 200 lives of our best citizens have been sacrificed." The *Leavenworth Dailey* reported that one hundred and ninety-three bodies were buried in Lawrence with seven additional remains being subsequently discovered.

As equally haunting was the *Dailey's* grim story of the survivors. There were an estimated 85 widows and 240 orphans made by Quantrill's slaughter. The property damage was assessed at one hundred and eighty-two buildings burned; eighty made of brick. The greatest devastation occurred on Massachusetts Street where 65 buildings were destroyed. A subscription of $100,000 was requested to rebuild the Free State Hotel, known as the Eldridge Hotel.[185]

On September 1, David Lee Campbell from Mapleton, Bourbon County, Kansas located south of Lawrence, wrote a letter to his brother James in Clayton, Adams County, Illinois. Campbell provided his brother a first-hand report of what he had observed of the devastation of Lawrence as he traveled through the town on his way to Leavenworth on business:

> *Dear Brother,*
>
> *. . . I came through Lawrence on Saturday. I [suppose] you have heard about Quantrill burning Lawrence and murdering the people last Friday week. Nearly all the good houses were burned. It is an awful sight to look at. They killed all the prominent men they could find. In fact they killed everything they saw except a few persons at the Whitney House, a few copperheads. Jim Lane fortunately got away and raised a few men in two hours and commenced to follow them.*
>
> *They stayed some two or three hours in town. They stole all the horses, money, and goods they could carry and burned the rest. They were*

---

[184] Leslie, 237.
[185] O.R., 22:2: 485-488; and Findlen, 63.

so hurried that they could not burn all the houses. And the women put out the fire when they could. But in the center of the town the rebels were too watchful to allow the fire put out. They were in the town before the people knew it.

They had their spies there for a week and some of the copperheads assisted them by pointing out houses to burn. There is no doubt but they intended to kill all the men of Lawrence and then go down through Kansas and kill and burn as they went . . .

The people of Kansas will be sure to avenge the slain of Lawrence. It looks hard to kill the wretches but if the Democrats of Clayton [Illinois] could see the women and children of Lawrence with horror and distress fixed in their faces. I think if they were not worse than brutes they would stop assisting Jeff Davis and his murdering thieves.

There are about 80 widows and over 300 orphaned children in Lawrence; all the work of these devils who all claim to be Democrats. But I must close....

Farewell,
D. L. Campbell[186]

Following the massacre, the excitement among the people of Kansas was expectantly bold, shocked, intense and furious. They demanded vengeance and insisted that extreme measures be imposed against western Missouri, and especially Jackson County. Three days after the invasion, Governor Carney wrote to General Schofield in St. Louis expressing the views of his constituents. The Governor described Quantrill and his band as the worst "fiends" known to mankind. Their actions had been nothing less than "savage barbarity." The Governor laid the responsibility for the successful "fearful, fiendish raid" at the feet of the people of western part of Missouri.

No body of men large as that commanded by Quantrill could have been gathered together without the people residing in Western Missouri knowing everything about it. Such people cannot be considered loyal, and should not be treated as loyal citizens.[187]

---

[186] Jackson, David W. "Order No. 11: The Lawlessness of Martial Law." *Jackson County Historical Society JOURNAL* (Autumn 2003), 17-20. The original of this letter is preserved at the Linn County Museum and Library, Pleasanton, Kansas (note: some punctuation and spelling corrected for clarity).
[187] O.R. 22:1: 489-490; and Goodrich, 239.

Schofield promptly responded to the Kansas Governor. He told Carney that nothing within his power would be "omitted to visit just vengeance upon all who are in any way guilty of the horrible crime, and to secure Kansas against anything of the kind in the future."[188] Schofield warned Carney not to allow the people of Kansas to retaliate, but allow the Kansas defense to be left to "United States troops to execute."[189] There was to be absolutely no retribution by Kansans. Schofield was determined to prohibit and prevent Kansas troops from entering Missouri to avenge the crime that had been perpetrated upon one of the oldest and "fairest towns" of the young state. He instructed Carney to adopt a system of organizing and arming the Kansas militia but not for invasion into Missouri. Rather the forces were to be assigned and garrisoned in every border town to defend against further guerrilla intrusion. Schofield insisted only he and Ewing had the authority to govern the situation, but he did provide Carney with some hope. He reassured the Governor that given the Lawrence massacre he and Ewing recognized that their previous military measures had been inadequate. He ensured Carney more sweeping steps would be forthcoming.[190]

As the last remnant of the guerrillas departed southeast from Lawrence, Lane conveniently slipped out of the cornfield and back into his ransacked, adopted hometown. He reappeared in the heart of town and resumed center stage. Lane was horrified, enraged and demanded retribution. He prayed and preached as if giving a fire and brimstone sermon. His impassioned homily raised an army of some fifty or more men. Following an impromptu counsel of war, his posse began its hot-pursuit of Quantrill and his thieves.

Lane and his cohorts had no problem locating the guerilla trail. They discovered burned houses along a ten-mile trek south of Lawrence. During Lane's forced march, he gathered more troops and eventually reached the rear-guard of Quantrill's retreating brigade. A skirmish ensued between the two opposing forces. Lane later asserted his men killed forty-five bushwhackers. He wanted more blood—all of it! Lane planned to organize more forces and return to Missouri, but federal troops prevented him from making any further intrusion into the state.

The stage had been set for a far-reaching campaign of retribution. Those targeted would be an entire civilian population. The consequences would be more exacting and terrifying than had ever been considered, prosecuted or implemented in American history to that date.

---

188 O.R. 22:1: 491.

189 O.R. 22:1: 491.

190 O.R. 22:1:15-16; Collins, 227; and Goodrich, 142.

CHAPTER X

Martial Law: General Order Number 11

As the enormity of the mass destruction spread like a wild fire across the nation, no one was more energized, excited, enraged and exercised than the senior United States Senator from Kansas. Senator Lane having freshly returned from his chase of Quantrill met with Ewing. Lane, never one hesitant to dominate, forcefully expressed to Ewing his frustration and dissatisfaction on the manner in which Ewing had handled the affairs within the District of the Border. Lane personally assailed and accused Ewing of directly contributing to the blood bath in Lawrence by acts and omissions of malfeasance and nonfeasance. During Lane's face-to-face encounter with Ewing he emphatically told the General the stark consequences if he did not issue General Order Number 11:

You are a dead dog if you fail to issue that order.[191]

Lane departed from his meeting with Ewing only when the General pledged his word that an order of total and complete depopulation would be issued.[192]

After Lawrence, Ewing faced several real dilemmas. Not only did he have to confront Lane's anger and venom, he had to resolve complaints from other political foes determined to further fan the flames. Ordinary Kansans, some of whom previously opposed to Lane, united behind the Grim Chieftain and demanded immediate retaliatory action. They did not hesitate to criticize and condemn Ewing. The excitement and fury had taken on a mob mentality clamoring to invoke the fullest measure of revenge. Specific charges were leveled against Ewing. He was ridiculed for having abandoned his own post at the time of the raid; failing to provide sufficient troops to defend Lawrence; and delaying his deployment of troops to hunt-down the invading rebels.

Ewing defended himself from these challenges. He justified his absence from his headquarters in Kansas City: he had to attend to official public business in Leavenworth; he was the only one who could personally address the troubles in Leavenworth; his departure to the Missouri River town had only occurred the day before the massacre; his implemented military measures included patrols in their saddles night and day for ninety miles; and he had utilized all the troops under his command in defense of the area. He

---

[191] The Devil's Knows How to Ride, 258; letter from Senator Jim Lane to Gen. Ewing dated August 22, 1863; and Brownlee, "Gray Ghosts," 125.
[192] O.R. 22:2:485-488.

was further dismayed that anyone would challenge him for not having permanently stationed troops to garrison Lawrence before the raid. No one had even previously suggested or requested this idea to Ewing. The General was willing to prove he had adhered to proper protocol and requested a military inquest to evaluate his conduct.[193] He was confident he would be vindicated by such an inquiry, and if not he would "quit the service at once" if "after candid investigation" it were determined he had been slightly negligent or had failed to exercise "average skill in the command" of his forces.[194]

Ewing realized the Lawrence disaster necessitated the implementation of prompt and sweeping measures. He also knew his political future required punctual retaliatory action. In evaluating his options, Ewing had to weigh and maintain a careful balance. He had to prevent Kansas' mobs from raiding into Missouri so as to avert further annihilation like that recently inflicted upon Lawrence. Ewing did not trust Lane or his radical supporters, but he was fully cognizant that Lane possessed political clout and a vindictive personality, and if there was anyone he had to pacify it was Lane.

Ewing was not the only political or military leader who recognized that strong measures had to be imposed in response to the raid. Provost Marshal Comingo knew that many members of the various terrorist organizations (on either side of the issues) were either common criminals or deserters. These men had become equally participatory in the violence and unrest. On August 24, Comingo issued an Order to encourage citizens to help rid the district of deserters. A reward of ten dollars plus "reasonable expenses incurred" would be given "to any person for the apprehension and delivery of any deserter at the headquarters of the nearest Provost Marshal."[195] All military methods to curb the violence needed to be utilized and controlling desertion would be an additional way to accomplish this end.

For a period of four days following the blood bath in Lawrence, Ewing listened to every argument and explanation on how to respond. In the end, he chose an aggressive alternative that would prove to be irrevocable to both his career and to the Missouri counties within his District. On August 25, Ewing issued his celebrated or infamous General Order No. 11. By this measure, Ewing was no longer going to wage war against rebels or their sympathizers. He was going to inflict total war against an entire civilian population. Except for a few carved exceptions, all residents of Jackson, Cass, Bates and Vernon County north of the Little Osage River were ordered to evacuate their property within fifteen days and report to the provost marshal at a military post within said time period. The areas exempted from the Order were located within one-mile radius of the military stations situated in Pleasant Hill and Harrisonville in Cass County and Hickman's Mill, Independence, Westport and Kansas City in Jackson County.

---

[193] O.R. 22:2:471-472.
[194] O.R. 22:2:473.
[195] *Liberty Tribune*, September 4, 1863.

All civilians in the impacted counties, regardless of their loyalty, were required to report to the aforementioned military stations. Those who established their loyalty to the satisfaction of the provost marshal or his designee were permitted to remain within the line of the military post, or the State of Kansas—but not those areas situated in the border counties of Kansas. All other persons were required to remove beyond the boundary lines of these military districts. All grain and hay found within the district was to be seized and turned over to the provost marshal. Crops not timely removed or used within fifteen days was to be destroyed. In some semblance of procedural due process, the name of all loyal owners and the amount of crops confiscated was to be properly certified. The purpose of this edict was designed for future reimbursement to loyal and guiltless persons left homeless by Order No. 11.[196]

Ewing's ultimate decision was based upon his deep seeded belief that citizens of Jackson County were disloyal. Their conduct to Ewing had included secreting and aiding robbers, brogues and murderers. He blamed these unscrupulous people for Quantrill's successes and cold-blooded invasion of Lawrence. Ewing reasoned that Order No. 11 would irrevocably terminate all the direct and indirect persistent assistance in which Quantrill and his followers had come to rely. It would rid the area from wanton bushwhacking as bushwhackers would no longer be able to seek comfort in the numerous hiding-places or other haunts from which they had been able to launch their terrorist attacks on federal troops, mail escorts or Kansas' towns. Ewing was steadfastly convinced that these disloyal folks had enabled Quantill's clandestine invasion of Lawrence—this would never happen again under his watch as such a calamity demanded drastic action. This meant total war and Ewing considered Order No. 11 as the only military measure in which he could curtail guerilla terrorist activities.

Ewing was fully mindful that the scope of Order No. 11 would impact blameless people. Although unfortunate, Ewing rationalized that the affect would be minimal because most loyal citizens had previously abandoned the region. The remaining loyalists had suffered at the hands of guerrillas, who had preyed, coerced, terrorized and threatened them for assistance, shelter and food. They were collateral consequences of the blood of some two hundred innocent victims that stained the streets of Lawrence. For Ewing, the burden placed upon these Union supporters was greatly outweighed by the benefit of evicting the vast majority of inhabitants who Ewing considered to be as culpable as Quantrill for the Lawrence massacre. Any and all pain or agony suffered by faithful Missourians was simply unavoidable and unfortunate, but he had no other option or alternative but total depopulation and banishment of everyone.

There were some forty thousand residents living within the four affected counties when Order No. 11 was issued. Within fifteen days, the population of Cass County

---

[196] Lambert Moon Printing Company. *Centennial Greeting Booklet, 1827-1927*, Independence, Missouri, 1927. Jackson Count (Mo.) Historical Society Archives, Document ID 76F28. An exact copy of Order No. 11 is provided in Appendix E.

decreased from about seven thousand to around eight hundred, and those who remained lived within a one-mile radius of the military posts at Harrisonville or Pleasant Hill. Bates and northern Vernon County became ghost towns (for Bates it became a ghost county). Neither Bates nor Vernon Counties had a military post in which its citizens could seek refuge. Jackson County with a population of about 23,000 suffered the most.[197] There was approximately 3,100 living in Independence, 4,400 in Kansas City and 1,200 in Westport. The total population of Jackson County evicted under Order No. 11 is estimated at about 14,000 citizens.[198]

Historians have written from both sides of the argument as to the "origin" of Order No. 11. The Order's nativity was certainly a natural evolution and outgrowth of Orders No. 9 and No. 10. To Ewing and his supporters, the new Order had been issued out of an exigency predicated upon the climate then existing in western Missouri. The ravages that had occurred in Lawrence justified the Order. To those adherents, Order No. 11 was the retaliatory war measure needed to curtail guerrilla activity and their untrammeled ruthlessness. Ewing had issued the edict out of military necessity.

The fundamental basis for Order No. 11 was the doctrine of military or exigent necessity. By 1863, this policy had become well established. Ewing was a lawyer and member of the judiciary long before he became a Union officer—he knew the law. His military duty was to ensure law and order. Ewing adhered to the political philosophy that the fundamental substantive laws of the United States were supreme to the rights of the citizens and states during a wanton and ruthless war.

Ewing had precedent on his side. Lincoln and later Congress had imposed drastic emergency military doctrines to quash the rebellion. Lincoln had abolished the writ of habeas corpus in the first year of the war causing the arrest and seizure of rebels or other disloyal citizens. Those so detained were deprived of any notice as to the reason for their confinement and were denied the right of trial by jury. Congress on July 17, 1862 further invaded the purview of the United States Constitution and the Bill of Rights when it

---

[197] The 1860 Census of Missouri provides the following total population in these counties. This list includes all white citizens, free blacks and slaves. The aggregate total of each county is as follows: Bates County, 7,215; Cass County, 9,794; Jackson County, 22, 896. Vernon County had a population of 4,850, but only few were impacted, as the bulk of the county was not located within the District of the Border. See, "Proceedings of the Missouri State Convention Held at Jefferson City." (St. Louis: George Knapp & Co., June 1862), 30-33. See also, The Devil's Knows How to Ride 260, McCullough, 31; Long E. B. with Barbara Long, "The Civil War Day By Day An Almanac 1861-1865." (New York: Da Capo Press, Inc., 1971), 401; and Larkin, Lew. "Missouri Heritage." (Columbia, Mo.: American Press, Inc., 1968), 13.

[198] Jackson County, Missouri, Census of 1860, abstracted by Hattie E. Poppino in 1964: Jackson County Historical Society Archives. This figure fourteen thousand is derived from deducting the number of residents living in these three cities (since they lived within one mile of the military posts) from the total population of Jackson County listed in the 1860 Census. This estimate does not take into account the large number of the seventy-five hundred residents living in those three cities who were banished by the provost marshal during Order No. 11 for having been southern supporters or agitators (e.g. Mrs. Johnston (Martha Ann "Mattie" Livingston) Lykins, later Mrs. George Caleb Bingham).

enacted the Treason Act or Confiscation Act. The Act declared those engaged in the rebellion as treasonous and subject to punishment. The declaration justified seizure of property out of military necessity.[199] The Emancipation Proclamation was nothing more than an extension of these war power policies. Lincoln in a confidential letter to Senator Orvill Hickman Browning described the practical application of military necessity.

> *If a commanding General finds a necessity to seize the farm of a private owner, for a pasture, an encampment, or a fortification, he has the right to do so, and to so hold it, as long as the necessity lasts; and this is within military law, because within military necessity.*[200]

For Ewing, Order No. 11 was consistent with these previously enacted war policies, all of which had been implemented to curb, destroy and end the rebellion to save the Union.

One week after Order No. 11 was decreed Schofield left St. Louis and traveled to Jackson County by steamboat. His mission was to investigate the situation developing on the western border. He first met with Ewing in Leavenworth and then held a conference with Kansas Governor Carney. Recognizing the political necessity of conferring with Senator Lane, Schofield held a private dialogue with him as well. Schofield attempted to mollify and assuage Lane's anger, animosity and apprehensions. Most significantly he cautioned Lane against sending Kansas forces to invade Missouri. He knew Lane and his adherents desired nothing less than to make western Missouri a desert and waste land. This Schofield could not and would not tolerate.

Schofield also listened to those who expressed open objections to Order No. 11. In Schofield's personal diary he wrote of his journey to the border and his inspection and discussion with men on both sides of the issue. He determined that it would be prudent "to modify Genl. Ewing's order . . . so that no property" would be destroyed as he deemed "the destruction of property unnecessary and useless." However, he was convinced that something drastic had to be done. He blamed the "chief evil" upon those "disloyal people" who had given support and aid to guerrillas or who had "conveyed" information or prepared "food for them." This wickedness would only be "removed" by enforcement of Order No. 11.[201]

In the end, Schofield gave his complete endorsement to Order No. 11. He expressly ordered Ewing to prevent any pretext by Lane or the other Jayhawkers from

---

[199] Kelly, Alfred H., and Winfred A. Harbison. "The American Constitution, It's Origin and Development." (New York: W. W. Norton & Company, Inc., 1975), 389; http://www.law.cornell.edu/wex/due_process.
[200] Neely, Mark E. Jr. "Lincoln and the Triumph of the Nation: Constitutional Conflict in the American Civil War." (The University of North Carolina Press, 2011), 123.
[201] John McAllister Schofield, *Diary of Events in Missouri, 1863*, Box 1, John McAllister Schofield Papers, Library of Congress, as set forth in "Missouri's War, The Civil War Documents," 160-161.

staging a Missouri raid. Lane had assembled over five hundred cavalry forces in Paola, Kansas to attack Jackson County. Ewing ultimately intervened and stopped the invasion but not without incurring the wrath and anger of Lane. The Grim Chieftain described Ewing as being nothing more than a traitor and coward.[202] Ewing was not deterred by Lane's oration. He had received Schofield's support to the war measures; for Ewing this had been a vindication of how he had handled the horrific events.

Ewing not only had to battle Kansans, and especially Lane, he faced detractors who insisted Order No. 11 was an illegal act waged upon the infirm and innocent. To the antagonists, the Order was in derogation of the natural order. It violated the fundamental substantive and procedural due process laws guaranteed under the United States Constitution. The most objectionable part of the Order was that a Kansan had imposed it upon private Missouri citizens. Provost Marshal Comingo and his advocates considered this the most devastating and humiliating part of the Order. Comingo saw Order No. 11 as an attempt by Ewing and his minions to prey upon individuals they had hated and despised for years—long before the official start date of the War.

A subordinate question has existed for years as to the exact location from which Ewing initially issued and executed Order No. 11. It is generally believed that Order No. 11 was decreed and signed at Ewing's headquarters at the Pacific House in Kansas City.

Others have maintained that Ewing declared and signed the document in Westport. A manuscript written by Adrienne Christopher following an interview she had with Mrs. August Rosa "Rose" (Becker) Hahn supports this conclusion. Hahn claimed that Ewing endorsed the Order over a dining room table in a sixteen-foot log cabin owned by Phillip Becker in Westport. According to this account, both Union forces and outlaws had utilized Becker's home for food, drink and shelter during the rebellion. She even insisted that Ewing was not in Leavenworth the night before the Lawrence raid as Ewing had maintained. Hahn said he was in the Becker house at the time of the attack. Washington H. Chick, a contemporary of Becker's, further supported this claim. He said he witnessed the reading of Order No. 11 in Westport and that upon hearing the Order he personally told Ewing that it was "the most outrageous thing he had ever heard in his life."[203]

---

[202] O.R., 22:2:577; and Smith, 203-206.

[203] Becker had been a pioneer shoemaker from Westport who died years later on November 27, 1913 at the age 81 in the home of his daughter, Mrs. August (Rose) Hahn, located 4205 West Prospect, Westport, Missouri, on November 27, 1913. Becker's log house stood on the corner of Broadway and Schafe for over 65 years. Becker had purchased the property from Joseph Glain, a bachelor, when he came to Westport from Germany in 1859. Prior to Mr. Glain, the cabin belonged to Charles Raber, another pioneer of Westport who was engaged in the freighting business and whose family had come to Westport from Switzerland in 1855. After purchasing the cabin Becker added two rooms and improved it by adding weatherboards. This cabin was later razed to extend Armour Boulevard to Westport Avenue. This 16-foot square cabin was a well-known landmark until it was destroyed. In the Jackson County (Mo.) Historical Society Archives Document ID 113F21, there are two articles that assert that document the signing of the

If the Westport execution is accurate, the location in that city was later disputed. Mrs. Joseph W. (Celeste Houk) Aiken disagreed that it had been signed at the Becker house. Mrs. Celeste insisted that Solomon Houk, a Santa Fe Freighter, had personally told her that he saw Ewing implement Order No. 11 at his residence in Westport located at 310-312 Westport Avenue. Houk told her that three Union officers had converted his home into their headquarters. Houk further related having seen Ewing drinking liberally until he became intoxicated. Ewing reportedly explained to Houk that his visibly inebriated state was because it was too terrible an edict to draft and sign while sober.[204]

Interestingly, Flora E. Stevens later supported both versions of the execution of Order No. 11—Westport and the Pacific House. Flora had been a "Southern" sympathizer, but her "father-in-law, Judge William Stevens, who lived in Kansas City" during the Civil War "was a Union man, and one of the Home Guards of the city." Flora claimed that Judge Stevens was a cousin of Ewing, both having been from Ohio. Flora's version was echoed by two Union men, Mr. John C. Gage, a leading Kansas City lawyer, and Mr. E. Stine. Mrs. James E. McHenry of Jefferson City, Missouri later sided with Flora's account. Flora's memorial of what Judge Stevens told her about the Order and the reason Ewing issued and executed it is as follows:

> [Judge Stevens] *detailed to me what I believe to be the true facts regarding this order. He stated that he saw Tom Ewing write at least the original draft of the order.* [There has been some discussion as to whether Ewing actually did write it.] *It was done at the old Pacific House in the city, still standing. This was in the presence of some of his staff and citizens he had invited to witness the occurrence. William H. Chick, an old citizen of Kansas City, yet living, told me he also witnessed the writing and heard Tom Ewing read it to the company.*

Order in the home Solomon Houk (or Houck) home in Westport. Jackson County (Mo.) Historical Society Archives, Document ID 82.06F12.

[204] *Kansas City Star*, August 21, 1912; Jackson County (Mo.) Historical Society Archives Document ID 82.06F12; Jackson County (Mo.) Historical Society Archives, Document ID 113F21; "Order No. 11: A New Story of its History--A Decanter of Whiskey Its Inspiration," reprinted from an article in the June 11, 1901 *Kansas City Times* and in the *Missouri Statesman*, Columbia, Missouri, June 21, 1901.

Original fascimilie of General Orders No. 11. Transcript in Appendix E.

Head Quarters District of the Border.
Kansas City, Mo. August 25th 1863.

General Orders)
No. 11.)

I. All persons living in Jackson, Cass and Bates Counties, Missouri, and in that part of Vernon included in this District, except those living within one mile of the limits of Independence, Hickman's Mills, Pleasant Hill and Harrisonville, and except those in that part of Kaw Township, Jackson County, north of Brush Creek and west of the Big Blue, are hereby ordered to remove from their present places of residence within fifteen days from the date hereof.

Those who within that time, establish their loyalty, to the satisfaction of the Commanding Officer of the military station nearest their present places of residence, will receive from him certificates stating the fact of their loyalty and the names of the witnesses by whom it can be shown. All who receive such certificates will be permitted to remove to any military station in this District, or to any part of the State of Kansas except the Counties on the eastern Border of the State. All others shall remove out of this District. Officers commanding Companies and detachments serving in the Counties named, will see that this Paragraph is promptly obeyed.

II. All grain and hay in the field or under shelter, in the District from which the inhabitants are required to remove, within reach of military stations, after the 9th day of September next will be taken to such stations, and turned over to the proper Officers there, and report of the amount so turned over made to District Head Quarters, specifying the names of all loyal owners and the amount of such produce taken from them. All grain and hay found in such District after the 9th day of September next, not convenient to such stations, will be destroyed.

III. The provisions of General Order No. 10 from these Head Quarters, will be at once vigorously executed, by Officers commanding in the parts of the District, and at the stations not subject to the operation of Paragraph I of this Order. — and especially in the towns of Independence, Westport and Kansas City.

IV. Paragraph 3, General Orders No. 10 is revoked as to all who have borne arms against the Government in this District since the 21st day of August 1863.

By Order of Brig. General Ewing
(Sgd.) H. Hannahs, A.A.A.G.

Recd
Dec. 10 '73.

Sept. 3/'75

This copy was made from the Newspaper slip with these features by a clerk this a.m. —
Jany 23 '73

> *He told Ewing 'it was the most outrageous thing he ever heard of....' Two of the staff expressed approval of the edict were Preston B. Plumb, later Senator from Kansas, and J. M. Hadley, father of ex-Gov. Hadley, who some years ago was the Republican executive of Missouri. Curiously enough, Hadley Jr., while previously Attorney General of Missouri, presented and secured settlement for a number of claimants against the United States for damages growing out of property destroyed by the Federal troops under this order which Hadley's father advocated and actively assisted in carrying out. Judge Stevens did not protest, since he did not think it wise to draw attention to himself at the time. Though Union, he was a kind-hearted man and had secreted in the cellar of his home--where the New England Bank now stands--a guerrilla of prominence, Col. Dick Chiles, of Independence, Missouri. He had been wounded and was in need of medical attendance. He was clandestinely visited by Dr. Ridge, a pioneer slave-holding physician. (It may be remarked that several times Dr. Ridge's own life was saved by sleeping in the same friendly cellar. Judge Stevens' silence may be readily explained. But he privately told Ewing, "it was a very tough order." Gen. Ewing answered that he was forced to issue this. It was conceived in the brain of Jim Lane.*[205]

Regardless of the true account as to where Ewing issued or executed Order No. 11, there is no dispute he declared, authorized and signed it. There is equally no challenge to the fact that the Order had a massive and everlasting impact upon Jackson County and the border counties to its south. It would eventually have a similar impact upon Ewing's career.

The Order was relaxed in November 1863 and was eventually rescinded by General Egbert E. Brown in March 1864. Brown's order allowed loyal citizens to return and attend to their farms. Brown's rescission order was the eleventh order he had issued after having taken over command from Ewing—ironically Brown's decree was known as General Order No. 11.[206]

---

[205] From an undated newspaper clipping from "The Sunday Republic Feature Section;" and see also, the Confederate Veteran, April 1917, Vol. 25, 154.

[206] Jackson County (Mo.) Historical Society Archives, Document ID 223.03F16. The football and basketball rivalry between the University of Kansas ("KU") and the University of Missouri ("MU") stood as the longest in sports history west of the Mississippi River until MU moved to the South East Conference ("SEC"). The collegiate animosity of KU and MU was pronounced along the border of Missouri and Kansas, becoming simply known as the "Border War." Sports fans probably do not realize today that the sporting tradition of the border-war between the two schools was an outgrowth of a tradition that dated from 1854.

CHAPTER XI

Responses to Order No. 11

The announcement of Order No. 11 created an outburst of protests from the majority of residents in Missouri and received ringing endorsements from others, especially citizens of Kansas. The *Kansas City Daily Journal* on August 29 proudly saluted the order and confidently stated, "families of the bushwhackers along the border are pulling up stakes and taking their way South. Drive these families all out, and the bushwhackers, if promptly followed up in the field, will soon leave, too."

Beleaguered Platte County, located to the immediate west of Clay County (the same county in which the editor of the *Liberty Tribune* had proclaimed its rigning endorsement of Ewing the day of the Lawrence raid) took immediate responsive and protective steps to avoid retribution. At least fifty "of the most noted secesh in that county" subscribed "from $1 to $10 each for the Lawrence fund." This action was taken in the hopes of eluding the anticipated devastation that Jackson County was to experience under Order No. 11.[207]

Senator Lane was only mildly pacified by the issuance of Order. He remained determined to see the total destruction of western Missouri. Lane was equally committed to utilize all of his political clout and devious means to accomplish this singular goal. He blamed Missouri Governor Gamble, Ewing and Schofield for preventing his retaliatory invasion into Missouri. Lane felt he had been rebuked by Schofield and thwarted by Ewing in bringing justice to Kansas. Lane was not one to be summarily dismissed or made to stand-down. On September 30, the Grim Chieftain led a contingency of supporters to the Executive Mansion to seek the President's intervention and support to their cause. Lane unleashed to Lincoln his fury by demanding the removal of Ewing and Schofield. Lane told the President that they had lost all confidence in the abilities of the two men. Lincoln listened to their complaints, not even the President could ignore the Grim Chieftain. Instead of dismissing Lane, Lincoln deferred the situation to General Halleck in Washington.[208]

No one could ever challenge Lincoln's western wit or his political talents. Perhaps the President's genius is illustrated by the manner in which he mollified Lane's wrath without agreeing to the Senator's dictates. Agreeing to investigate the situation without taking any affirmative steps to adopt Lane's insistent demands was Lincoln at his

---

[207] O.R., 22:2:486.
[208] Collins, 237.

diplomatic best. Although the President did not bow to Lane's insistent pleas, the Senator retained Lincoln's confidence and ear. Several factors account for this. Lincoln held a certain admiration for Lane. He had come to rely upon Lane for political assistance on Capitol Hill, and the Grim Chieftain had been a staunch supporter of the President. Lincoln's own political views were more aligned with Lane's than other leading Republicans. Neither man was part of the abolitionist wing of the party. They were westerners with similar philosophies. Lane and his "Frontier Guard" had defended the Executive Mansion in the opening month of the War; the Grim Chieftain provided the Administration a dependable voting block in Congress, and Lincoln needed to retain his relationship with Lane and his cohorts. Lincoln was aware of Lane's "lawlessness" and devilish ways, but he felt Lane's heart and soul faced "Zionwards."[209]

Numerous individuals and organizations commended and approved Ewing's Order. As stated above, Schofield was the first to publicly uphold and endorse it. After a protracted visit to the District of the Border examining the situation, Schofield acknowledged that the Order was severe but justified "the wisdom of it" as being "necessary at the time."[210] Historian, Joanne Chiles Eakin, in her book, "Tears and Turmoil," provides a copy of a letter Lt. Colonel Bazel F. Lazear of the 12th Mo. State Militia wrote to his wife on September 17, 1863 in which he expressed his opinion and validation of Order No. 11:

> *A great many think that was a very unjust and cruel order of General Ewings but I think that was one of the best orders that has been issued and I think will have a good effect. It may prove hard on some few Union people but very few as they . . . have had to leave a long time ago. And if we can't get the Bushwhackers out of here soon this region will be much the same way. I intend to send all the families of Bushwhackers out of this and Johnson county just as fast as I can give them notice to leave.*[211]

Toward the end of the Civil War, Major W. C. Ransom, who had been stationed in Independence when the Order was issued, sent the following missive to new Missouri Governor Thomas C. Fletcher on March 17, 1865, ratifying his former endorsement of Order No. 11.

> *Since Order No. 11, no Union man has dared to go beyond the limits of the military posts without incurring great risk and . . . no Union man has returned to his farm with the exception of a few in the immediate vicinity*

---

[209] Collins, 237.

[210] *Kansas City Star*, August 25, 1913.

[211] Eakin, Joanne Webb Chiles. "Tears and Turmoil: Order No. 11." (Independence, Mo.: Two Trails Publishing, 1996), 49.

of Kansas City and Independence, whereas on the other hand a majority of the rebels have returned . . . and are surely in league with the bushwhackers . . . we dare not go to our farms for fear of being devoured by the bushwhackers and their kindred spirits . . . Give us good and true Union men for our civil offices.[212]

Ransom was not the only observer who spoke favorably about Order No. 11 during or after the War. In an article entitled, "A Half Century Since Order No. 11," Schofield was still being quoted as to the sound reasoning behind Order No. 11:

I directed General Ewing to remove the families of all guerrillas and all those who were known to aid them and also the slaves of all disloyal persons living in those counties.

Such was the purposes of Orders No. 9 and No. 10 dated August 18. Even before Order No. 11 was officially decreed, Schofield wrote the following letter to Ewing:

A country inhabited almost solely by rebels could not be permitted to remain a hiding place for robbers and murderers from which to sally forth on their errands of rapine and death. It is sincerely hoped that it will not be necessary to apply this remedy to any other portion of Missouri. But if the people of disloyal districts wish to avoid it they must unite to prevent its necessity, which is clearly in their power to do. This order will be executed by Brigadier General Ewing, commanding District of the Border, and such officers as he may specially detail for the purpose.

Later, Schofield in his official report upheld the drastic measure of Order No. 11 because the two prior orders had not been "adequate," and "total depopulation of the district" was essential to eradicate the "chief" haunts "of the guerrillas."[213]

Other supporters claimed it was warranted, not expressly in terms of military necessity but under the doctrine of exigent circumstances. To these proponents the Order was an essential and necessary response to an emergency situation that had grown out of control. They compared the condition along the border to the "Great Fire" in St. Louis. In the spring of 1849 the *White Cloud* steamboat was docked at port on the Mississippi River. For unknown reasons, fire erupted on the steamer and soon spread to some twenty-two other ships tied to the wharf. It did not take long before the entire riverfront was engulfed in flames and the conflagration swept uncontrollably, consuming block after

---

[212] Missouri State Archives, Box F 14495.

[213] O.R. 22:1:573.

block of the city's downtown. The only remedy available to the volunteer fire department was to detonate kegs of black powder to destroy several businesses and city blocks. The plan was successful as once the flames reached the destroyed area it died out—the fire had no place in which it could further extend. The area destroyed by the detonation had created a fire-block or firewall against what had seemed as an insatiable inferno.[214] Hence the desolation of the counties in western Missouri had become the firewall to prevent further extension and incursion of guerrilla activity into Kansas.

Years after the war, there remained support for the Order and Ewing's actions. Although the comparison to the St. Louis fire was not raised in the "History of Bates County," the analogy is apparent. Bates County, located two counties south of Jackson, lost its entire population of some seven thousand residents. It was the only county of the four in which absolutely everyone was forced to vacate the area—even the county seat in Pappinsvile was destroyed. Despite this background, the author considered Order No. 11 as the only action in which guerrillas and bushwhackers had been successfully thwarted as it had deprived them of their base of operation. In other words, the conflagration could no longer spread because the entire land between the engulfed fires in Lawrence had been destroyed.[215] Even Captain Trow, a Quantrillian raider at Lawrence later claimed that Ewing had been unjustly censured for the promulgation of Order No. 11.[216]

Despite the contemporary and subsequent endorsements of Ewing's decree, those in opposition were instantly more vocal (and have remained so to this day). First and foremost was General Frank Blair of St. Louis. Blair was from one of the most influential families in America. His father Preston, a former Democrat and outspoken proponent of the late former United States Missouri Senator Thomas Hart Benton, had been a major political force on the national level. Preston Blair had also played a leading role in the establishment of the Republican Party. Lincoln could not ignore the influence of the Blair Family. In 1861 Lincoln had appointed the elder Blair's son Montgomery to his cabinet as Post Master General.

General Frank Blair's influence fit his family's pedigree. He represented St. Louis in Congress at the outbreak of hostilities and was a steadfast Union man. He had staunchly stood beside General Lyon during the early days of the rebellion in 1861 to ensure Missouri remained within the Union. By the time of Order No. 11, Blair served as a general under Grant and Sherman. When the Blair's spoke, the President listened! After General Blair heard the announcement of Order No. 11, he promptly described it as a "damnable atrocity" and an act of "imbecility." Brigadier General Odon Gutiar who commanded the district to the north of the Missouri River equally shared Blair's

[214] www.stlouis.com/history.

[215] National Historical Company. "History of Cass and Bates Counties, Missouri." (St. Joseph, Mo.: National Historical Company, 1883).

[216] Burch, 156-157.

sentiments. Guitar spoke of Order No. 11 as being one of the most atrocious edicts ever issued.[217]

In a letter written by Guitar to Bingham only a few days before the latter's death, Guitar questioned the validity of the Order and voiced his objections to it. Guitar described the Order as "unnecessary, unwise, cowardly and barbarous" and proclaimed it was "one of the monuments of military despotism and infamy." To Guitar there was ample evidence that Ewing issued the Order for no other reason than to retaliate against western Missouri. To prove his theory, Guitar presented rhetorical arguments.

> *First, if a measure of necessity and its object the lives and property of the depopulated counties of Missouri against the incursions of "rebel marauders," with whom a majority of their inhabitants were supposed to be in sympathy, why was not the same generous protection extended to the citizens of the contiguous border counties of Kansas, for whom these marauders entertained the bitterest of hatred? It seems "sauce for the goose" was not "sauce for the gander."*
>
> *Second—How did it happen that, immediately upon the promulgation of the order (if not before) the "red leg banditti" of Kansas swarmed over the line into Missouri and commenced the track of devastation and plundering; the coincidence of these events was too marked to be accidental.*[218]

Other public figures came forward to protest the Order. Three leading citizens voiced their outrage and opposition in a letter to President Lincoln. Included among these dissenters were Provost Marshal Comingo and Richard C. Vaughn of Lexington—whom Bingham considered to be "a gentleman of high standing, culture and impeachable veracity." On August 27 these men sent a vehement protest to Lincoln articulating their disdain about the Order and their objections of Ewing as commander of the District of the Border. These "memoralists" were adamant that the counties in Missouri should be immediately detached from the District of the Border and that no one from Kansas, especially Ewing, should hold dominion over Missouri; to them Ewing had to be removed.

[217] "Bingham Letters," 532.

[218] Letter Odon Guitar to Bingham dated July 4, 1879, "Bingham Letters," 537.

His Excellency ABRAHAM LINCOLN,

President of the United States:

Your memorialists would respectfully state that they are loyal citizens of the United States and of the State of Missouri, and, having been such at all times, they regard it as their right and duty to represent to Your Excellency the unhappy condition of affairs now existing along the western border of their State, and to pray Your Excellency's interposition in behalf of a suffering people. Your memorialists feel that justice and humanity demand at least this much at their hands. They therefore beg Your Excellency's attention to the facts hereinafter appearing . . .

Your memorialists further beg leave to say that one of the most fruitful sources of trouble in Western Missouri is the attachment of a part of her territory to the District of the Border . . . Old animosities existing between the people of Missouri and Kansas . . . embittered and intensified by the recent barbarous acts of a guerrilla band perpetrated upon the citizens of Lawrence, in the latter State, will develop themselves, and will seek gratification in retaliatory acts upon the citizens of the former . . . A fearful retribution no doubt awaits Missouri.

In view of these facts, your memorialists respectfully, but most earnestly, pray Your Excellency to rescind the order by which a part of Missouri is attached to the District of the Border, and to order that it be reattached to the Central District of Missouri, or to any other district in our State . . . that some tried and faithful officer may be placed in command over the soldiers and people in the counties of the border--some officer whose sense of duty and of love to his country rises far above his political aspirations and party ties and prejudices . . .

This is all that the masses of the people desire, and for this your memorialists will ever pray, &c.[219]

A. A. KING.
R. C. VAUGHAN.
A. COMINGO

[219] O.R. 22:2:485-488. The complete text of Comingo's letter is set forth in Appendix F below.

The response provided by the President to the missive from the "memorialists" was dispatched on August 31; it was short, terse and bluntly pragmatic:

> *It is not improbable that retaliation for the recent great outrage at Lawrence, in Kansas, may extend to indiscriminate slaughter on the Missouri border, unless averted by very judicious action. I shall be obliged if the General-in-Chief can make any suggestions to General Schofield upon the subject.*[220]
>
> *A. LINCOLN.*

In other words, just deal with it! Lincoln could not ignore the tragedy of Lawrence and would not interfere with the new war policy; Ewing would remain in power. The President was also not willing to be placed in a position to further the wrath of the Senator from Kansas, Jim Lane.

Members of the civilian population came forward to express opposition to Order No. 11. Rev. George Miller, a Presbyterian minister in Kansas City, insisted that the area impacted by the Order was no longer important to Quantrill. Rev. Miller insisted that the eviction Order would cause great hardship because only about one in five families even had a mode of transportation. The minister was further concerned about the hardships that would follow by forcing the relocation of so many people as the fall and winter seasons were approaching.[221]

The prayer for relief by Provost Marshal Comingo and his "memorialists" along with other leading citizens of Jackson County fell on deaf ears. Their appeal may have been ignored, but there was yet another voice to be heard from the State Capitol. He had his own political connections in the Missouri and Washington. He also held firmly to the same positions articulated by Blair, the "memoralists" and Rev. Miller. Word came out of mid-Missouri that the State Provisional Treasurer George Caleb Bingham was incensed and was not going to sit back; he would take immediate action.

Bingham had always been known for his fighting demeanor and willingness to attack any adversary. Order No. 11 would become his most climatic propagandist fight—becoming a lifelong assault and personal vendetta against the character of its author, General Ewing. Bingham also possessed a strong sense of political knowledge, opinion and skill, but his impulsive temper and impatient demeanor often caused him trouble. The summer of 1863 would illustrate this dichotomy. Throughout the war, Bingham held back much of his venom and disdain towards political leaders from Kansas, but when Ewing had confiscated and converted his late father-in-law's property into a woman's prison Bingham's limited patience was tested. When the "prison" collapsed killing five women

---

[220] O.R. 22:2:488.

[221] Niepman, Ann Davis. "General Orders No. 11 and Border Warfare during the Civil War." *Missouri Historical Review*. (January 1972): 66: 2: 185-210.

and injuring many others, Bingham placed the blame squarely upon Ewing.[222] Finally, with the issuance of Order No. 11, Bingham's anger reached a boiling point and exploded.

When word reached Bingham in Jefferson City of the issuance of Order No. 11, he became exercised and enraged. Loyal Missourians advised the State Treasurer to go to Kansas City to demand the rescission of the Order and request the arrest of Lane as a seditionist.[223] It has become part of the Bingham lore that he traveled from the Capitol by steamboat to Kansas City where he personally challenged Ewing and his Order. No extant letters or official contemporary records exist of Bingham having made the trip or meeting with Ewing at his headquarters in the Pacific House. Bingham, however, proclaimed years after the War that he had in fact done so.

According to Bingham, he met with Ewing at his office in the Pacific House. Bingham claimed that when he met Ewing, he promptly demanded the immediate rescission and vacation of the Order. Bingham reminded Ewing that the Order would "damn" him before the world. Ewing's response was emphatically curt and pointedly explanatory: "I am a soldier and have to obey orders."[224] It is not certain if Ewing was expressing that the Order came from himself or from somewhere higher up in the chain-of-command, such as Schofield.

In the end, the meeting did not go well for Bingham. Ewing was not moved by Bingham's insistence. This was neither the time nor the hour to exchange political niceties with another politician. Ewing promptly refused Bingham's demand and summarily ordered the State Treasurer out of his office. With this rebuke and discharge, Bingham spun on his heels allegedly proclaiming, "I will make you infamous with my pen and brush as far as I am able." Another translation of Bingham's blistering statement was that if God spared his life "with my and pencil, I will make this order infamous in history."[225] Regardless of the quote or whether any meeting was held, Order No. 11 marked the beginning of Bingham's "war" against Ewing. Bingham desired retribution and it would be waged against Ewing the rest of his life and beyond—literally.

Ewing was not deterred by Bingham's threat. He remained steadfast and stood firm to the Order. He was not inclined or willing to give-in to anyone, including Bingham. To Ewing, the outrages on the border had become so uncivil. He was firmly convinced regardless of the potential hardships to innocent civilians or to him personally

---

[222] Letter from Bingham to James S. Rollins dated December 21, 1863, "Bingham Letters," 283; and Nagel, 103. Bingham would spend the rest of his life in a failed attempt to be compensated by the federal government for the $5,000 loss his family experienced by the collapse of the property.

[223] Smith, 204.

[224] Jackson County (Mo.) Historical Society Archives Document ID 82.06F12 and Dcoument ID 113F21.

[225] Nagel, 112; and Rash, 191. Rash contends that the only documentation to the encounter between Ewing and Bingham came from a Kansas City newspaper story in 1877. However, Bingham's letter to the editor of the Republican Newspaper dated February 22, 1877 ratifies the account of him going to see Ewing in an attempt to have the Order rescinded or modified, "Bingham Letters," 472.

Order No. 11 had to remain in effect and be aggressively enforced. He could not and would not rescind the Order; it was issued out of military necessity. Bingham's pledge and the impact of Ewing's firm rejection of the State Treasurer's demand would in the end prove to haunt both men the rest of their lives.[226]

[226] *Kansas City Genealogist*. Vol. 23, No. 1 & 2, 7.

CHAPTER XII

Smoke in the Sky

Order No. 11 had a devastating impact upon the lives and property of all citizens of Jackson, Cass, Bates and northern Vernon Counties.[227] Many were irrevocably touched, affected, compromised and destroyed by the Order.

Artwork by Bill Wehner

[227] A roster, biographical data, and stories of those known to have been banished specifically under Order No. 11 are delineated in Appendices G and H below.

Mrs. Emily Steele summarized the horrible account in a letter she wrote from Lexington to her son George on September 20.

> *I think my troubles are more than I can bear – our family is scattered to the four winds . . . [E]verybody, man, woman and child had to leave the border counties. Those counties are entirely depopulated except for Bushwhackers, and Scouts of Federal Troops and Kansas Jayhawkers driving off stock and hauling off grain which the citizens had not time to get away and dare not go back for. We got away some few head of horses and beds and some of our best furniture, we have it in Lexington. Your Pa is trying to sell the stock, but prices are low.*[228]

Mrs. Steele's saga was similar to those suffered by other residents of Jackson County. They all experienced the pain and agony of being "scattered to the four winds" under Order No. 11. The Order was compelling and complete in its totalitarian effect, and it subjugated the citizens to the omnipotent dominion of the provost marshal.

> *We learn that quite a number of individuals and some entire families, popularly known as 'sympathizers,' have been ordered to leave this District. We do not understand that these parties are accused of aiding or abetting the guerrilla movements, hereabouts, but that they are removed in accordance with the same policy which has induced the Commanding General to order ALL of the inhabitants out of certain portions of the District. We are informed that the Commanding General has ordered away such parties only as, upon the best evidence he could find, were publicly regarded either as positively disloyal in speech or sentiment, or as in sympathy with the rebellion. He regards the presence of such individuals as highly prejudicial to the peace of the border, and as liable to result in extending of aid and comfort to the guerrillas whom he is endeavoring to exterminate.*[229]

The extermination was described as a "devastating sight." Broken wagons, lame horses and hordes of long-trains of families were seen for miles and miles. Two of the "memorialists" who had written to President Lincoln, Vaughan and Provost Marshal Comingo had the opportunity to witness first-hand the impact in which Order No. 11 had upon ordinary citizens. Vaughn described the horror "as the cause of more suffering than

---

[228] Eakin, "Tears and Turmoil," 26. See also, Eakin, Joanne Chiles. "Our Far Away Boys: The Civil War Letters of the Emily Steele Family." (Independence, Mo.: Two Trails Publishing, 2004).

[229] Eakin and Hale, 54, quoting, *Daily Journal of Commerce*, 1 Sept. 1863.

all the other incidents of the war combined.[230] Comingo in a letter he wrote to the State Provost Marshal General James O. Broadhurst on August 30 described the affect of the Order in a chilling and descriptive manner. The "people are crazy from fear and terror with which their lives are filled . . . weeping and wailing like children." Comingo saw citizens crowded into "dilapidated outhouses," tents or in huts "constructed from the boughs of tree" and "huddled together in the little villages in which posts have been established and are suffering from the want of food and other necessities of life."[231]

Lt. Colonel Bazel F. Lazear, of the 12th Missouri State Militia despite having been on the hunt for Quantrill for fifteen days and an advocate of Order No. 11, was emotionally moved by what he had witnessed. Lazear in a letter written from Lexington to his wife on September 10 described what he saw.

> [T]*here is hundreds of people leaving their homes from this country, and God knows what it to become of them. It is heart sickening to see what I have seen since I have been back here. A desolated country and women & children, some of them almost naked. Some on foot and some in old wagons. Oh, God. What a sight to see in this once happy and peaceable county.*[232]

For miles there could be seen long, moving trains led by sorrowful exiles. Teams of every sort, except good ones, drew wagons and vehicles of every shape, size and all other types. Some could be seen driving their "flocks and herd or leading a cow or a skeleton horse, with a bundle or pack fastened upon it, or a packhorse, on which the feebler members of the family rode by turns." One solitary woman was seen leading a team of oxen while her three children rode inside the cart. Another lady was observed having two cows hitched to a wagon with her little boy leading the team.[233] The majority was only able to remove provisions they could carry or pack into their makeshift wagon or wheel barrels. Most of the victims were women and children who were "bare-footed and bareheaded . . . stripped of every article of clothing except a scant covering for their bodies" who were "exposed to the heat of an August sun" and compelled to march quietly and dejectedly "through the dust on foot" carrying in their arms little bundles or their children.[234] Clouds of smoke encircled the sky from the multitude of fires emitting from the burning buildings, and dust that filtered into the air from the dirt roads that were

---

[230] "Bingham Letters," 529-530.
[231] Fellman, 78.
[232] Eakin, "Tears and Turmoil," 48; and McCullough, 32.
[233] Eakin, "Tears and Turmoil," 97.
[234] Bingham's letter to the Republican Newspaper dated February 22, 1877, "Bingham Letters," 427.

stirred by the column of wretched fugitives departing from Jackson County.[235] The day and night blended together.

There refuges were filled with great confusion on where to relocate. Some sought comfort with families inside Missouri or somewhere else far removed from the state. Others who made their departure had no idea where to seek an asylum. Many were precluded from resettlement by strict military prohibitions. In Clay County, just north of Jackson County on the northern bank of the Missouri River, the commanding provost, Major John R. Green, issued on September 9, 1863, the following special order from his headquarters in Liberty.

> *SPECIAL ORDER*
>
> *All persons who are leaving Gen. Ewing's district in compliance with his order No. 11 are hereby prohibited from stopping in this county to reside. All those failing to comply with this order will be escorted beyond the lines of the county.*
>
> *John R. Green*
> *Major Commanding Post*
>
> *By Robt. W. Flemming, Act. Post Adjt.*

Officials in Ray County (to the immediate east of Clay) posted similar orders' as did other northern Missouri River counties.

The victims of Order No. 11 had become fugitives in flight. Confederate troops encountered many traveling through Arkansas. They were described as women and children and old men, in rickety wagons, drawn by teams too shabby for army service. The late summer of 1863 in Jackson County was "truly awful."[236] Loyal, disloyal and neutral citizens experienced the wrath and vengeance from Missouri Militia, Kansas Red Legs, Jayhawkers and rebel guerrillas. Years after the war, Bingham described what he saw from the ravages that had been rout upon Jackson County by the Order:

> *Families were peremptorily ejected from their dwellings. Robbery and murder sought no concealment. Men were halted in thoroughfares and compelled to surrender their purses regardless of the presence of spectators. Large sums of money were extorted from the wealthy to glut*

---

[235] Martin Rice, Rural Rhymes, and Poems from the Farm, Millett & Hudson Kansas City, Missouri 1877; Martin Rice, Rural Rhymes, and Tales of Olden Times Ramsey, Millett & Hudson, Kansas City, Missouri 1882, Martin Rice Remembers: What I saw of Order Number Eleven, Missouri Historical Review, Volume 106, No. 3, April 2012; Union, 940-941; and http://www.historiclonejack.org/whatisaw.html.

[236] Webb, 184-185.

the avarice or support the profligacy of officers. Stores were entered and the goods of the proprietors taken from the shelves, boxed up and shipped in Government wagons to the State of Kansas, and there sold for the benefit of the plunderers. Carriages, buggies, wagons and horses were alike unscrupulously appropriated. Citizens were incarcerated in dungeons, and there held until they made to their persecutors a bill of sale of the property, as the price of their release. Many were arrested and executed without a hearing or a trial, and their sad fate made known to their friends only by the discovery of their lifeless bodies suspended from the limbs of trees by the wayside.[237]

No one was safe during the enforcement of Order No. 11, regardless of their sympathies.

Of the many personal hardships, the story of Dr. Lydall W. Twyman and his wife Frances "Fannie" (Fristoe) Twyman and their family is commonplace. They were only given a few hours to vacate their property. Most items of their property had been previously taken or destroyed. They were compelled to load all that remained into a damaged wagon and an old buggy. They first left Jackson County and traveled to Howard County. Later they relocated to Missouri City in Ray County and did not return to their home until after the War. When they returned, they discovered their property greatly damaged and their house in a dilapidated condition.

The following stories are selected to represent the experiences of thousands of Jackson County, Missouri, victims of Order No. 11, best told from their own personal sufferings and recorded accounts.

Additional, authentic, historical accounts, and a complete roster (as we know it to-date 150 years later, in 2013) of Jackson, Cass, Bates and Vernon Countians affected by Order No. 11, may be found in Appendices G and H below.

[237] From a pamphlet published by Bingham in late 1871, "Bingham Letters," 322. In an oral history provided by Howard Lay an attorney in K.C. to the author on June 12, 2013, Mr. Lay stated that his mother told him that her uncle who lived outside of Independence traded his depreciated real property holdings for a mule and escaped to Kentucky under Order No. 11.

Saga of the Luttrell Family

Like many residents of Jackson County the Luttrell family had deep southern roots. The Luttrell patriarch, Samuel, was born in 1813 in Patrick County, Virginia. His parents, Willis and Sarah Luttrell, had moved them to Missouri in 1830 where Willis Luttrell purchased more than a quarter section of land. Samuel's wife, Armilda Dalton, was born June 27, 1831 in Russellville, Logan County, Kentucky. Her parents, Lewis and Matilda Rabourne Dalton had likewise migrated to Missouri when Armilda was a young child. Both the Luttrell and Dalton families lived near each other in eastern Jackson County.

After Samuel and Armilda were married, they built a log cabin about three miles southeast of Independence. During those trying Civil War years, the Luttrell's were known for their southern sympathies. They had family and friends within the ranks of noted guerrillas including Cole Younger and Frank James. On one occasion, they provided shelter to Quantrill after he had barely escaped capture by Kansas Red Legs.[238]

[238] Cole Younger's family was distant cousins to the Daltons. Moore, LaVonne Belew. "Missouri Town-1855: A Program in Architectural Preservation." (Kansas City, Mo.: Jackson County Parks and Recreation in cooperation with Lowell Press, Inc., 1987), 44-45.

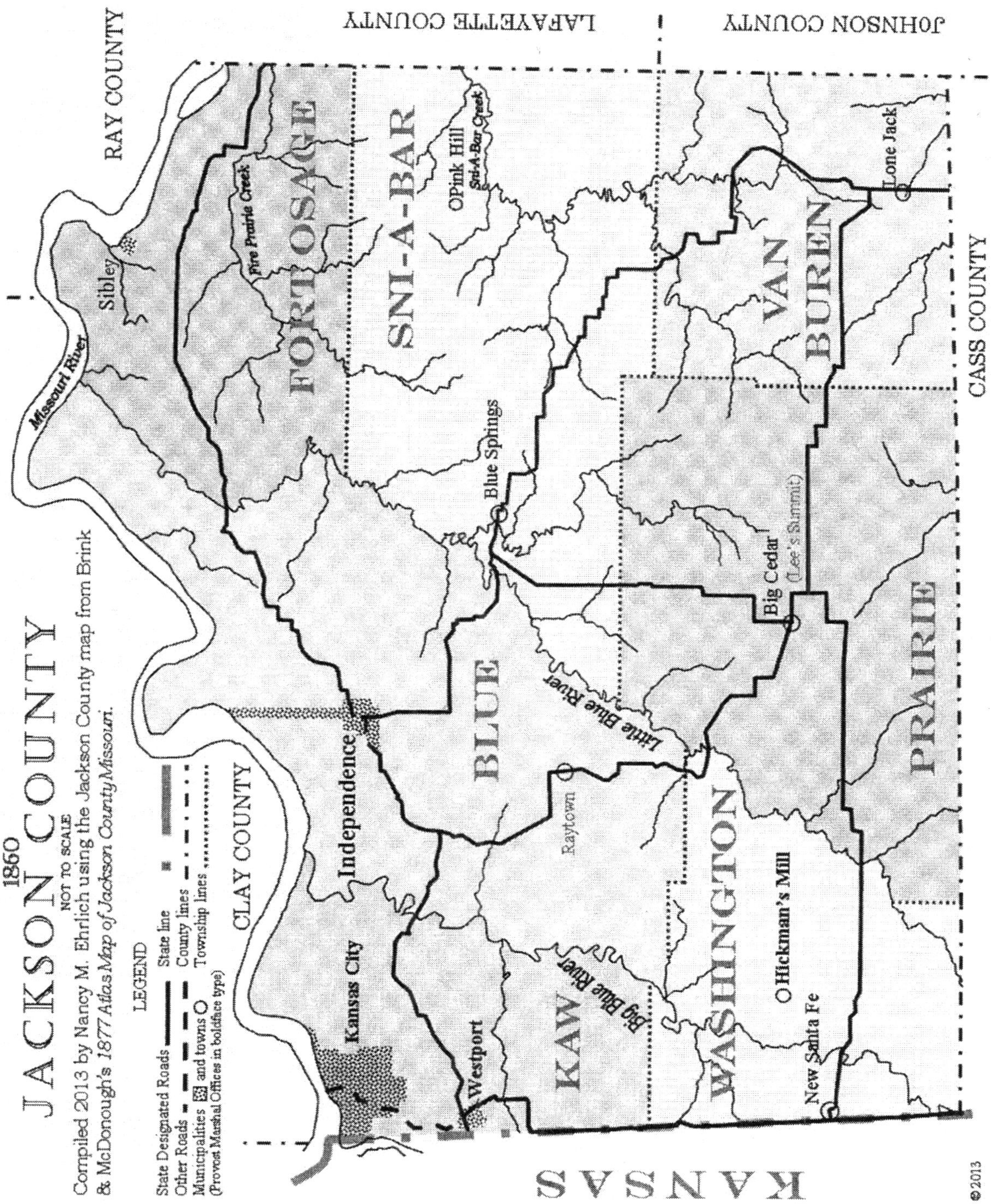
1860
JACKSON COUNTY
NOT TO SCALE
Compiled 2013 by Nancy M. Ehrlich using the Jackson County map from Brink & McDonough's 1877 Atlas Map of Jackson County Missouri.
LEGEND
State Designated Roads
Other Roads
Municipalities and towns
(Provost Marshal Offices in boldface type)
State line
County lines
Township lines
RAY COUNTY
LAFAYETTE COUNTY
JOHNSON COUNTY
CASS COUNTY
CLAY COUNTY
KANSAS
Missouri River
Sibley
Fire Prairie Creek
FORT OSAGE
SNI-A-BAR
Pink Hill
Sni-A-Bar Creek
VAN BUREN
Lone Jack
Blue Springs
Big Cedar
(Lee's Summit)
PRAIRIE
BLUE
Little Blue River
Independence
Raytown
WASHINGTON
Hickman's Mill
New Santa Fe
Kansas City
Westport
KAW
Big Blue River
©2013

The Luttrell and Dalton families lived on adjacent properties when the 1860 Census was enumerated. Matriarch Sarah Luttrell died shortly thereafter on October 13, 1860 at age 72. By the time Order No. 11, the Luttrell's and their surrounding neighbors and friends were compelled to report to the provost marshal at the military post in Independence. Before they departed, they took painstaking measures to conceal some property they were unable to take with them. With limited time to spare, the Luttrell's dug trenches and buried their most precious belongings, including their feather beds.

> *Deep trenches were dug to bury their feather beds, especially, for they must be kept dry and could not be replaced until they raised geese to yield the feathers to make new ones. For in those early days the settlers and farm people did not have factory-made mattresses. Even to the turn of the century we filled our unbleached muslin ticks each summer from the fresh straw at the straw tacks after the wheat was thrashed. Our feather beds which were used on top of these straw mattresses were treasures in the household. It is not known where the Luttrell family went for shelter but they left on horseback and could only take with them what they could carry on their horses. It was almost two years before the War ended. Let us hope the buried feather beds were not ruined! No wonder Mrs. Luttrell always had an intense hatred for Abraham Lincoln.*[239]

The Luttrell's did not return to Jackson County until after the Civil War. There are no extant records that reflect whether the treasures buried in those trenches were ever found, but they were more fortunate than most other neighbors as they found their home still standing when they returned.

Impact Upon the Family of a Future President

The political make-up of the household of Solomon Young and his wife Harriet "Louisa" (Gregg) paralleled many other residents in Jackson County at the outbreak of

---

[239]Union, 915; Manuscript written by Ethel Milton Hughes, 1963, for the Jackson County Historical Society Contest, entitled "Some Interesting Ancestors Who Lived in Jackson County." Jackson County (Mo.) Historical Society Archives, Document ID 77.04F6. See also, Hughes' account in the February 1962 edition of the Jackson County Historical Society JOURNAL, entitled "Some Interesting Ancestors Who Lived in Jackson County." Jackson County (Mo.) Historical Society Archives, Document ID 77.04F6. Also, Jackson County Census, Blue Township, June 20, 1860, page 178. The Samuel Luttrell log cabin was located about today's intersection of 39th Street and Lee's Summit Road, near present day St. Mark's Catholic Church. The house survived Order No. 11. In February 1962 the structure was saved from destruction and was physically relocated to the living history museum, Missouri Town 1855 near Lake Jacomo in Lee's Summit, where it remains part of the living history village. Missouri Town is owned by Jackson County and is operated by Jackson County Parks + Rec.

the rebellion—one of divided loyalties. Solomon, whose grandfather had served in the American Revolution, was firmly an avowed Union man; his wife's family, the Gregg's was southern sympathizers. The Young's eldest son William had enlisted in General Sterling Price's Missouri State Guard in 1861, and their third child Sarah Ann was married to the notorious guerilla Jim "Crow" Chiles.[240] Solomon and Louisa were natives of Shelby County, Kentucky; he was born there on April 24, 1815, and Louisa (Gregg) on October 15, 1818. The couple was married in the county of their nativity on January 11, 1838, and three years later moved to Jackson County.

By 1861, the Young's lived on about 2,000 acres along the Blue Ridge in Washington Township in the southern part of Jackson County. Seven children, Susan M., William A., Sarah Ann, Harrison, Laura, Martha Ellen, and Ada and their slaves made up the household. Solomon and his sons engaged in farming and stock dealing, including twenty year history of conducting cattle drives and leading wagon trains out west. When the Civil War erupted, Solomon was once again on one of his numerous western business ventures.[241]

The Young family would experience the horrific drama that played out along the border of Missouri throughout the war. Their first personal encounter came during the first year when Solomon was in the west and son William was serving under General Sterling Price. As Jim Lane and his Jayhawkers were raiding through Jackson County whether in route to ransack, pillage and burn the town of Osceola in St. Clair County or for some other despicable purpose, he and his men paid an uninvited and unwelcome visit to the Young farm. According to family accounts, Lane and his cohorts entered their property and proceeded to kill four hundred head of Hampshire hogs while the family matriarch was ordered to bake biscuits to fill the mouths of the hungry troops. Lane, being aware that William Young was fighting for the south, directed his men to ascertain the whereabouts of Solomon and William. They seized Harrison Young and tightened a noose securely around his neck. Harrison was repeatedly strung-up but each time he insisted he did not know the whereabouts of his father or brother. Eventually, Harrison was released, but it did not end Lane's reign of terror upon the Young farm.

Lane ordered the Young hay burned and their silver and feather beds seized.[242] This proved not to be the first or last time in which Kansas' troops inflicted their wrath on the Young's. On separate occasions Jayhawkers returned to confiscate horses, hogs,

---

[240] About ten years later, Jim Crow who had by then become well known as the town bully was killed on the streets of Independence when he was shot dead in a shoot-out with Deputy Marshal James Peacock. Sarah Ann not only lost her husband that day, but during the gunfight, their son, Elijah, was hit and died the next day. The Young family was not saddened by the death of their nefarious son-in-law. After his death, no one was permitted to mention his name in the home.

[241] The Young property was located near present day Grandview. Truman, Margaret. "Harry S. Truman." (New York: William Morrow & Company, Inc., 1972), 45; McCullough, 29; Union, 358 and 987; and Wilcox, 311.

[242] McCullough, 31.

wagons, corn, hay and countless other items, including large sums of gold that Solomon had earned out west.[243]

The trauma felt by the Young's during the first couple of years of the war paled in comparison to their plight after the issuance of Order No. 11. Of their family possessions, they were only able to remove that which could be loaded into one wagon. Little Martha Ellen would later recall how she had trudged behind their rickety wagon and traveled northward down a hot, humid and dust-filled road. The exile from their family home while not forever, would prove to be irreparable. Solomon led his family to Kansas City where they remained until they were authorized to return to their property.

Solomon Young's favorite grandson, Harry S Truman

When the Young's returned to their farm, they rebuilt and re-established their former holdings, but the war's pain never subsided. Mrs. Young's bitterness towards Kansas Jayhawkers and Red Legs even extended to the color blue, as it had been the dye of the Union uniforms—a coloring that had persecuted and accosted her family throughout those four-long suffering years on the border of western Missouri.[244]

For years Solomon pursued federal claims of compensation for the property damage and losses the Young's had suffered during the War. Mrs. Young persisted in pursuing restitution even after her husband's death. She reportedly submitted her final application in 1902. In all, the Young's sought damages in excess of twenty thousand dollars (into the millions today).[245]

Around 1904, Mrs. Young and her daughter Mattie were visiting one evening at their Grandview farmhouse. They were anxiously awaiting the arrival of Mattie's 21- year-old son. The young man had grown-up working on the farm of his

[243] Eakin, "Tears and Turmoil," 53.
[244] McCullough, 32.
[245] McCullough, 31.

maternal grandparents. The boy's parents and even his paternal grandfather had all resided together in the Solomon household. Growing-up, he had been a faithful farmhand—his late grandfather Solomon just adored the boy.[246]

When Solomon's favorite grandson entered the residence that evening, his maternal grandmother and mother were engaged in conversation. He had recently enlisted as a private in the Light Artillery Battery B, First Brigade of the Missouri National Guard, and his new uniform was blue with red striped trousers and piping red cuffs. The young man was so excited to model-off his military dress-blues. Almost instantly upon having entered the room, he realized something was dreadfully wrong. The colors he wore conjured in the mind of his aged grandmother those hateful memories dating back well over forty years. She loathed the thought her grandson could be wearing those colors. Outraged, she promptly objected to his clothing and ordered them instantly removed.[247] Certainly, she reminded him of those haunting days of the Civil War, and in particular the forcible removal from their property under Order No. 11. The lad's impression of his grandmother's reaction followed him the rest of his life, and even into the White House. Harry S. Truman of Jackson County, Missouri, would never forget!

Tales of Laura (Harris) Flanery and Family

Of the numerous stories of long trains, privations and sufferings by those banished under Order No. 11, perhaps one of the most troubling is that of Laura (Harris) Flanery and her family. Her saga and exile began on August 29, but it would certainly prove neither to be her first nor last encounter with the horror and ravage of the War.

Laura Harris was born May 14, 1838, on a farm about two miles southeast of Independence (on land that is today known as Drumm Farm). Her father Samuel Harris and her mother Jane (Hall) Harris were natives of the State of Virginia. In the 1820's her parents settled in Jackson County where her father in 1827 purchased property from the Federal government at $1.25 per acre. Laura was one of eleven children. Her father raised mules and hemp, which were

Laura (Harris) Flanery

[246] Solomon Young died in 1892 at the age of seventy-seven. McCullough, 39.

[247] Truman, Harry S. "Autobiography of Harry S. Truman." (Boulder, Co.: Colorado Associated University Press, 1980). The story also appears in the first volume of Truman's memoirs and whose recollection of the family story told by President Truman to Robert Ferrell who published the account in Truman's autobiography; and, Eakin, "Tears and Turmoil," 52.

shipped from Wayne City Landing on the Missouri River. She affectionately referred to her father as "Pappy". Tragically, he was killed when a mule struck him in 1854. His death would not be the last misfortune Laura would experience over the next ten years.

On February 17, 1859, Laura married Martin Van Buren Flanery. She and her husband were living on a farm located on the high road from Westport to Lone Jack (near Lee's Summit today) with their toddler William V. when the rebellion erupted. Martin soon enlisted in the Missouri State Guard under General Sterling Price. He was discharged three months later suffering from rheumatism. After Martin's return, their family home, nearby church and twenty-seven other houses was burned by Kansas' troops on October 27, 1861.

> *Before the church was fired, the soldiers took the Bible out and placed it securely on a stump, away from the flames. After the battle eleven of us girls gave first aid to the wounded and dying soldiers. Southern people after the battle of White Oak drove the federal troops into Kansas City, killing all but 23. Captain Washington Wells and Captain John Williams were the only two rebels killed in this engagement.*

After experiencing the destruction of their property and surrounding community, the family rented a farm near present day Lee's Summit. Martin resumed farming until one day while plowing corn "he was seized by the White Rags," a band of federal "marauders" from Northern Missouri. They forced him "to direct them on a trip through the southern part of the state on a thieving expedition." After the raid, the "White Rags" intended to return to Jackson County and hang Martin in the square of Independence. He was saved from the lynching when his father Zion Flanery garnered assistance from William Hagan, a Union sympathizer.

Following this incident Martin "hastened" to the woods and enlisted with "Quantrill and his men, as an act of safety to himself . . ." When word reached federal troops that Laura's husband had joined Quantrill, a gang of Jayhawkers "set upon" the home and seized her "baby" by his heels" and "threatened" to kill him if she "did not divulge the secret where Quantrill was hiding." Laura told the soldiers she "did not know" while begging and pleading with them to spare the life of little William V. "After some time they left and no one was further harmed" that day.

Laura's next horrifying experience came with the issuance of Order No. 11. She described it as the "worst blow of all." She and other women were forced into a state of panic not knowing which way to turn.

> *We had led sheltered lives, busy in our own households . . . all important decisions were made by the men of the family. There was nothing of value left in our formerly well furnished and spacious homes. The only stock that*

was left on our farm was one crippled horse and a blind mare. But the women—including my sister, Mrs. Julia Irwin—with indomitable courage made preparations for departure."

Twenty-three families whose fathers, husbands and sons were in the rebel army "set out eight days after the order was issued." They gathered at the farm of Alvis Powell located near Lone Jack. Laura loaded into one wagon one hoarded feather bed, two pair of pillows and one small trunk that she described as "a prized object brought from Virginia" by her parents. Inside the trunk was just "a few bare necessities in the way of cooking utensils, etc., for myself and small son." She placed her boy in a wagon and they started towards Clinton in Henry County, Missouri under an escort provided by Laura's 84-year-old neighbor, Buckner Muir, and his son, Sam Muir. There were some sixty-one women and children in the trek to Clinton, including Laura, her son William V., her sister Mrs. Julian Irwin, who Laura described as "possessing indomitable courage" and Mrs. Walton and her three-year old son Thomas. At the time of the Order, Mrs. Walton's husband (little Thomas' father) was a lieutenant colonel in the Trans-Mississippi Department of the Confederate Army.

When the caravan arrived in Clinton, they encountered more heartbreak. Some 300 Kansas Red Legs stole all their stock, save a single team of horses, and scattered what food they did not carry-off with them. They were left in "sorry circumstances" after this episode, and soon "began to look around for some means of locomotion." Finding no further assistance, they began their arduous journey toward Texas on foot and under the care of Buckner and his son. The group did find some temporary relief outside of Clinton:

About a mile away, to my joy, I ran across my old crippled, blind mare, which the robbers had taken, but left along the road. I suppose because she was blind and maimed. Some of the party found a farmer who had a team of oxen. We traded what few belongings we had left for the team and several other teams until we had enough to hook our wagons. Try as we would we could not drive the oxen since the lead oxen were hitched to my wagon, I was compelled to lead them every step.

The group made about five miles per day until they reached Van Buren, Arkansas where they stopped for a rest. They continued their long journey south through the Indian Country (Oklahoma today). Their wandering was difficult and filled with danger and disease. By the time they had traveled through the Cherokee Nation their water supply was exhausted. On the third day out, the oxen began to "bawl and run."

We knew this meant there was water close at hand. God must have been with us for we had gone scarcely a mile when we stumbled on a large

> *shallow hole in the earth filled with water. Millions of flies and mosquitoes were hovering over the water, which had a green scum over it, but at that, nothing has ever looked so good to me.*

Despite the discovery of water, their misfortunes did not end. Laura's sister, Julia became stricken with fever. Three-year old Thomas Walton also grew gravely ill with membranous croup. A Native American woman attempted to administer treatment to the little boy, but her efforts as well as those of the other women was of no avail, and Tommy soon died. His mother was prostrated with grief. Laura placed his body into her wagon where she and Miss Maggie Baker kept the deathwatch over him. They were determined not to bury the youngster in a desolate wilderness.[248]

It took nearly two painstaking months before the beleaguered group of women and children crossed into Texarkana, Texas. When word reached Laura's husband she and little William V. had arrived in Texas, he along with Lieutenant Richard Yeager were granted a leave of absence to find their families. Martin Flanery found his wife, son, family and friends along the trail. He promptly "procured a mule" for Laura and William V. to ride into Sherman, Texas. They stayed about one week at the home of Jim "Crow" Chiles before settling some ninety miles away at the house owned by Alexander Chiles. Laura and her son remained with the Chiles' family until the end of the Civil War.

Although Laura and her boy had arrived safely into Texas, it did not bring to an end her sufferings. Her husband Martin was killed during an ambush on June 29, 1864; his body was buried in Johnson Station, Tarrant County, Texas. After the end of hostilities Laura, young William V, her sister Julia and her husband and several others returned to Jackson County. She graphically described their journey as having been "made under even worse conditions than the trip out."[249]

Laura (Harris) Flanery, in later life as Mrs. Bridges, grandmother of Missouri Congressman, William J. Randall.

The tragedies experienced by Laura (Harris) Flanery and her family during Order No. 11 paralleled the story of her husband's family. Laura's in-laws were Zion and

---

[248] When the party reached Boggy Depot in Texas, they met Colonel Harris and a detachment of about one thousand five hundred Confederate troops. The men were sickened over the death of Little Thomas. They also knew his father was an officer in the Confederate Army. In proper military fashion, Col. Harris and his soldiers provided a coffin for Tommy, and gave him a full military funeral and burial out of respect for the boy's father.

[249] Leslie, 264-265.

Lucinda Flanery. The Flanery's had eight children: Martin Van Buren (Laura's husband), James S., Rebecca J., Melvina W., John W., Mac P., George Z., and Horace W., and all of whom were southern sympathizers, two of their sons, including Laura's husband Martin, were away serving in the War.

Their teenaged daughter, Rebecca Flanery, had her own personal problems with federal forces because of her southern proclivities. She not only had two brothers serving for the south, but she had an uncle who served under Quantrill. Rebecca on one occasion reportedly protected Captain Dick Yeager, son of Judge Yeager, by giving aid to him and his guerilla friends. She served two weeks in the Knobnoster jail for her sin, living on bread and water for her crime, until she signed the loyalty oath and was released.

Knowing full well the dangers his family faced in Jackson County, especially because of their southern sentiments, Zion purchased property in Saline County and safely relocated his family to their new home during Order No. 11. Once his family had been established in Saline County, Zion returned to his property in Jackson County to gather needed corn for his livestock; he never returned. His widow, Lucinda, went to her grave without knowing her husband's fate. Two men on horseback had chased Zion out of Jackson County and into Lafayette County where they shot and killed him. Jimmy Calloway saw the murder and asked the men why they killed Zion. They shrugged merely stating because he was a southern man. Calloway then made a coffin of walnut and buried Zion under a sycamore tree near the spot where he had been killed.

It was not until some twenty-five years later that the fate of Zion was made known to his family. The unraveling of the tragedy was made under ironic circumstances. Zion's daughters Rebecca and Melvina had gone to Arrow Rock (once the home of George Caleb Bingham) while they were living in exile in Saline County under Order No. 11. A romance soon developed between Rebecca and James Gault. According to family lore, Gault after having first laid eyes on Rebecca told his friend John Duggen (later a well-known attorney in Saline County): "John, there's the girl I'm going to marry!" John inquired how he could be so certain since he had not previously met the young woman; James boldly announced he would have someone introduce them. It was not long before James Gault's prediction became reality and he and Rebecca were soon married. Still Zion's plight remained a mystery.

One of the children of Rebecca and James, Anna Lou later married G. L. Park in 1888 and the year after their union, they moved to a livestock farm in Henry County. It was there that the fate of Zion was revealed. Jim "Uncle Jimmy" Calloway had become the Deputy Sheriff of the county. As the story goes, Uncle Jimmy told Anna Lou about how had seen the murder of her maternal grandfather and how he had buried Zion. Anna

Lou then shared the story with her mother, Rebecca—a story that had been buried for twenty-five years, and since the waning days of Order No. 11 in 1863.[250]

Family Ties Near Lone Jack

On Sunday August 30, John Roupe, owner of a blacksmith shop south of Lone Jack, traveled to the homestead of Rile Geam who lived in Pleasant Hill. To comply with Order No. 11, Roupe sought Geam's assistance to help move his family to the military post at Pleasant Hill. Roupe did not make it to Geam's property before encountering serious trouble from Mr. M. S. and other bushwhackers. Despite his southern sympathies, the guerrillas immediately recognized Roupe for his chosen profession. The rebel leader summarily ordered Roupe to return to his forge and produce five hundred horse nails before sundown "that very day or he would be killed." Roupe knew the threats were no mere words as he had experienced privations from both sides during those embattled years. Roupe quickly returned to his shop and with the help of his son John Henry they went to work and manufactured the requisite number of horse nails within the allotted timeframe. Although the guerrillas did not return at the assigned hour, they did arrive two or three days later to collect their nails.

Shortly after this event, one of Roupe's neighbors, Benjamin Potter, made a visit to the Roupe farmstead to ascertain whether the family intended to report to Pleasant Hill. The blacksmith assured Potter he had suffered enough and was making plans to depart as soon as possible. He also told Potter that his latest terrorist encounter had been inflicted not by federal troops, but by Quantrill's men. Roupe prudently decided to abandon their farm and he and his family left. The 9th Kansas Regiment promptly converted the vacated Roupe property into their military campsite and their field command post.[251]

Benjamin Potter was seventy-five years old at the time Order No. 11 was issued. He had been born in Surry County, Tennessee on May 17, 1788, and served his country in the War of 1812. He and his wife Mary "Polly" (Hays) had eleven children; Polly had two brothers living in Lexington and Jackson County. After Polly's death, Potter moved to Jackson County with seven of his children where they settled on property near Lone Jack around 1854. When Potter's son William died on a western journey, Potter took on the responsibility of raising his son's three small children, Marian, Amanda and Thomas.

---

[250] "Independence Woman's Heroic Walk to Texas is Still Vivid Memory," *Kansas City* (Mo) *Journal-Post*, 16 May 1926; "Mrs. Laura Harris Bridges, 90, Believes in Present Generation," *Independence* (Mo) *Examiner*, 26 Apr. 1929; and, "Mrs. J. P. Bridges's Life a Story of Hardy Pioneers," *Kansas City* (Mo.) *Star*, September 1930; as located in the Jackson County (Mo.) Historical Society Archives, Document ID Vertical File Newspaper Subject Files: Biography: "Bridges." Also, Laura Elizabeth Bridges Missouri Death Certificate, 24 Sept. 1930; Eakin, "Tears and Turmoil," 46; and Thruston, Ethylene Ballard. "Echoes of the Past." (Kansas City, Mo.: The Lowell Press, 1973), 19-20.

[251] Eakin, "Tears and Turmoil," 11-12.

Throughout the war, Potter had been known for his southern sympathizes. He was once asked by Jennison warriors his sentiments and Potter gave a blistering reply: *"I was for the Old Union, I fought for it in the War of 1812; I am not for the sort of Union it has become."* His advanced age precluded him from taking an active role in the war, but three of his sons did. His son, Steven, was killed in 1862, shot in the back, and two other sons, John and Melchert, served in the Confederate Army. The family always attributed Steven's death to the war.

Consistent with the Potter family's southern loyalties, they were known for having provided sustenance to guerrillas. The last time the Potter family gave them aid was when Quantrill and his 300 mounted men were in route to Lawrence. Two days before Black Friday, Potter fed the large assembly and even introduced two of his daughters to Quantrill. The chief guerilla ate supper with Potter in his house.[252] At the time, no one in the Potter household knew the purpose of the mass number of troops—it would not take long before they learned the reason.

Potter's sizeable family and extended family lived near his property. His daughter, Amanda, had married John Cave and they had a young daughter named Lillian. Potter's other daughter, Juda, was married to John D. Hunter. The Potter, Cave, Hunter, Bennett and Bynum families were either related directly or by marriage.

The parents of John D. Hunter were John and Elizabeth (Jeans) Hunter who were in their seventies, had nine children, including John D., Matilda, Nancy, Berrilla, and William. David Ousley, the husband of their daughter Matilda died fighting for the Confederacy requiring Matilda to rely upon her 17-year-old son, Andrew Ousley, to raise her small children. Another daughter, Nancy married John S. Cave on September 9, 1836.[253]

MARTIN RICE AND HIS FAMILY in front of the Rice home in Lone Jack. Fabius Martin Butler, grandson of Martin Rice, has written a book about the Civil war in Jackson county and the volume also includes some poetry by his grandfather.

[252] Eakin, "Tears and Turmoil," 3-10.

[253] Union, 935-936. John S. Cave was born in Virginia in 1818.

By 1863, John and Nancy (Hunter) Cave had the following children all born in Missouri: Mary J., age 22; Allguis, age 20; Susan F., age 17; Ella E., age 14; Martha "Matilda" A., age 12; Berilda T., age 7; and John, age 4. According to the 1860 census, the Cave's owned one slave. The Caves and Hunters were among the original charter members of the Lone Jack Baptist Church.[254]

Another well known personality that lived near Roupe's farm was Martin Rice, the same man who had delivered the committee's resolution opposing Orders No. 9 and No. 10 to the attention of General Ewing about ten days earlier. Martin Rice was born November 22, 1814 in Campbell County (now Union County, in east Tennessee). He was the oldest of eleven children born to Enoch and Mary Rice. In his youth Martin did not have the advantage of a scholastic education. He did not attend public school or an academy and barely attended a subscription school. Before he was fourteen, he quit school and never saw the inside of a classroom. These limitations did not deter Martin from being self-educated. He studied Webster's American Spelling Book, the Bible and Pike's Arithmetic, and while he labored behind a ploy, he carried Lindley Murray in his hat to further his own personal education. Despite Martin's limited education, he taught the first free school in the district, Claiborne County, Tennessee in the fall and winter of 1832.

The family moved to Jackson County, Missouri, in the fall of 1833 and by October of that year, his father purchased a farm near Lone Jack in Van Buren Township in southeast Jackson County when Martin was nineteen years old. In the winter of 1833, while making rails and grubbing by day to open up a farm on the new Missouri home, Martin studied Gibson's Surveying at night, and continued the study until he had mastered the trade. He assisted his father to open and fence a small farm in the brush. A neighbor hired Rice during crop season for ten dollars per month before Rice resumed teaching in a small house with very little pay. In 1835, he helped plant crop with his uncle Absalom Powell, who lived near the road between Independence and Blue Mills and near where the Salem Church was later established. In the fall of that year, he molded and burned the first brick that was made in Van Buren Township. The brick he made became popular as many chimneys in the region came from the brick. Later in the fall he bought land near his father's property. Rice soon sold the property within a few months,

---

[254] Union, 330; United States Census, Van Buren Township, Jackson County, 1860. Nancy (Hunter) Cave was born in 1819 and died in 1905. The Cave's daughter Matilda later married on October 23, 1870 a former Union soilder, J.K. Reid. Reid had been born in Delaware County, Ohio on July 26, 1843. He enlisted in Company D, 20th Ohio Infantry, but became ill just after the Battle of Shiloh, and was discharged from the army in the fall of 1862. He went to Illinois in the spring of 1864, and farmed one season in Starke County, but reenlisted on October 12, 1864 into Company A, 31st Illinois, and served until the close of the War. In the spring of 1866 he went to Colorado, and engaged in trading with Indians. He settled in the Sni-A-Bar Township of Jackson County in the fall of 1868, and returned to farming and raising stock. Matilda and her husband had four children: Mary V., Lucy O., Ella W. and George B.

and in March 1836 he purchased property in Cass County (then known as Van Buren County).

The following month on April 3, 1836, Rice married sixteen-year old Mary Lynch of Lafayette County near Green Town. Martin Rice moved Mary to his farm on the Shawnee Trail located on the road toward Clinton some twenty miles away. The nearest neighbor lived about two miles from their house. Martin and Mary had nine children (seven survived to adulthood), including Isaac, Spencer, Elizabeth, Nancy, Martha, and James. Their children were among the first pupils to attend the first school near Lone Jack in the late 1830's.[255]

Rice soon became active in political endeavors. In August 1836, he was elected county surveyor, and in April 1837, he surveyed, or laid out the town of Harrisonville. In 1841 Rice and his wife became members of the Baptist Church where he continued to serve as the clerk of the Lone Jack congregation for well over twenty-five years. After his father's death, Rice sold his Cass County property and purchased his late father's property near Lone Jack. Politically Martin Rice had been a prominent member of the old-line Whig Party, but by 1860 he fully supported the Republican Party, and had become a strong Unionist. These sentiments placed him in the minority in his community. In December 1855, Rice's wife, Mary, died at age thirty-six.

Thirty-two-year-old William Calvin Tate had also become an integral part of the Lone Jack community. He had been born in North Carolina in 1832 and had married into the Rice clan in 1856 by marrying Rice's daughter, Martha J. in Lafayette County. When Order No. 11 was issued, the Tate's had three young children, John age 5, Mary age 3 and infant Nancy Rosalie.

Everyone in the area was either friends or relatives, and principally engaged in farming. On September 6, during the heat of Order No. 11, the lives of these intermingled neighbors would be irrevocably shocked and destroyed. Martin Rice would be at the forefront of the disaster that day, and he would be questioned throughout the remainder of his life about his involvement and his loyalty or the lack thereof by all these neighbors, friends and family members.

[255] Union, 332; *Daily Missouri Republican* (Saint Louis, Missouri) 11 Sept. 1863.

Tragedy and Loyalty: September 6

Martin Rice's first knowledge of Order No. 11 was when he read about it in the newspaper on August 30—the same day that guerrillas had accosted Roupe and had ordered him to manufacture the five hundred horse nails. Rice years after the war, wrote of his initial observations of Order No. 11.

> *I thought I had witnessed and felt the hardships and privations of civil war and martial law before, but it was reserved for this, the last week in August and the first ones in September, 1863, to teach me and others how much the human body and mind can bear up under and still survive. The 30th of August, instead of being a Sabbath of rest, was to all a busy day of care and labor, preparing to obey the stern mandate, and abandon the homes procured by many years of toil and labor, followed too, by other days of care, toil, and anxiety.*
>
> *Previous to this, if one were brought into a strait, or got into trouble or difficulty, he could appeal to some friend or neighbor for help, and the appeal was seldom made in vain. But now all were in the same strait; the same weight of sorrow and distress was pressing upon all; there was no exception, and none in our part of the district were exempt from the general hardship. Though none were well prepared to obey the order, some were much better prepared than others. But whether well or ill prepared, there was no help for it; all must go. On Monday, the last of August, a number of citizens, myself amongst the number, repaired to Pleasant Hill, in order to prove loyalty and get certificates or permits to remove to the military posts, or other parts of the district outside the doomed or proscribe territory.*[256]

Learning of Order No. 11, Rice took a load of family provisions and reported to Captain John Ballinger of the 1st Regiment of the Missouri State Militia, who commanded the post in Pleasant Hill. Rice duly took the loyalty oath and Ballinger provided him the appropriate certificate. Rice then secured for his family a house in the "suburbs" of Pleasant Hill. The following day when Rice was nearly half-way home, he was overtaken by a messenger who warned him that to relocate to Pleasant Hill would be unwise as "bushwhackers would not permit" him to safely move to that town. Rice then took some of his property to his brother-in-law, William F. Snow in Johnson County. He

---

[256] Martin Rice, *Rural Rhymes, and Poems from the Farm, Millett & Hudson* Kansas City, Missouri 1877; Martin Rice, *Rural Rhymes, and Tales of Olden Times Ramsey, Millett & Hudson, Kansas City, Missouri 1882, Martin Rice Remembers: What I saw of Order Number Eleven*, Missouri Historical Review, Volume 106, No. 3, April 2012; Union, 940-941; and http://www.historiclonejack.org/whatisaw.html.

took another haul to Mr. Galloway in Lafayette County (near Odessa today) where he had decided to temporarily resettle. His brother E. N. Rice of Cass County also changed his plan on resettlement and chose to seek refuge in Clay County.[257]

Rice was not the only resident living in the area of Lone Jack who looked east towards Lafayette County. John Cave, John D. Hunter, and Thomas Bradley also took articles to the yard of Mr. Galloway. Other neighbors and friends throughout the remainder of that week were generally "removing necessary articles out of the county to places where they would be in some degree safe, until they could find a temporary home to which they could be removed . . . away from the ill-fated county."

On Saturday, September 5, Rice did venture back to the military post at Pleasant Hill, where he had no difficulty in getting another certificate of loyalty, which would authorize him to go to any part of the country, outside the three counties of Jackson, Cass, and Bates. While at the provost marshal's headquarters, Rice secured certificates of good character for his family. He returned home "in better spirits than I had enjoyed for several days, and had a better night's sleep than I had had for a week before."[258] This would become the last pleasant night of his life.

On the morning of September 6, Rice and other neighbors and family members were making their final arrangements to leave. While so engaged, a squad of the Kansas 9th Volunteers under the command of Captain Charles F. Coleman came suddenly upon Rice, his son Isaac, and his son-in-law William Tate. Rice despite having presented his certificates was arrested with his son and Tate. They were ordered to follow the soldiers to the Roupe farm, about a mile in distance. The Kansas force already had in custody William J. Hunter. As they neared "the residence of old Mr. Hunter," Hunter's 17-year-old "grandson, Andrew Ousley rode up to see about arrangements for moving, and he, too, was taken into custody." Andrew's 75-year-old grandfather, John Hunter, was not molested by the troops.

Having now five men in custody, the Kansas troops led them a short distance when they came upon the house of John S. Cave and his brother-in-law John D. Hunter. Both men were promptly arrested. They had only proceeded some one hundred yards when they met seventy-five year old Benjamin Potter; he marked the eighth man taken under charge. The 9th Kansas then led their captives about three-quarters of a mile until they reached their encampment at the Roupe Farm.

---

[257] Ibid. Although Rice's brother had been a strong Union man, some federal scouts had represented to him that they were bushwhackers. E. N. Rice was terrorized and frightened. He became so alarmed that to escape capture he said something that in the eyes of the Union soldiers was disloyal. The men considered taking his life, but instead they carried him to Kansas City where he was incarcerated for several months.

[258] Martin Rice, *Rural Rhymes, and Poems from the Farm, Millett & Hudson* Kansas City, Missouri 1877; Martin Rice, *Rural Rhymes, and Tales of Olden Times Ramsey, Millett & Hudson, Kansas City, Missouri 1882, Martin Rice Remembers: What I saw of Order Number Eleven*, Missouri Historical Review, Volume 106, No. 3, April 2012; Union, 940-941; and http://www.historiclonejack.org/whatisaw.html.

The hostages were escorted to Colonel Charles S. Clark, regiment commander who had been scouring the countryside ever since the Lawrence massacre. Clark took down the names of the eight prisoners and Rice's paperwork and retired while the prisoners sat on a fence awaiting Clark's return. Rice and his son were the only members of the eight who held certification as to their loyalty.

Clark was going to determine the fate of his prisoners. He did not consume much time before returning to the captives. Clark asked Rice to identify his son. Rice pointed towards Isaac. Rice boldly even complained to the Colonel that one of his men had taken his son's hat. The soldier was ordered to restore Isaac's hat to him without delay. Thereafter Clark retired again. Barney Dempsy, an acquaintance of all the men held under arms, who had been acting as pilot for the Kansas 9th, came and visited with them in "friendly words" and then left. During the entire ordeal, "neither the officers nor any of the men spoke a harsh or menacing word" to any of them. Captain Coleman who had first taken the eight men as prisoners then approached Rice and said: "You will take your son and travel." Rice grew apprehensive over the manner in which Coleman had spoken to him, and he grew alarmed that the other members of the party may have been in "real danger." However, Rice did not delay departing with his son and "immediately left as commanded," leaving the other six in the fate of the 9th Kansas Regiment.[259]

Rice and Isaac journeyed back to their home. They had not been on their property for any length of time before they heard "the report of several guns in quick succession." Rice was alarmed but held out the futile hope that the soldiers had been shooting fowl for breakfast on the Roupe farm. Family members were not as naïve as Rice.

Miss Jane Cave rushed towards the area in which the gunfire had been heard. She discovered that the Kansas troops had left, but six men had not—all of them were dead, some of the bodies had been "pierced with many balls." When Rice and others arrived at the crime scene, they found the victims near the spot where he and his son had left them. Two of them appeared to have been shot sitting by the fence, and the others but a few feet away. According to Joanne Chiles Eakin, in her well-researched work, "Tears and Turmoil: Order Number 11," before Andrew was killed, he had attempted to escape imprisonment. As he fled, one member of Col. Clark's regiment yelled "*Five dollars to the man who drops the boy.*" Andrew's body was later discovered several feet from the other five slain men.[260]

[259] Ibid. Col. Clark in his report to Ewing dated September 8 made absolutely no mention of the massacre. O.R. 22:1:610.
[260] Eakin, "Tears and Turmoil," 16.

6-Man Cemetery Memorial Marker

There was nothing left to do but the sad affair of burying the fallen dead. The burial detail included Rice, his son and their 75-year-old aged friend, John Hunter. John had the great misfortune as he assisted the Rice's in laying to rest his only two sons, John D. and William J., his grandson, Andrew, a son-in-law, John S. Cave, and two other relatives—one of whom was Rice's own son-in-law, William C. Tate. With their hands and crude instruments they dug a wide but shallow grave near the spot where the fallen men had been found on the Roupe Farm. They laid the six "side by side, in their bloody clothes; blankets spread over them and covered with earth." When the last clod of dirt and soil had covered the remains, the men said a prayer and with "heavy hearts" departed in silence.[261]

Late in the afternoon on the day of disaster the survivors made preparations as best they could to journey eastward. Rice and his family and neighbors left under those piteous conditions. The despondent party consisted of John Hunter, and his surviving relatives, the surviving heirs of Benjamin Potter and Martin Rice and his family.

The Hunter's were a devastated clan. The shocked family included John Hunter's two daughter-in-laws, the widows of his two murdered sons and their fatherless children, a single daughter, and his daughter Nancy (Hunter) Cave whose husband had died that morning. The group also consisted of Matilda (Hunter) Ousley and her five surviving children whose husband, Dana, had recently been killed in the War and who left behind her oldest son Andrew in a shallow grave on the Roupe farm. Among the mournful travelers of the late patriarch Benjamin Potter were his three grandchildren, his daughter, now the widow Juda (Potter) Cave and her two fatherless girls along with Mrs. Jacob Bennett whose husband was safely in Ohio.

On December 14, 1864, Martha Jane "Mattie" (Rice) Tate wrote a letter in which she expressed the anguish and turmoil she and her children had experienced during Order No. 11. The letter also provides additional insight to the background of her late husband and family (original style is preserved for authenticity).

> *I am the wife of Calvin Tate who was in Indian Terr*[itory] *ten years ago when you was quite small I have often hear him speak of you and your Pa when he was there. But alas I never shall hear his lovely voice again on earth for he is gone gone from me forever and I am left*

---

[261] Ibid.

with three small children to take care of and not able a great part of my time to take care of myself and what is to become of them I cannot tell But the Lord has said I will be a husband to the widow and a father to the fatherless and I hope he is such to me and mine for if ever a poor family needed his help it surely is mine alone for there are many these war times that need a great deal of help from above such is and has been the [case] with me and six other familys all related to my husband save one and all living within one mile of each other and although I say it myself a better set of men never live or die then they were in September 1863. "Ewing made the Order No 11 to devastate the Countys of Jackson Cass Bates and part of Vernon and people had but nine days to get out of those Countys. [A]*nd the men would have started in one hour on Sunday Morning. But before they got off some soldiers from Kansas came and taken My Father Brother and my Husband and five others two Hunters Brothers and one by the name of Cave Brother in law to the Hunters and one boy 17 years old named Owsley Nephew to the Hunters and one old man 60 years old by the name of Potter Father in law to one of the Hunters and taken them something over half mile from home and shot them all But father and brother they were released and came home and had just got home when the guns were heard that made four widows and 26 fatherless children and Oh what a sight to behold they were shot all to pieces* [most?] *and left on the ground by them to stay there Forever for what they* [--?--] *there was but three men left to bury them my Father Brother and Uncle John Hunter 70 years old the Father of the two that was killed everybody was gone but those familys and one other woman and we had to just dig a hole and put them all six in without any coffins or boxes nothing but blankets and right where they was Massacred by a band of Murders and a set of Blood thirsty__________ None of those men had ever been in arms on either side except David Hunter who was in the Molisha at Kansas City about two weeks they was all at home and had been and was minding their own Business they were killed in the morning about 10 o'clock and we left home about 4 in the evening that home that was to be home no more for me we went to Ray Co[unty] on the North side of the River and stayed all winter and came back in the spring when we left home we all had to leave all our stock and left with one wagon a piece and what could be put in them. Mr. Hunters family and Caves have not come home yet they live near Dover Lafayette Co[unty] there is four families of them and but one man among them and that is Uncle John Hunter the women has all the work to do in and out of the house...." Mattie Jane Tate P.S. address me at*

Pleasant Hill Cass Co, Mo, (my childrens' names John Martin / Mary Matilda / Nancy Rosalie).[262]

After their exodus the families found temporary places in which to reside. Mr. Hunter and his family, Mrs. Cave and her children and Mrs. Ousley and surviving children settled in the eastern part of Lafayette County. David Hunter's wife and her late father's family went on to Indiana.

Accompanying Martin Rice was his elderly mother, his six children, including his son Isaac, and his widowed daughter and her heartbroken three children. They crossed the Missouri River at Lexington, but when they arrived in Ray County they were not allowed to stay. Notices or proclamations were posted along the roadside, forbidding all persons who had been banished from the counties south of the river from stopping in the county without appropriate certification from a provost marshal. This marked the second time within a two week time-period that Rice's Union loyalty and certificate proved invaluable—the first on September 6 when it had saved he and his son's life; this time by allowing his family refuge. They remained in Ray County before returning to their home in Jackson County. Upon arriving to their property, Rice was fortunate to find his farm property and buildings still standing, and his fencing "much less damaged than" he had expected.[263]

The property of neighbors and relatives who had departed together on the sad and tragic day back in September did discover their land in the same condition as had Rice. Their homes and properties as were hundreds of other farms, structures and buildings had been destroyed by intentional arson or consumed by prairie fires—there was nothing except naked land where once beautiful homes and outbuildings had formerly stood.

Post-script of Order No. 11

After the atrocious massacre on September 6, Clark dispatched Coleman and his detachment of the 9th Kansas Regiment into the woods to slaughter Quantrill and his bushwhackers. Coleman's ferocious efforts to intensify the war against the enemy began when he received word that Quantrill boys' had burned down the flower mill in Lone Jack. Coleman scoured the woods until he came upon a trail that led him directly to a rebel encampment near the Sni-A-Bar River in eastern Jackson County. During their raid,

[262] The original Tate letter was donated to and is preserved by the Lone Jack Historical Society by Mr. and Mrs. Jane Foltz (Independence, Wisconsin). A copy of the letter, plus the transcript by Sharon Snyder and edited by David W. Jackson are filed in the Jackson County (Mo.) Historical Society Archives Document ID 140F16.
[263] Ibid, and Eakin, "Tears and Turmoil," 17.

they killed two "sesch," captured forty-seven horses, secured ammunition and weapons, and other supplies.[264]

The Federal and Kansas militia operations against the invaders of Lawrence were relentless but not overwhelmingly successful. For a period of ten days, the Federal reports commonly exaggerated the number of bushwhackers slain, asserting that not less than one hundred to one hundred fifty outlaws had been killed since the raid on Lawrence. They were ordered to continue their "same energy of pursuit" and they became naively convinced that guerrilla forces were "now very much dispersed and broken."

There was no question that Ewing was determined never to allow another Lawrence. He saw total war as his only remedy to quash and vanquish those responsible for the attack, regardless if they were a border ruffian or civilian, innocent or guilty:

> *The enforcement of Order No. 11, which is going on with great activity, will soon leave the guerrillas no place of rendezvous--no recruiting places--no resident spies. They must depend solely upon themselves and the brush, and if an active campaign is kept up against them, they must soon be driven out or exterminated. They will not yield readily or quietly. On the contrary, we look to see them put forth the most desperate efforts to retain possession of their old haunts, and to continue their ravages upon the border.*
>
> *The only safety against their concentration, and their making another raid, is in hunting them constantly in the field. Both infantry and cavalry should scour them down, should be kept at the work. Thus, and thus alone, may we hope to see this border warfare cease. There will be no virtue in proclamations and orders, unless backed up by energy in the field. Bushwhacking families may be driven out, and sympathizing citizens compelled to leave, but the guerrillas will remain to carry death and desolation to the border, unless they are hunted like wolves in their dens and fastnesses. There is no patent method of getting along with this job.*
>
> *One great difficulty, hitherto, has been that there were not sufficient troops sent to the border to perform the work necessary to be done. This error is being rectified. With the reinforcements promised him from Gen. Schofield, and with the fresh troops now being raised in Kansas, we trust that Gen. Ewing, before the fall months are over, will have the work substantially completed.*[265]

---

[264] *New York Times*, September 24, 1863; and O.R. 22:1:620.
[265] Eakin and Hale, 54, quoting, *Daily Journal of Commerce*, 1 Sept. 1863.

The guerrillas responded against Federal troops inflicting similar retaliatory measures of their own. The persistent back-and-forth between the Kansas troops along the prairies and the rebel bushwhackers from the brush became a game of cat-and-mouse. It persisted until fall's frost had overtaken Jackson County. By late fall, some peace had been restored to the region. There were two obvious reasons:[266] Quantrill and his guerrillas had retired south for the winter, and both Schofield and Ewing had kept Kansas raiders from making further excursions into Missouri. On November 3, Schofield wrote to Ewing relating the practicality of relaxing Order No. 11.

> *Now that the border counties of Missouri are free from guerrillas, and are likely to remain so during the winter at least, it appears to me that the loyal people of those counties might be permitted with safety to return to their homes. If they return now they will be able to gather their corn and some other crops, and make themselves comfortable for the winter. If this policy be adopted, I think the test of loyalty should be rather liberal than severe, the object being to permit those, and only those, to return who will hereafter be faithful to the Government. Under the reign of terror which has so long existed on the border, active loyalty could not be expected. All who return should be enrolled, and their names registered at the nearest military post. Those who are able to bear arms, of whatever age, should be organized into companies and armed. They should be required to give the most active and unqualified support to the Government. Nobody, especially in those counties, can now be permitted to occupy a neutral or equivocal position.*[267]

With some stability restored and with Schofield's recommendation, Ewing issued General Order No. 20 on November 20. The new order partially rescinded Order No. 11 to the extent that it authorized loyal citizens to return to their homes.[268] Some did, but most did not until after the rebellion ended—some never did.

As the calendar turned to January 1, 1864, the most tumultuous year in Jackson County history had unceremoniously ended. The evil rules of the chess game of terror in Jackson County remained as they had been on the first day of 1863. Kansans still trembled by even the whisper of the name "Quantrill;" children in Missouri continued to be punished by the threat of the word "Lane." Quantrill had returned to Texas and Lane to Washington. Comingo held his discomfited position as the 6th District Provost Marshal

---

[266] O.R. 22.2.732.
[267] O.R. 22:2:693-694.
[268] O.R. 22:2:713-714. By March 1864, General Egbert Brown pursuant to General Order No. 11, fully rescinded Order No. 11. Ironically, Order No. 11 vacated Order No. 11 in its entirety.

hearing complaints, filing reports and reporting troop sizes.[269] Ewing retained his rank, but not on the Border. He had been transferred to Southeast Missouri, a venue considered at the time to be less controversial than Jackson County.[270] Schofield had been relieved from the agony and drama of overseeing the most infested, guerilla-plagued district in the country. His replacement was the man he had succeeded earlier in the year, General Sam Curtis. Bingham remained ensconced in his political fiefdom at the State Capitol as Curtis regained dominion from his headquarters in St. Louis, the banking capitol of the state.

The fateful summer of 1863 marked the climax of ten long, suffering years of struggle and horror along the border of two states. It had been the longest and most terror-stricken nightmare of terror in America's history. As the clock struck midnight on December 31, the most horrifying, appalling, shocking, unsettling, sickening and terrifying year in Jackson County had finally come to an end. The year 1864 loomed in the horizon, and while the ravages of the guerilla war would cry out for one more "curtain call" nothing would ever compare to the thunder on the border during the summer of 1863.

---

[269] *Liberty Tribune*, December 24, 1863.
[270] Smith, 211.

CHAPTER XIII

Bingham's War Against Ewing

When Bingham's term as State Treasurer ended in 1865, he moved his family to Independence. They purchased a stately home at Pacific and Osage located about five blocks south of the county courthouse on the city square, and two blocks west of the Gudgell Place, the home of the former provost marshal and friend Abram Comingo.[271] The Bingham household included his second wife, Eliza, and their four-year old son James Rollins. Of his children from this marriage, Horace's stay in Independence would be short-termed, and Clara who had married Thomas Benton King on June 2, 1864, lived north of the Missouri River in Ray County.[272]

[271] Bingham's Independence residence built circa 1856 remains as an historic landmark located on Pacific Street, and it is on the National Register of Historic Places, listed under the name "Bingham-Waggoner Estate." The Bingham home is owned today by the City of Independence and operated by the "Bingham-Waggoner Historical Society." Jackson County (Mo.) Historical Society Archives *Abstract of Title* Collection. Comingo lived at 6th and South Main in Gudgell Place; a church is located there today.
[272] Bingham's first wife, Sarah, died on November 21, 1848. Horace was born in 1841, only hours after the death of the couple's three-year-old son, Newton. Horace died in 1869. Bingham and Sarah had another son, Joseph; but he died in infancy shortly after his mother's death. Clara was born in 1844 and she and her husband would eventually settle in Texas and have a large family. Clara's father-in-law, Austin A. King,

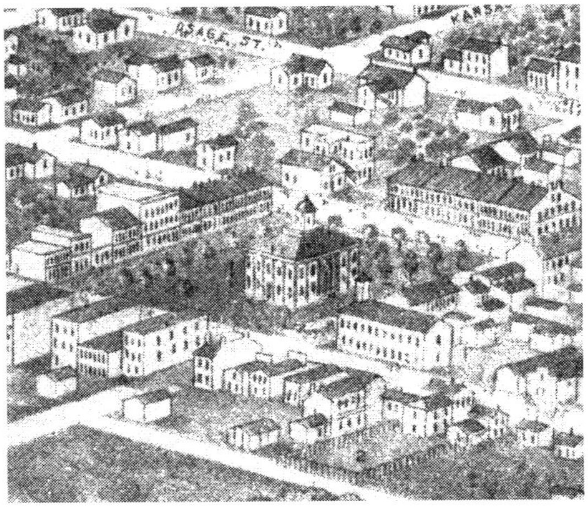

Bingham's home was 5 blocks south of Independence Square

The Civil War may have been over but the former state treasurer was not going to remain silent. Bingham was determined to ensure that the tragedy inflicted upon Jackson County, and in particular Order No. 11, would never be forgotten. He was determined to use his political rhetoric and artistic wrath to vilify several individuals he considered culpable for the havoc and pain they had wrought upon the county during the rebellion.

He first launched a speaking and writing campaign against Lane and Jennison. This was followed by his denouncement of the atrocities and criminality of the Jayhawkers and Red Legs. He reserved his largest indictment against Ewing and Order No. 11. Bingham described Ewing's Order as a pure act of "arbitrary power, directed against a disarmed and defenseless population" and had been an "exhibition of cowardice in its most odious and repulsive form."[273] To document his complaints against Ewing, Bingham needed a more permanent forum. Mere words or a verbal lashing would not describe Ewing's "criminality" and dictatorship. Bingham soon began to pour-out his anger on canvass with his pen and paintbrush.

Bingham's public scorn was not limited to Lane, Jennison or Ewing. His first post-war rage was to denounce the Drake Constitution enacted on June 6, 1865.[274] The Constitution was an aggressive, reconstruction post-war policy implemented by Missouri Radical Republicans designed to punish those who had been disloyal during the late rebellion. The Constitution included a "Loyalty Oath," also called the "Iron Clad Oath." It disqualified any person from the right of franchise who had given "aid, comfort,

---

was Governor of Missouri from 1848-1852, and he had been a long-time close associate of Senator Thomas Hart Benton. James, Bingham's only son with his second wife, was born in Kansas City on September 21, 1861. "Bingham Letters," 3, 8, 84 and 332n6, "Bingham Letters," 8 and 44n2, and letter from Bingham to James S. Rollins dated November 21, 1871, Letters 327, and "Bingham Letters," 253n6.

[273] Bingham's letter to the Editor of the Republican Newspaper dated February 22, 1877, "Bingham Letters," 427.

[274] The name "Drake Constitution" was derived from the last name of the sponsor of the legislation, Republican State Senator Charles Drake of St. Louis (later U.S. Senator in 1867). The Drake Constitution barely received voter approval. Kansas City Journal of Commerce, 1866; *Missouri Historical Review* (April 1924): 18: 3: 365 note 80; Dean, Abner Holton (Reverend). "Reminiscences of Half A Century." Jackson County (Mo.) Historical Society Archives Document ID 3F12 and 269F9; and "Bingham Letters," 296, note 3.

countenance, or support to persons engaged in any such hostility . . . or been in the service, of the so-called 'Confederate States of America'; or has ever been engaged in guerrilla warfare . . . known as "bushwhacking . . ." Not only were such persons disenfranchised, it further provided:

> [N]*or shall any such person be capable of holding, in this State, holding any office of honor, trust, or profit... including . . . an officer, councilman, director, trustee, or other manager of any corporation, public or private, now existing or hereafter established by its authority; or of acting as a professor or teacher in any educational institution, or in any common or other school; or of holding any real estate, or other property, in trust for the use of any church, religious society, or congregation.*[275]

With strong convictions and an argumentative personality, Bingham's anger and disdain reached fever pitch in 1866. He detested the Iron Clad provision, but most egregious to Bingham was the constitutional prohibition that precluded members of the cloth from preaching the Word of God without submitting to the loyalty oath. With his pin and brush, he chose to attack the Drake Constitution as he was in the process of doing against Ewing. During the Conservative Union Convention, Bingham publicly condemned "Radical Republicans" and damned their Constitution as not "the will of Heaven" but rather the "will and rule of Hell."[276] He further provided strong testimonial evidence against the objectionable law by reporting the tragic story of Reverend Abner Dean—a tale he had illustrated not in just words but by his brush.

Rev. Dean, a Baptist minister whose wife had been murdered by Kansas' troops during the War, had been a loyal Union man. He had held the rank of major in the Union Army, but when the Drake Constitution was enacted, the preacher protested the "Loyalty

---

[275] http://www.civilwarstlouis.com/History/Oathofloyalty.htm; Section 3, Missouri Constitution 1865; and The Missouri Oath of Loyalty of 1865, Anderson, Galusha, Excerpted and Introduced by G. E. Rule from "The Story of a Border City During the Civil War" (1908). Loyalty Oath: Section 6 of the Missouri Constitution 1865: "I, A. B., do solemnly swear, that I am well acquainted with the terms of the third section of the second Article of the Constitution of the State of Missouri, adopted in the year eighteen hundred and sixty-five, and have carefully considered the same; that I have never, directly or indirectly, done any of the acts in said section specified; that I have always been truly and loyally on the side of the United States against all enemies thereof, foreign and domestic; that I will bear true faith and allegiance to the United States, and will support the Constitution and laws thereof, as the supreme law of the land, any law or ordinance of any State to the contrary notwithstanding; that I will, to the best of my ability, protect and defend the Union of the United States, and not allow the same to be broken up and dissolved, or the Government thereof to be destroyed or overthrown, under any circumstances, if in my power to prevent it; that I will support the Constitution of the State of Missouri; and that I make this oath without any mental reservation or evasion."

[276] Rash, 210.

Oath" and refused to submit to it for religious reasons. His position was steadfast and resolute:

> *I did not receive my commission from the state to preach, and the state had no right to take it from me. I never believed in mixing matters of church and state, and could not, in any sense endorse the state's disposition to meddle with spiritual affairs nor to curb soul liberty.*[277]

Rev. Dean continued to preach the word of God in his church in Harrisonville without submission to the oath. Dean despite his Union history was prosecuted and convicted in Cass County for his crime. After his conviction, the parson was transferred to the Jackson County Jail in Independence.

Jackson County Jail in Independence

The warring painter visited the incarcerated preacher. Bingham had previously started a mural that summer to memorialize Ewing and Order No. 11, but he placed the painting on hold and diverted his attention to the plight of Dean under the Drake Constitution. According to Dean, "Captain Bingham came several times . . . and brought me books, and once he brought his camera, and took my photograph, from which he afterward painted my portrait."[278] Bingham took the photograph of Dean, and then painted the photograph in oil (an unusual process in the mid 1800's to touch a photograph with paint). The painting was designed to provide a 'canvass protest' of the Loyalty Oath. The painting also gave Bingham a forum in which to artistically vent his anger towards the "Radical Republicans."

The painting depicts Dean sitting quietly in his jail cell reading the bible by the bare stream of light emitting through the bar window. Bingham's painting, "Major Dean

[277] Dean, 17; See also, Barclay, Thomas. "The Test Oath for the Clergy in Missouri." *Missouri Historical Review*. (April 1924), 18: 3.
[278] "Réminiscences, 17.

in Jail," received endorsements from various newspapers throughout Missouri. The Columbia *Missouri Statesman*, quoting the *St. Louis Dispatch*, said the painting of Dean would "illustrate to future generations an ignoble period of our State's history, and to be a memento of the reign of Radicalism." The *Independence Sentinel* celebrated the image:

> *It will be a picture that every family should possess, and it should bear the name of 'Missouri under the Radical rule in the middle of the nineteenth century.' We suggest that upon the back of each picture [i.e., photograph], there will be the likeness of Charles D. Drake and the devil, if it be not too much of a disgrace to his satanic majesty to put him in such bad company.*[279]

The heat over the Drake Constitution became even hotter during a late, humid night in Independence. On the evening of June 13, 1866, Rev. Dean was still being held captive in the Jackson County Jail. Among the other prisoners was Bill Faley, a horse thief, and Joab Perry, a guerilla crook who had been named as a suspect in the Liberty bank robbery in January of that year—the first day-time bank robbery, and the first laid at the feet of Frank and Jesse James. Faley and Perry had been convicted of horse stealing, and they were to be transferred the following day by the Jackson County Sheriff Charles Daugherty to the state penitentiary in Jefferson City.[280] It was said that the jail cells that evening were full of inmates.

Rev. Abner Holton Dean in the Jackson County Jail, by George Caleb Bingham, and copied by Charles Goslin for exhibit at the Jail Museum

The town's people had retired for the evening, as had the county's jailor and his family. Henry Bugler, an Irish immigrant, had served as the jailor or

[279] Rash, 209-211. Cummings v. State of Missouri, 71 U.S. 277 (December Term 1866). Father Cummings, a Catholic priest, had filed a constitutional challenge to the Loyalty Oath. The U.S. Supreme Court declared the loyalty oath unconstitutional. For Bingham and other anti-Radical Republicans the high court's ruling was vindication to their cause.

[280] Dean, 17.

marshal of the Jackson County Jail in Independence—a post he had held since 1859. Bugler and his pregnant wife Mary and their six children all resided in the house that was part of the jail.

Around ten o'clock on the night of June 13, the tragedy of the late war returned to Jackson County. A band of former guerrillas, about six or seven in number assailed the jail. As Bugler opened the front door of the jail, the bushwhackers demanded the release of Perry and Faley. Despite the insistence of the armed gunmen, Bugler refused to discharge the two prisoners. The gang warned Bugler they had over five hundred men backing them in their demands, and he would not be harmed if the two men were turned over to them.[281] Bugler not acquiescing remained steadfast in opposition to the demand of the guerrillas. At that moment, one of the criminals still mounted on his horse shouted to Bugler: "We are here for Perry and Faley!" Bugler pointed his gun declaring, "You're not getting anybody!" Bugler turned his back to the ruffians and started to close the door behind him when gunfire erupted, and Bugler immediately fell to the ground dead—shot through the heart. One of the many errant shots went wild and a bullet went through the upstairs bedroom window inflicting a wound in the wrist of Bugler's four-year old son, John.

Jailor Henry Bugler postmortem, laid out in the 'Marshal's Office'

Following Bugler's death and his son's injury, the guerrillas attempted to set the jail on fire. Dean "heard the leader of the mob say, with an oath 'strike a match and stick it in that bed and burn down the whole thing.'" Dean reported that the hearts of those inside the jail sank in the fear that if the structure "was burned, the flames and smoke would come through our grates and we" would be "suffocated or burned up without any hope of escape." As the roof of the building began to ignite, a number of

[281] *Liberty Tribune*, Liberty Missouri June 22, 1866 (The *Liberty Tribune* dates the murder of Bugler as June 14, 1866); *Warrensburg Standard*, June 15, 1866. However, see, Wilcox, 438. Wilcox date's the raid and Bugler's death as June 13, and she further identifies the inscription on Bugler's tomb stone at Woodlawn Cemetery, Independence, Missouri which reads as follow: "Memory of Henry Bugler, who was assassinated while in charge of his duty as jailer, Jackson County, Missouri, June 13, 1866, age 41 years. See photograph of Henry Bugler's monument at Woodlawn Cemetery depicted in Hansen, Duncan, "A Reunion in Death: Gravesites of Quantraill's Men and the James Gang." (Two Trails Publishing, 2002), 48; and, Jackson, David W. "LOCK DOWN." (Jackson County Historical Society, 2009), 54. Based upon this documentation, it is the author's opinion that the murder of Bugler occurred on June 13 and not June 14. Yeatman, 87; Stiles, T. J. "Jesse James Last Rebel of the Civil War." (Alfred A. Knopf, 2002), 175.

Independence citizens rushed to the jail and extinguished the flames and drove "off the mob."[282]

Townspeople who had gathered outside the jail were angry and enraged. Memories of the war's drama returned and sentiments reached a boiling point desirous of revenge and retaliation against the attackers. The following day several resolutions were enacted designed to destroy rebellion and bushwhacking. A committee of twelve men was named to carry out the declarations. A "company of 160 loyal men" was "organized," armed and a posse was formed "to enforce the law, and bring those of the murderous crew to punishment, which dare to remain."[283]

As the resolutions were being adopted, Bingham along with attorneys including, Abram Comingo and William Chrisman went to court. They filed a writ of habeas corpus and secured the release of Reverend Dean. After his discharge, Dean returned to his home in Cass County.[284] Dean's release was a vindication for Bingham and further popularized his painting of "Rev. Dean in Jail."

Word of the rebel raid into Independence soon reached the State Capitol. Missouri Governor Thomas C. Fletcher responded with what Bingham described as an illegal and outrageous order inflicted upon "suffering" people. The Governor called forth the militia to garrison and defend Independence. Bingham objected to this summons as being illegal. His anger intensified to madness when he learned the militia commander was Captain Ridgeway, a former Jayhawker. For Bingham the use of a Kansas Jayhawker was unpardonable. Bingham compared Ridgeway's invasion into Jackson County in 1866 to those Kansas invaders who had robbed, stolen and committed other privations and

---

[282] CDV Revenue Stamp –Regan & Parker Photograph of Henry Bugler (Jackson County (Mo.) Historical Society Archives); *Warrensburg Standard*, June 15, 1866, "The Marshal of Independence Murdered—His Child Shot; Dean, 17; and Jackson County Records Center, Jackson County Court Record Book No. 13, July Term, 1866. Some 15 years later, young John Bugler was accused of being a member of the James' Gang charged with having robbed the Alton Chicago Train in Jackson County, Missouri on 7 Sept. 1881. The Bugler trial commenced in April 1882; but, when Jesse James was murdered on April 3, the Jackson County Prosecutor, William Wallace, requested a recess of the trial to investigate the assassination in St. Joseph. Upon his return to Independence, Wallace dismissed the case against Bugler claiming that he had obtained information while visiting the remains of Jesse James that young Bugler had not been involved in the train robbery.

[283] *Warrensburg Standard*, 15 June, quoting *Kansas City Journal of Commerce*, 14 June 1866. The newspaper accounts provide that the citizens of Independence were greatly "excited over the inhuman outrage" and adopted resolutions "prohibiting rebels and bushwhackers coming into the city, armed, under penalty" of punishment." The story further told of another resolution adopted which declared that "no man should be permitted to make a speech of publication in the city inciting resistance to the laws of the State, and that every rebel or supposed bushwhacker having no visible signs of support should be notified to leave the city within twenty four hours."

[284] Dean, 17.

offenses against property and people of the county during the Civil War.[285] Despite Bingham's intense arguments and scathing cries of radicalism, the militia remained in the town for nearly three months. The scars, pains and sorrows of the rebellion were not being allowed to heal in Jackson County.

Later in the year, Bingham became embroiled in another controversy. The subject of Bingham's wrath and political rhetoric was U.S. Army Major Wyllys C. Ransom who had recently been elected Clerk of the Jackson County Circuit Court. Bingham leveled blistering accusations against Ransom asserting that the Major had committed acts of thievery and destruction while he and his men were stationed and serving in Jackson County during the War. Ransom responded to the allegations by filing suit against Bingham for slander and attempting to impugn his reputation. On March 26, 1867, Bingham wrote to his friend James Rollins that Ransom's suit had been postponed because one of his important witnesses was absent. Bingham proclaimed that he was "ready for trial and would have proved all that I had charged against him." It appears that the lawsuit was eventually dismissed.[286]

Of all the post-war controversies faced by Bingham, the most sensational was his aggressive assault and personal war against Ewing. Around 1865 from his home in Independence, Bingham made the decision to fulfill the threat he had made to Ewing at the Pacific House after the general had refused to rescind Order No. 11. It became Bingham's lifetime cause and duty to illustrate Ewing and his Kansas radicals in a manner that would elicit disdain for the deeds they had inflicted upon the "innocent people" of Jackson County during Order No. 11. While Bingham was painting a life size portrait of the martyred Union General Nathaniel Lyon and Rev. Dean in Jail, he contemporaneously began to villainies Ewing on another canvas.[287]

Bingham's painting of Order No. 11 or "Martial Law," and originally entitled "Civil War" or "The War of Desolation" was intended for historical presentation and preservation. Bingham did not want the crimes of Ewing, Lane, Jennison or other Kansas evildoers to be lost to antiquity. Bingham's purpose was to provide a "due warning to posterity" and to guarantee that the "Red Legs' of Kansas and their equally demoniac associates known only as thieves and assassins" would be criminalized in perpetuity.[288]

---

[285] Bingham's letter to the editor of *Jefferson City People's Tribune* dated July 4, 1866 and published on July 11, "Bingham Letters," 285-287; Rash, 208. In 1866, Bingham ran for Congress out of Independence, but he failed to gain the nomination of the Conservative Union Party.

[286] Bingham's letter to James S. Rollins dated March 26, 1867 and dated March 8, 1870, "Bingham Letters," 288-289; and "Bingham Letters," 299.

[287] Rash, 192. The painting of Lyon was eventually hung in the Capitol in February 1867, but it was lost in the fire that destroyed the Capitol building on February 5, 1911.

[288] Bingham's letter to the editor of St. Louis Missouri Republican dated March 12, 1869, 296-299; and Bingham's Address to the Public Vindicating a Work of Art published in pamphlet form and dated 1871, "Bingham Letters," 315-326. By the 1870s, Bingham entitled the painting "Desolation of Border Counties

In an article written by Bingham after he completed Order No. 11, he outlined the purpose of the work.

> *Art being the most efficient hand-maid of history, in its power to perpetuate a record of events with a clearness second only to that which springs from actual observation. I sometime since became impressed with the conviction, that, as one of its professors, I could not find a nobler employment of my pencil, than in giving to the future, through its delineations, truthful representations of extraordinary transactions indicative of the character of military rule which oppressed and impoverished large numbers of the best citizens of our State during our late sectional war.*[289]

"Martial Law," or "Order No. 11" by George Caleb Bingham

---

by Military Orders Issued by Brig. General Thomas Ewing of the Federal Army, Aug. 25, 1863 (Today it is simply known as "Order No. 11").

[289] Rash, 202.

Bingham's Independence studio where he painted, "Martial Law," or "Order No. 11"

Bingham's painting of Order No. 11 has become an iconic depiction of the ravages of a war along the border of Missouri and Kansas and in particular under "martial law." Bingham for definitional and posterity purposes provided a full description and meaning of his own painting. Bingham further insisted that the scene in the picture was true and accurate, and his artwork was designed to exhibit the worst of the War and the illegal despotism of those who instigated, promulgated, issued and enforced Order No. 11.

The noted author and editor of the *Kansas City Times* John N. Edwards poignantly and prophetically described the future impact of the picture. He proclaimed "Bingham's painting of burning and plundered houses of a sky made awful with mingled flame and smoke, of a long train of helpless fugitives . . . will live longer than the strife."[290]

The scene portrayed in the image is illustrative of Bingham's hatred and resentment of Order No. 11. Bingham himself described the painting. The house on the left of the picture is located about twenty-eight miles west of Lexington in the eastern portion of Jackson County and is being pillaged and ransacked by Red Legs. Bingham affirmed that the family being ejected was loyal Union supporters. The "band of scoundrels" in the picture Bingham insisted were "no more the people of Kansas than were the Border Ruffians" representative of the "great Commonwealth of the State of Missouri."[291] He prominently displayed Ewing perched on his horse—possessing a cavalier, unconcerned, dispassionate demeanor, passively and purposely ignorant of the scene of death, bloodshed and waste. Standing to Ewing's left and in the center of the painting is Jennison adorned with his red-leathered leggings in the act of either pulling or

[290] Noted Guerrillas, 206.

[291] Bingham's letter to Rev. R. S. Johnson dated December 23, 1868, "Bingham Letters," 292-295.

returning his revolver to his holster. Death had reached the house of the poor victims, as the son of the family patriarch lies dead on the ground in the clutched arms of his weeping widow. Bingham's own words provide the details and meaning of his artwork.

> *The principal group in the foreground of the picture, chiefly consists of a venerable patriarch and his family, who have just been ejected from their dwelling, which is about to be committed to the flames.*
>
> *A daughter clings to the defiant old man, imploring him to temper his language so as not to incur the vengeance of the brutal assassin, who, in the act of drawing a pistol, threatens him in front. Another daughter is on her knees before this wretch vainly endeavoring to awaken some emotion of humanity in his callous breast. A married son lies weltering in his blood, his young wife bending in agony over his lifeless body. His murderer is seen in the scowling ruffian nearby with a discharged pistol in his hand. The aged mother has fallen in a swoon, and is supported in the arms of a faithful negro woman. A negro man retires weeping from the scene, accompanied by a negro lad, whose face bears the unmistakable marks of fright and horror.*
>
> *Immediately in the rear of the outraged family the myrmidons of Kansas, aided by their criminal allies in Federal uniform, are busily engaged in the work of pillage. Some of them, on horseback, have already encumbered themselves with spoil. Wearing apparel, household furniture, and everything portable is being placed in wagons, a long train of which well freighted with plunder, is seen in the distance, wending its way westward, while a melancholy procession of dejected and impoverished refugees, fleeing from their desolated homes, file off to the right in an opposite direction.*[292]

The backdrop of the painting consists of smoke rising into the sky as caravans of refugees are shown aimlessly marching from their burned out homes to destinations of which they are uncertain.

> *The scene in the picture is but one of hundreds of similar character and the charred remains of broken walls and solitary chimneys yet to be seen all over the fearfully desolated region attest the truth of my delineations, and would cause any other than the art critic of the Democrat to blush in attempting to discredit them.*

---

[292] Bingham's Address to the Public Vindicating a Work of Art published in pamphlet form and dated 1871, "Bingham Letters," 325; and Nagel, 113.

> *To classify the brutal actors in such scenes as "Union military men" is an insult to every honest soldier who periled his life in defense of his flag. They were the "Red Legs" of Kansas, and their equally demoniac associates known only as thieves and assassins, in whose judgment all men owning property were sufficiently rebels to be robbed and murdered.*[293]

Bingham created two versions of his legendary art piece. The first version he sent to the famous steel engraver, John Sartain in Philadelphia. Bingham hired Sartain to create engraved prints of the painting for sale. The other painting he created on a tablecloth in 1869. This version accompanied Bingham during his long speaking tours in which he journeyed by railroad to "towns of any importance on these roads" denouncing Ewing and taking subscription orders for his soon to be released steel engraving of his painting.[294]

"Martial Law," or "Order No. 11" when on display in the Colonel Joseph W. Mercer home, Independence, Missouri

[293] Bingham letter to the editor of the St. Louis Missouri Republican dated March 12, 1869, 296-299.
[294] Bingham letter to James S. Rollins dated March 8, 1870, "Bingham Letters," 300.

Ewing had returned to Ohio after the war and slowly crept back into politics. Throughout the 1860's Ewing had returned into the fold and ranks of the Democrat Party. He realized that the Republican Party of his Kansas days had grown too radical and was not consistent with the political climate of his home state of Ohio. Even in Kansas, Ewing had been a party centrist and with radicalism dominating the Republicans he abandoned their cause and soon began espousing the Democrat philosophy.

By 1870, Ewing was seeking public office.[295] To shore up his support within his new party, he recognized that his Catholic Faith would negatively impact his political future. He chose to leave the Church and embrace Protestantism. His conversion to the Protestant Faith and Democrat Party failed to gain the trust of the party's leadership.

Bingham continued to follow Ewing's career and political aspirations. When Bingham learned Ewing had designs to run for Congress, he chose to take decisive and concrete action. He knew his war against Ewing would no longer be waged by his tongue or on parchment paper. Rather he had to illustrate his bitterness against Ewing through his artistic talents. Bingham's first open political challenge to Ewing took place in 1870 as Ewing sought the Democrat nomination for Congress. Several Ohio Democrats who opposed Ewing requested Bingham to provide them copies of Order No. 11. Bingham was disappointed when he could not fill the order to produce "250 copies of our print to be used in defeating" Ewing's "nomination." Bingham "regretted" he could not comply with their order, and he blamed it on Sartain for not having timely completed the engravings. Ultimately, Ewing lost the election, and Bingham felt he had played a major role in ensuring that Ewing was "defeated in his aspirations in Ohio."[296]

By the end of 1871, Bingham grew more and more alarmed over Sartain's long delays in finalizing the proof of Order No. 11. To avoid further setbacks in the completion of the work, Bingham traveled to Sartain's studio in Philedelphia where he supervised and corrected the final proof. As final arrangements were made to release the prints for sale, Bingham recognized that his depiction of Order No. 11 would "likely be attacked on political grounds" and he knew that he would have to be prepared "to meet such attacks." Given these concerns, Bingham felt that the initial release of the Sartain's prints should "begin in Kansas City in the midst of the witnesses who can verify the statements of our art record." He also submitted articles to the local press as preemptory measures against these suspected challenges.[297] Bingham had faced negative comments about the painting when it was first released, and he was not going to be caught off guard

[295] Smith, 315-316.

[296] Bingham letter to James S. Rollins dated June 4, 1871, "Bingham Letters," 304. Bingham's use of the word "our" in his letter to Rollins is because Bingham, Rollins and Mr. R. B. Price had acquired an interest in the painting. Rollins and Price had advanced $5,000 for the cost of the engraving. "Bingham Letters," 329n1.

[297] Bingham letter to James S. Rollins dated January 23, 1872, "Bingham Letters," 335-336.

this time. When his print was available for sale, Bingham aggressively addressed every comment expressed in opposition to his painting.

Bingham also renewed his commitment to launch another war and campaign to thwart Ewing's political ambitions. Bingham's painting propaganda against Ewing had not always been successful. In 1876 Ewing was elected to Congress and two years later he was re-elected. During Ewing's second term in Congress he began to set his sights on the governor's mansion in Ohio.

Ewing knew his run for governor would would reignite Bingham's war and crusade against him. Bingham had zealously utilized his painting of Order No. 11 in an attempt to embarrass and defeat him in the past and Ewing had to soften the anticipated attack. He chose to implement a preemptive strike against Bingham's venomous voice by soliciting the defense of his former commander. General Schofield sent a letter to Ewing that was published in newspapers in Ohio and Washington (and even made its way into Missouri newspapers) in which Schofield defended Ewing and Order No. 11. He insisted that the Order impacted very few people as the counties had been previously depopulated by "savage guerrilla warfare." Schofield further provided an itemized list of historical precedents that supported the issuance of Order No. 11 including the various military orders issued out of necessity during wartime. He insisted that those directives had paralleled Order No. 11.[298] In other words, Schofield emphasized Ewing had been justified in issuing the Order.

Schofield's correspondence initiated a prompt critical response from Bingham. He swiftly responded to Schofield's letter and Ewing's effort to hide behind his former general. His criticism of Schofield was succinct:

> *The general has exercised caution, characteristic of all great military commanders, in allowing nearly fourteen years to transpire before venturing upon the defense of a measure which, for heartless atrocity, has no parallel in modern annals. He will be apt to discover, however, that there are those yet surviving who will be able to confront him in this prudently delayed effort to subordinate history to the service of tyranny.*

Bingham after having dismissed Schofield's apology, returned to yet another chorus of his crusading vendetta against Ewing:

> *General Ewing has doubtless discovered that this, his crowning military achievement of 1863, was not of a nature as well calculated to secure the favor of the Democracy with whom he is now associated, as it was to win*

[298]Bingham's letter to editor of the St. Louis Republican dated February 22, 1877 and published in the newspaper on February 27; "Bingham Letters," 426-430.

> *to his support the jay-hawkers and corrupt rabble of Kansas through whose aid there is reason to believe, he then looked for political preferment, and hence his effort, arising from the necessities of his shifted aspirations, to secure it a gloss, which his associate, in responsibility therefore, has endeavored to put upon it, at the sacrifice alike of justice and truth.*[299]

Ewing never found solace or safety from the insults, castigation and ridicule of the Missouri artist. By 1879, Ewing in his second term in Congress received the Democrat nomination for Governor of Ohio. Bingham's retaliatory convictions against Ewing never subsided and when Bingham learned of Ewing's nomination, the old warrior renewed his steadfast commitment to disclose to the world the unconstitutional measures Ewing had imposed upon an "innocent and helpless people." Bingham's persistent and unmerciful campaign against Ewing and his infamous Order No. 11 never wavered. With the governor's race at stake, Bingham's determination to villainies Ewing, Lane, Jennison and their radical supporters by his artwork of Order No. 11 remained resolute. Bingham sent letters to newspapers and political leaders arguing against Ewing's demagoguery and tyranny. Bingham's logical arguments against Ewing resonated in Ohio and especially in Jackson County. The *Independence Sentinel* strongly endorsed Bingham and published the painter's scathing commentary of Ewing and his Order.

> *The truth of his rehearsal of the facts of his portrayal of the infamies of Order No. 11, in this communication, can no more be denied now, after the lapse of sixteen years, than at the time of their occurrence. None dare to deny it.*[300]

The article demonstrated that Bingham's continuous reprisals against Ewing continued to garner the support from a citizenry who had suffered under Order No. 11.

Bingham's propagandist painting and personal crusade against Ewing followed the artist well into his grave. Bingham died before Ewing's gubernatorial campaign ended, but this did not stop efforts to derail Ewing's candidacy.[301] Following Bingham's death, his son James Rollins picked up the baton and the cause on behalf of his late father. He traveled to Ohio where he campaigned against Ewing displaying the Sartain prints of his late father's painting. James Rollins also ensured that the last known letter written by his father prior to his death would be published. Bingham's posthumous words

---

[299] Bingham's letter to editor of the *St. Louis Republican* dated February 22, 1877 and published in the newspaper on February 27; "Bingham Letters," 426-430.

[300] "Bingham Letters," 528-529.

[301] Nash, 214-215.

were entitled, "A Voice From the Grave." It marked the final opinion and epilogue of Bingham's warfare against Ewing.

> *I have always thought and said, and still entertain the opinion, that the measure in question, in any light in which it can be viewed was unnecessary, unwise, cowardly and barbarous, and I have no hesitation in expressing my sympathy with your efforts to canonize it as one of the monuments of military despotism and infamy, which stand out now and then as beacons on the pages of history, to keep us off the rocks on which others have split.*
>
> *With this further evidence thus presented, the merits of this controversy are submitted to the candid judgment of every honest an enlightened citizen.*[302]

Whether Bingham's political rhetoric or his painting of "Martial Law" or, "Order No. 11" impacted the Ohio gubernatorial race will probably never be definitively determined. There is no question that Ewing did not win the governor's race that year, and it is equally unquestioned that Ewing later failed in his ambition to become United States from Ohio.[303] Bingham would certainly have felt vindicated by Ewing's political defeats. Most assuredly, he would have boasted and taken credit claiming his paintings depicting Ewing as demonic tyrant caused him to lose the races.

George Caleb Bingham, self-portrait, ca. 1877

[302] Bingham's letter published in *St. Louis Post-Dispatch* on July 23, 1879, "Bingham Letters," 526-538; Smith, 336; and Nash, 214-215.

[303] "Famous Personages of the Civil War in Missouri, Thomas Ewing," *Missouri Historical Review* (July 1962), LVI: 4.

CHAPTER XIV

The Fates of Major Players: Lane, Bingham, Quantrill, Schofield, Comingo and Ewing

Senator Jim Lane
(June 22, 1814-July 11, 1866)

By the winter of 1863-1864, Lane had returned to the Capitol and resumed his senatorial duties. He continued to be haunted by the Black Friday tragedy. Quantrill's raid seemed to have compromised Lane's dominion over Kansas, and he began to lose his omnipotent power. He constantly battled competing factions within the state. Lane's adversaries had grown weary of the Grim Chieftain's wrath and relentless determination to prevail regardless of the cost. An adversarial relationship is part of the business of politics, but for Lane it bordered on the maniacal. He was ruthless and cunning. Eventually, he found himself at odds with his political opponents and well as those who had been aligned with him or had benefited by his influences. Governor Carney would prove to be no exception.

Carney's gubernatorial success had been assured through Lane's efforts. However, by 1864, Carney and a mounting number of Lane-men had grown exhausted and drained by Lane's controlling nature. The Governor also had a personal motive; he wanted to supplant Lane in the United States Senate. Carney a political disciple of the Grim Chieftain utilized Lane-stye schemes and plots against his teacher in an effort to squeeze Lane from office. To defeat a man of Lane's ilk, Carney knew it would take cunning measures. The Governor chose a method that would have made Lane proud of his protégé, except when the master was on the receiving end of the plot.

Carney called for a special election for Lane's senate seat. On February 9, 1864, the Kansas Legislature elected Carney to the United States Senator—or so they thought. Lane's power may have been waning in Kansas, but his clout in Washington remained undeterred because of his strong relationship with the President. The United States Senate disallowed the vote of the special election and would not unseat Senator Lane. There had been no basis for a special election, and Carney would have to wait to re-challenge Lane after the fall elections.[304] The Governor violated the first principal in politics: when challenging an incumbent, do not lose. Lane held his power base, and from his bully pulpit was able to turn public opinion against Carney and within a few months, Carney's political clout was in free-fall.

Carney's special election challenge was not the only difficulty Lane faced in 1864. Although he retained his senate seat, he recognized his re-election chances would be problematic. He needed a pro-Lane Kansas Legislature to be elected that fall, and this was in serious doubt. He was less concerned about the internal political wrangling in Kansas than the fact he needed a favorable legislature to be elected during a presidential election year. Everyone in Kansas knew of Lane's long-term relationship with the President and Lane was fully cognizant of the nation's declining support for Lincoln. The Civil War continued to drag on with increasing death tolls. Copperheads blamed Lincoln for not ending the war and many within the Union accused Lincoln of failing to have won the war earlier. Speculation even grew that the President might not even be re-nominated. In the end, Lincoln was nominated and ran for re-election against the Democrat candidate General George B. McClellan.

Lane had to decide which presidential candidate to endorse. Lane could not afford to waiver on the issue, as it would make him look weak. Yet, if he chose the losing candidate, his likelihood for prevailing would also be doubtful. After some reflection, Lane decided to remain loyal to Lincoln and publicly endorsed the President. Lane knew there could be political risk in supporting Lincoln, but once Lane made up his mind, he was unflinching. He promptly returned to the stump traveling the country campaigning for Lincoln.

The campaign circuit did not overcome the ongoing and persistent obstacles Lane experienced during the election year in Kansas. A corruption scandal erupted causing the reputation of Lane and his friend General Blunt to be questioned. As the November elections approached, it appeared that Lane was destined to defeat unless something cataclysmic occurred. The war gods soon began to smile upon the Grim Chieftain. Lane was stumping for Lincoln in St. Louis when word reached him that General Sterling Price had re-entered southeast Missouri from northeast Arkansas to recapture Missouri for the Confederacy.

Lane promptly left canvassing for a vote and returned to battling an invasion. Lane left St. Louis and rushed westward to Kansas. He wanted to defend his state from Price's invading army (a cynic might claim he offered his services for political gain). Price once again entered Missouri with the same misconception he had during the first year of the war: the men of Missouri would flock and enlist in the Confederate Army and he would recapture the state for the "cause."

---

[304] Collins, 240.

Price's military plan was to capture St. Louis and then move westward and seize the State Capitol. Marching northward out of Arkansas, Price's invasion strategy was thwarted by General Ewing at Pilot Knob, located south of St. Louis. Ewing's defense of the earthworks at Ft. Davidson caused Price to change his course of action. Price no longer could risk further advancement towards St. Louis and instead chose to march through the heart of mid-Missouri and on towards Independence, Kansas City and even Kansas. Ewing's success at Pilot Knob became a blessing for Lane as it allowed Ewing's former client the opportunity to regain the spotlight in the defense of the Union.

By the time Price was advancing through Independence towards Kansas on October 22, 1864, Lane had returned to the war effort. He served on the staff of General Samuel Curtis, the Union commanding general stationed at Westport in Jackson County. In what became known as the Battle of Westport, the largest land battle fought west of the Mississippi River during the Civil War, Curtis (with Lane's assistance) destroyed Price's Confederate Army. The overwhelming defeat of Price's military campaign through Missouri proved to be a fortuitous turn of events for Lane's political fortunes. A favorable Kansas Legislature was elected within two weeks of the Union victory at Westport; Lincoln was re-elected to a second-term; and Lane was returned to the United States Senate.

Lane had retained his senatorial hold on Kansas, but it would fade as quickly as it had been rejuvenated by the Battle of Westport. His epilogue would last only one year longer than that of his enemy, William C. Quantrill. Lane's misfortunes began when his friend Abraham Lincoln was assassinated. By the President's death, Lane faced concrete problems in Washington. He was no longer part of the Radical Republicans. Lane had been distancing himself from their radicalism for some time, but his opposition had widened after the radicals abandoned Lincoln's reconstruction policy of atonement and had chosen a course of retribution.

Lane had other disagreements with the radicals, but it was his decision to back Lincoln's successor Andrew Johnson (as would General Ewing) that caused him the greatest difficulty. It served as the precursor to the unraveling of his political career. He fought the Radical agenda in Washington as he attempted to support the policies of the Johnson Administration. Lane was further compromised by his inability to forge the same rapport with Johnson as he formerly held with Lincoln. Lane's separation from radicals in Washington led to accusations he intended to destroy the Republican Party in Kansas.

By 1866, the political infighting and persistent battling in Washington and Kansas compromised Lane's mental and physical health. When the session ended that spring, he left Washington gravely and visibly ill. Lane stopped in St. Louis where a physician examined him and gave him a prescription and recommendation for rest. Lane's condition did not appreciably improve and the doctor advised him to immediately return to his home in Kansas. Lane traveled by steamboat up the Missouri River from St. Louis to Ft. Leavenworth. In route, Lane's health continued to decline and when he reached

Leavenworth he was physically unable to make the journey to his home in Lawrence. Arrangements were made for Lane to remain near Leavenworth at a farm operated by his brother-in-law. Lane reportedly told a friend he expected to die as his life was at its end.[305]

Lane's life was in fact near its end. His mind had suffered; years of political turmoil had drained him; he suffered melancholy and depression; he no longer dominated Washington; his political power in Kansas had plummeted; Lane had lost his reign—and he knew it. Lane considered death as his only salvation.

James "Jim" Lane tombstone

On Sunday, July 1, 1866, Lane was gripped in the hold of one of his many morose moods. His friends faintly hoped a buggy ride in the fresh air would help improve the spirits of the Grim Chieftain. When the prairie ride ended, they returned to the house. Lane jumped out of the carriage and pulled out a revolver he had concealed under his frock coat. He looked towards the men and across the plains of his adopted state. He then ejaculated "in a loud and cheerful voice, 'Good-bye, gentlemen!'" Lane then put the pistol in his mouth and pulled the trigger.[306]

Even in death, Lane was a hard man to kill! While the self-inflicted gunshot wound was mortal, he hung onto life for another ten days before finally giving up his ghost on July 11. He was buried in Lawrence, Kansas, two days later.[307] The life of the Grim Chieftain had expired, but his legend did not.

---

[305] Collins, 271-273

[306] Castel, "Frontier State," 232; and Collins, 274-276.

[307] Yeatman, 87.

George Caleb Bingham
(March 20, 1811-July 7, 1879)

Bingham remained politically active throughout his years of protest and war against Ewing. He served on the Independence School Board until 1870 at which time he sold his house on Pacific Street—the same residence from which he had painted both murals of Order No. 11. He, his wife and son James Rollins moved to Kansas City. The decision to relocate was certainly motivated by the population growth and economic development in Kansas City. In 1869 the first expansion railroad bridge over the Missouri River, known as the "Immortal Bridge" or "Hannibal Bridge," opened in the City. The completion of this project almost overnight turned Kansas City into a thriving industrial, railroad and transportation metropolis. Bingham knew he would be financially more secure by relocating to Kansas City, and by 1871, he and his son James established the firm "George C. Bingham and Company."[308] A more dense populated community would mean "more heads" to paint and more income for the family.

The move also furthered Bingham's political advancement. On May 11, 1874 he was appointed President of the Kansas City Police Commission. The night after his appointment Bingham demonstrated the law and order direction he intended to follow. He issued an order that caused "from two to three hundred gamblers" to be "turned out on the streets, and every gambling house in the city suppressed." No dram shop or drinking establishment would be permitted to open on "any Sunday."[309] These new laws were not well received by the local merchants, but the criticism did not dissuade Bingham's determination to implement these tougher policies.

The following year in January 1875 Bingham received the state's highest law enforcement appointment when Governor Charles H. Hardin named him Adjutant General of Missouri. From that day forward, the Missouri artist was known as General Bingham. After assuming his new duties, Bingham promptly became embroiled in a criminal case that has remained steeped in controversy and almost overnight turned public sentiment and sympathy towards the James Gang.

On the evening of January 26, 1875, a number of men alleged to be members of the Pinkerton Detective Agency from Chicago surrounded the James farm near Kearney, Clay County, Missouri. Reports that the bandit brothers Frank and Jesse James were inside the home had led them to the homestead. Someone within the group threw a "hand-grenade or explosive instrument" or other device "saturated with turpentine or oil" through "the kitchen" window to flush the outlaws into the open night air. The result and consequence was complete disaster. The accelerant began to burn and then exploded mortally injuring the half-brother of Frank and Jesse James, eight-year old Archie

---

[308] Nagel, 116.

[309] Bingham's letter to James S. Rollins dated May 26, 1874, "Bingham Letters," 363-364, and 365n4; Bingham's letter to James S. Rollins dated June 7, 1874, "Bingham Letters," 365-366.

Samuel. The mother of the boys Zeralda James Samuel was permanently maimed in the explosion, as her husband Dr. Reuben Samuel was forced to amputate her "shattered" right arm below the elbow. The tragedy produced a bitter outcry of protests and a staunch demand for a criminal inquiry. Newspapers throughout the state carried the story and pointed blame squarely upon the Pinkertons'.[310]

Governor Hardin on the morning of January 29 dispatched "General Bingham to proceed to Clay county and investigate as far as he might the facts of the assault upon the family of Mr. Samuel." Hardin was determined to have his new Adjutant General fully investigate the disaster so that those responsible for the outrage "upon the peace and quiet of society" answered for their crime. Hardin in a report to the Missouri General Assembly made his intentions certain. He would do everything within his power to see the "offenders" from outside the state who had "undertaken, in a wanton, brutal and wicked way" the attack during the blackness of night brought "to trial."[311]

Bingham left the Capitol and traveled to the James property to scrutinize the situation. Bingham's focus was "to co-operate with the local authorities" in Clay County "to bring the perpetrators to justice." He conducted interviews of several witnesses, including Mrs. Samuel. Bingham by design "omitted" identifying "the names of citizens from whom" he had spoken "believing it best for the ends of justice that they should not be given to the public." This was in part to shield them from harm as some still wanted "the James brothers . . . captured and brought to justice."[312]

On February 4, 1875 Bingham submitted his official report to Governor Hardin. Bingham announced at the outset that his inquest was hampered because he lacked the "power to compel witnesses to testify under oath." He was certain "several unknown" invaders had ridden in a "caboose attached" to a special train that had "stopped in the woods about two miles north of Kearney." These men Bingham concluded had been responsible for the calamity. The Adjutant General took possession of a pistol found in the woods that had "marks upon it . . . known to be on pistols of a well-known band of detectives." Bingham was convinced that neither Frank nor Jesse James was inside the dwelling at the time of the clandestine raid and that "the James brothers" had not "been seen in the vicinity since last April." Bingham was satisfied that only through "the inquisitorial powers of a grand jury" would evidence be elicited in order to "lead to the identification of the parties engaged" in the "transaction."[313]

---

[310] *The State Journal*, Jefferson City, Missouri, 5 Feb. 1875.

[311] *The State Journal*, Jefferson City, Missouri, 12 Feb. 1875.

[312] *Ibid.*

[313] *Ibid*, *Liberty Tribune*, 12 Feb. 1875; Yeatman, 140-142 and Appendix D, 360-362; Ventimiglia, Jack "Miles." "Jesse James in the County of Clay," *The Friends of the James Farm* (2001), 47, 85-86. Manslaughter indictments were eventually returned in Clay County in March 1875 against eight men including Allan Pinkerton, the head of the Agency. They never stood trial for their crime.

Nothing during Bingham's two years as Adjutant General ever became as noteworthy as the James investigation. The bulk of his tenure was spent negotiating claims of military veterans. He spent a considerable amount of time in Washington where he lobbied Congress to ensure benefits were paid to former soldiers. He also used the opportunity to press for reimbursement for his property loss on August 13, 1863. Bingham was relentless in the pursuit of his casualty claim against the federal government. He insisted Ewing's seizure of the property and its destruction while under Ewing's command entitled him to just compensation (there is no record found that Bingham ever received compensation for the loss of his property). His monetary damages furthered his fury against Ewing.

Bingham experienced many personal family tragedies after the War. His son Horace died in 1869. His second wife Elizabeth began suffering from mental illness sometime in 1876. Bingham became alarmed at her condition and even considered taking her to warmer climate in the south hoping it would help her recover.[314] By the middle of October 1876, Bingham realized his wife's affliction would "require all the fortitude" he could "summon" in order "to bear" her mental condition. Her "deranged" state "compelled" Bingham "to take her to an asylum." He told his friend Rollins that Eliza's "derangement exhibits itself in brilliant visions of heaven, and her departed relatives and friends who are inhabitants thereof." She said "her Savior" visited and conversed with her and had given her instructions on "what to do."

Eliza's infirmity left Bingham no alternative and he had his wife institutionalized at the insane asylum in Fulton. Her confinement at the facility began on October 31 and she shortly thereafter died on November 3.[315] Eliza's death had a profound impact upon Bingham. He entered a spirit world in which he claimed his late wife frequently visited him. He also experienced a religious revelation leading to more of a spiritual calling.[316] Only time would heal the pain.

Within five months of Eliza's death, Bingham found the solace he needed to overcome his grief. He began courting a long time friend. When word of the new relationship reached his son James R., Bingham encountered yet another troubling ordeal and crises. His son protested the thought of his father remarrying. James ridiculed his father and soon broke off all communication with his father. Bingham was disheartened by his inability to assuage his son's fears and concerns. Once again, Bingham turned to his life-long friend, James Rollins, to help heal the wounds between father and son.

---

[314] Bingham's letter to Governor Hardin dated October 9, 1876, "Bingham Letters," 414.

[315] Bingham's letter to James S. Rollins dated October 15, 1876, "Bingham Letters," 414-416; and Bingham's letters to James S. Rollins dated October 29, 1876, "Bingham Letters," 416-417. In 1876, the institution was named, The Missouri State Lunatic Asylum, now Fulton State Hospital. The facility had opened in December 1851 to provide care for the mentally ill; See, "Bingham Letters," 417n1.

[316] Nagel, 143.

Rollins did intervene and he facilitated a resolution of the rupture to the great satisfaction of Bingham.

Bingham's affections had turned to the widow, Martha Ann "Mattie" (Livingston) Lykins. Mattie's late husband, Dr. Johnston Lykins, had been a pro-union mayor of Kansas City. Bingham considered Mattie "in the highest sense of the term" as "no ordinary woman."[317] The Bingham's and Lykins' had been family friends before the deaths of Dr. Lykins and Bingham's second wife. Before Eliza had become mentally infirm, she worked with Mattie on various civic projects and on behalf of Mattie's orphanage in Kansas City.[318] After the deaths of their respective spouses, the widower and widow began to court, and on June 18, 1878, they were married in Kansas City.

Bingham's Kansas City studio. "Order No. 11" is in the background; the young lady in the foreground is Mrs. Roma Wornall.

Although Bingham held strong Union proclivities during the war and his new wife, Mattie, had been an ardent southern sympathizer, they shared certain commonalities. Bingham had lost his family property on that fateful afternoon in August 1863, and Mattie had personally witnessed the massive destruction of the building. Bingham detested Order No. 11, and Mattie personally experienced the pain and agony of banishment under the Order.[319] Most in common was their shared distaste and determination to mark Ewing's evilness through the promotion of Bingham's paintings of Order No. 11.

The marriage of Bingham and Mattie would not last long. On July 7, 1879, and only some thirteen months after their union, Bingham died at their residence in Kansas City, known as the Lykins Institute—a home for homeless and orphaned girls that Mattie had established. At the

[317] Mattie Livingston married Dr. Lykins on October 12, 1851, in Washington and the following year they moved to Kansas City. Bingham's son James Rollins died of pneumonia in 1910 at the age of forty-seven. Findlen, 110-111.

[318] Bingham's letter to James S. Rollins dated January 17, 1872, "Bingham Letters," 334.

[319] Despite Dr. Lykins strong Union loyalties, his wife Mattie had been banished from Jackson County during Order No. 11 and was forced to relocate to the northern side of the Missouri River into Clay County—see biography of Martha Ann "Mattie" (Livingston) Lykins in Appendix H.

time of his death, General Bingham was also serving as the first professor of art at the University of Missouri in Columbia.

The day following Bingham's passing the *Kansas City Daily Journal* eulogized the artist. They gave the history of Bingham's nativity, settlement in Franklin then in the Missouri Territory, his dedication to the Whig Party and his artistic talents as a portrait and genre artist. The article lauded Bingham's "flatboatmen, the county election," his "landscapes" which had given the painter "great eminence" as had his full-size images of Clay, Jackson and Blair. The *Journal* wrote of Bingham as opinionated, firmly with the Union during the rebellion, seized with a tenacious "often aggressive" personality. He was considered a "controversial writer" but his writings were described as "clear, simple, seldom ornate but incisive to the last degree." As to his politics, he had converted to the Democrat Party during the 1864 presidential election and refused to support Lincoln. The *Journal* returned to the most political of Bingham's artwork—Order No. 11, and prophetically opined:

> *From that time* [1864] *until his death he was a Democrat. It is the opinion of many that it was the issue of Order No. 11 that dated his alienation from the more radical Union men of the State. Be that as it may, his most recent prominence, so far as the general public is concerned, was in connection with that celebrated order. He has immortalized it both by pen and pencil, and it is from this fact, as much as any other, that it is destined to an immortality that will live even after the incidents it represents are forgotten . . . In his death Missouri has lost a distinguished citizen, Kansas City has lost one of her best and most widely known men, and society has lost a man who, whatever may been his connection with public affairs, has left behind that noblest of all characters—that of an honest man. He was true to friends, but, above all, true to his convictions—he valued friends, but preferred what he thought right even to friendship. He had many and often bitter controversies, but even those who most keenly felt the sharp thrust of his pen or tongue, will be among the first to do honor to the integrity of his motives and the purity of his private and public life.*[320]

Bingham's funeral was held on July 9, 1879, and it was described as "one of the largest" seen in years in Kansas City. The "ceremonies were impressive and in some respects remarkable." Speakers included President Sam Laws from the University of Missouri and Bingham's "life-long and most intimate of all his friends, Hon. James S. Rollins." Rollins provided a "noble tribute" and "eloquent" commentary on the life of his friend. Those who heard Rollins' eulogy according to the *Journal* would remember it as

[320] *Kansas City Daily Journal*, 8 July 1879; "Bingham Letters," 523-525.

"an event of a lifetime."[321] Mattie, now twice widowed, made arrangements to have Bingham buried at Union Cemetery in Kansas City and adjacent to the grave of her first husband Dr. Lykins.[322] The second-timed widow had a special tribute inscribed on Bingham's tombstone that read:

> *Eminently gifted, almost unaided he won such distinction in his profession that he is known as The Missouri Artist.*

Mattie spent the remaining years of her life working for charitable causes and condemning Ewing and his Order No. 11. On September 20, 1890, Mattie Livingston Lykins Bingham passed away from complications of cancer. She was laid to rest next to her two late husbands.[323] Under the terms of Mattie's Last Will and Testament, Bingham's art was to be sold with proceeds being donated to the Confederate Veterans Home and cemetery in Higginsville, Missouri. The auction was held at the W. W. Finley Art Gallery in Kansas City in March 1893. The Bingham art sold that day was a combination of his political ideology and his genre settings. It symbolized Bingham as an artist and statesman.[324] The proceeds per Mattie's instructions were then given to the veterans' home—which today is a Federal Cemetery for military veterans of all wars, including the Civil War.

George Caleb Bingham tombstone

Over twenty years after Bingham's death, Frank James the former Quantrill guerrilla and leader of the post-war James-Younger Gang gave a press interview. The article appeared in the *Louisville Kentucky*

[321] *Kansas City Daily Journal*, 10 July 1879; "Bingham Letters," 525-526.
[322] Findlen, 106.
[323] Findlen, 108-109.
[324] Rash, 222; and Findlen, 109.

Journal on September 29, 1901, and quotes Frank's tribute to Bingham and his painting of Order No. 11:

> *I am glad General Bingham put that on immortal canvas. That is a picture that talks. That order simply ruined hundreds of peaceful homes in western Missouri. I know one man up in Jackson County who made a fortune going around picking up cattle that had been abandoned, a high toned cattle thief.*[325]

Even today, Bingham and his painting of Order No. 11 remain an artistic indictment of military martial law. Was it truth or propaganda? Bingham would have insisted it had been his finest moment as a political artist. To his antagonists, the painting had been the worst the Missouri artist had ever created. Regardless of one's perspective on this particular artwork, there is no debate that Bingham was Missouri's Nineteenth Century Artist: a painter, politician and propagandist.

---

[325] Hale, 112.

William Clarke Quantrill

(July 31, 1837-June 6, 1865)

After the raid on Kansas, Quantrill and his guerrillas regrouped and began their long trek south towards Texas. The bloodletting and murderous atrocities in Lawrence was not the guerilla chief's last notorious act in 1863. On October 6 Major-General James G. Blunt with an escort, including a military band, was in route to the fort at Baxter Springs in southeastern Kansas. As Blunt advanced he saw about one hundred men "all dressed in Federal uniform." Blunt thought the troops were part of a Union detachment "on drill." He did not suspect they were Quantrill's enemy forces until confusion broke out in his ranks.[326] By then it was too late and more carnage followed as the bushwhackers, each possessing "two to six six-shooters apiece" annihilated the federal force. Blunt ironically was able to escape the massacre by leaping from his buggy, mounting a horse and out distancing the pursuing guerrillas.

Baxter Springs had become a slaughterhouse for the federal troops. Schofield gave the official grim totals: fifty-four killed.[327] In returning to the battlefield, Blunt discovered that every member of the "brigade band, teamsters and all headquarters' clerks who were first captured were murdered"—shot through the head.[328] Schofield's casualty total was inconsistent with those provided by Confederate officer Peter Hardeman who encountered Quantrill's raiders within days of the attack. While Hardeman and his forces did not take part in the violence, he gave a substantially higher death toll. Hardeman in a letter to his wife dated October 12, 1863, stated the federal losses totaled "about one hundred men" while Quantrill only "lost four killed and three wounded." Following the total annihilation and guerilla victory at Baxter Springs, Quantrill and the remainder of his men slipped off to the winter quarters in Mineral Creek some fifteen miles northwest of Sherman, Texas—the same location they had spent the previous winter.[329]

During the winter of 1863-1864, Quantrill experienced significant internal and external military dilemmas that would ultimately prove to undermine his leadership. His organization and authority began to be subverted by George Todd and Bloody Bill Anderson. This power struggle severally compromised Quantrill. Some of his followers joined the mutineers, deserted or like William Gregg enlisted in the Confederate Army.[330]

---

[326] O.R. 22:1:688-689.

[327] O.R. 22:2:622; Hale, 59; and, Hardeman, Nicholas P. "The Bloody Battle That Almost Happended: William Clarke Quantrill and Peter Hardeman on the Western Border." *Civil War History: A Journal of the Middle Period* (Sept. 1977), 23: 3: 255.

[328] O.R. 22:1:689.

[329] Leslie, 284.

[330] In his memoirs Gregg's says he enlisted in Jo Shelby's Confederate cavalry unit and elected captain of Company H. Leslie, 289.

The Confederate brass had grown weary and detested the manner in which Quantrill and his bushwhackers waged war. These officers publicly expressed outrage over Quantrill's actions along the border of Missouri and Kansas during 1863. Privately, they recognized the value of Quantrill's guerilla operations, but only if he and his men operated within military protocol and strategic plans. Quantrill was disinterested in obeying orders. His narcissistic personality precluded him from surrendering authority to anyone, including the Confederate Government or those dissenting members within his ranks.

Quantrill returned to Missouri sometime around April 1864, but his command had disintegrated around him. His military actions that year became marginalized at best. He spent much of his time cavorting with his mistress Kate King. The bulk of the guerilla bloodletting during the summer and fall of 1864 were masterminded by the ruthlessness of Anderson and Todd.

Quantrill did meet with General Sterling Price in Boonville, on October 11, 1864 as Price made his final march through Missouri. During their meeting, Price ordered Quantrill to destroy telegraph lines, burn bridges and as historian Donald R. Hale described, "to carry on warfare more or less in their own manner and doing as they chose."[331] Quantrill did not play any significant role in Price's campaign as his support had waned and the atrocities of Anderson and Todd even disturbed the guerilla chieftain. Price's raid proved disastrous to the Confederate Army and resulted in the deaths of George Todd on October 21 near Independence and Bloody Bill Anderson five days later near Richmond in Ray County.

Price's defeat and the deaths of Todd and Anderson would serve as the final Civil War days for the guerrilla raiders in Jackson County. By January 1865, Quantrill and his remaining loyal followers traveled east into Kentucky. Among the guerrilla veterans was Dick Burns, Sylvester "Vest" Akers, Clark Howkinsmith, J. D. Hylton, John McCorkle, John Koger, Jim Little, "Ole" Shepard, Bill Bassham, "Babe" Hudspeth, Allen Parmer and Frank James.[332]

On April 18, "Captain" Edwin Terrell, a 20-year-old with a scoundrel's reputation, had been placed on the Union "Secret Service" payroll. Terrell's directive was to pursue, capture or kill "Quantrill and his band of about sixty men."[333] The choice of Terrell was fulfillment of the old maxim: "it takes a thief to capture a thief." It would take Terrell and his men a few weeks to accomplish the mission, but in the end, it proved successful for Terrell and disastrous for Quantrill.

Quantrill and some twenty-one men had taken refuge on a farm located near Taylorsville, Spencer County, Kentucky and owned by James H. Wakefield, a Confederate loyalist. On May 10, the guerrilla force was in the Wakefield barn loafing,

---

[331] Hale, 64.
[332] Allen Parmer later married Susan James, the younger sister of Frank and Jesse James. Leslie, 343.
[333] O.R. 49:2:406; and Leslie, 361.

sleeping, playing cards or throwing corncobs at each other when their peace was disturbed.[334] Terrell and his men had quickly galloped upon the property and wholly surprised the bushwhackers. As the rebels attempted to flee on horseback, Quantrill and Vest were hit. Vest's bullet wound was to his shoulder, while Quantrill had been shot in the back.

Vest later recounted his experience at the Wakefield farm to his grandson. He said that the only reason he had not been killed was simply good fortune. He just happened to detect a Union solider taking dead aim at him. Vest was hit in the shoulder and not in the heart because when he spotted the sniper, he threw his pistol down and shot; the bullet struck the post in which his assailant had been resting his rifle causing splinters to strike the man's eyes. While Vest was not severely injured, he along with many other bushwhackers was taken prisoner. They were held in an upstairs room in one of the buildings. Around 11 or 12 o'clock that night the guard came around to make final rounds. Vest and the other captives took blankets and sheets off the beds and tied them together and escaped through the window. Vest hid in the corn shucks until he was able to find a sympathetic family where he stayed until he recovered.[335]

Quantrill was not as fortunate or lucky as Vest. The bullet had penetrated his spinal column leaving him paralyzed. Terrell approached Quantrill as he lay in the mud. After an interrogation by Terrell, Quantrill identified himself as Captain Clarke of the 4th Missouri Confederate Calvary. When "Captain Clarke" assured Terrell that he would not attempt to escape, Terrell resumed his chase for "Quantrill." The fallen warrior was removed from the yard and taken into the Wakefield house. Frank James reportedly arrived that evening to "free" Quantrill, but the former guerilla leader told Frank it was no use.[336]

Two days later, Terrell returned to the Wakefield's having learned that the felled officer was none other than the hunted guerilla, William Quantrill. Terrell seized Quantrill and took him to Louisville.[337] Physicians concluded that the guerilla chieftain's wound was mortal; he was taken to a Catholic convent where nuns provided him medical care and attention. Before Quantrill died on June 6, he was baptized and given the last rights of the Catholic Church by Father Michael Power; Quantrill died a Catholic.[338]

Quantrill may have been dead, but the story of what happened to his remains has become legendry. He was originally buried in the "Portland Cemetery" in Louisville, Kentucky. To prevent thieves from stealing the body, Father Power refused to allow a

---

[334] Leslie, 365.

[335] Jackson County Historical Society Archives: Box 10 F15 Akers, Joseph Harold, from tape-recorded interview of Joseph Harold Akers on 08-27-1985. He was born in 1902 in West Central Missouri and was the son of Joseph Henry Akers and the grandson of Sylvester "Vest" Akers; and Union, 953.

[336] Noted Guerrillas, 436; and The Devil Know How to Ride, 365-366.

[337] Hale, 130-131.

[338] Noted Guerrillas, 438-440; Castel, "Frontier State," 229-230; and Burch, 251.

head stone to mark the grave. Despite this precaution, the resting place of the fallen rebel did not stay undisturbed. Some pilfering did take place, but not by common "relic" hunters. In December 1887, Quantrill's mother and his childhood friend, William W. Scott, editor of the *Iron Valley Reporter*, journeyed together to the cemetery. They convinced the sexton to open the grave so they could rebury the bones in a zinc-lined box. Through some slight-of-hand, Scott secured the bones and the skull leaving the grave virtually empty.

With the bones removed, it would take over one hundred years before they would be finally re-interred—and even then not in one place. On December 17, 1888, Scott wrote Franklin G. Adams, secretary of the Kansas State Historical Society, seeking to sell Quantrill's skull. Scott insisted he was "not speculating in dead men's bones" but only wanted to recoup some of his expenses. Scott enclosed in the letter a lock of Quantrill's hair.[339]

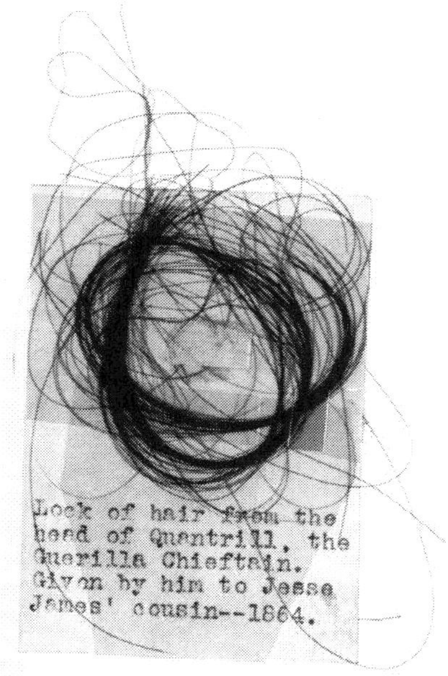

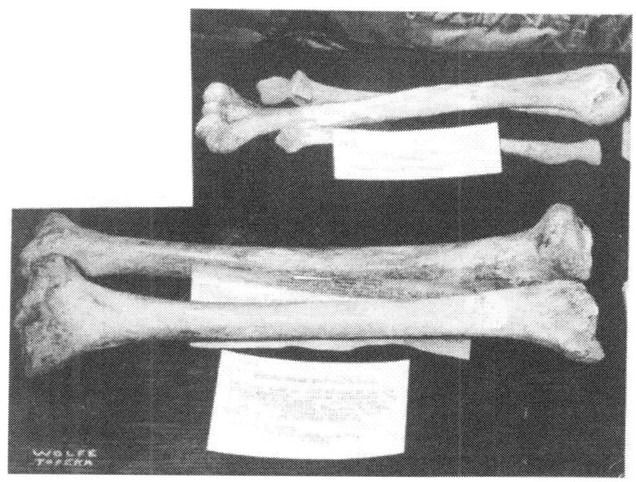

William Clarke Quantrill hair and arm and leg bones in years prior to re-internment.

[339] Leslie, 413; "Quantrill's Bones," Edward E. Leslie, *American Heritage* (July-August 1995); wesclark.com/jw/quantril.html.

In the spring of 1889, Quantrill's mother returned to Dover and in a private ceremony in which only five people attended, she had her son's bones entombed in an unmarked grave. The number of bones buried was never fully ascertained.[340]

William Clarke Quantrill skull when on display at the Dover (Ohio) Historical Society

From that point forward, Quantrill's remaining bones and skull became like a vaudevillian touring show or a disappearing act. Scott had previously donated shinbones to the Kansas State Historical Society and he retained his own collection, this was just the beginning of the saga. A wax model of the skull was made and then sculptured into a life-like image of the dead bushwhacker. After Scott's death in 1902, the bones and skull were displayed in glass cases, donated to historic societies, traveled from state to state,

[340] "A Piece of Quantrill: But Which Piece?" *Kansas City* (Mo.) *Times*, 21 Dec. 1977. Adams told this story to his son Zu Adams in a statement dated August 7, 1901. The leg bones were purportedly donated to the Kansas State Historical Society. William E. Connelley said he donated Quantrill's right arm bones to the Kansas State Historical Society (how he obtained possession of these bones remains a mystery). A lock of Quantrill's was reportedly given to an unknown cousin of Frank and Jesse James and somehow purportedly ended up in possession Kansas State Historical Society. Fraternities and other similar organizations were said to have used some of the leg bones during their rituals.

viewed in museums, used in fraternity rituals, tested scientifically, offered for sale, bartered and handled by historians. They were also discarded, overlooked, forgotten, lost, discovered, separated and rediscovered. An entire book could be told on the legend of the chronicle of the skull and those 'ole' bones.

Eventually, the scattered bones, skull and hair were given a final resting place—or places. On October 24, 1992, Quantrill's leg bones, arm bones and lock of hair were wrapped and placed in a red-plastic cooler and laid inside a replica 19th Century oak coffin draped under a Confederate flag. Nearly 1,000 were in attendance in Higginsville, Missouri, as Father Hugh Behan read the Roman Catholic funeral liturgy. A Confederate organization under the direction of Robert L. Hawkins III provided the final tribute. Hawkins gave memorial remarks that were followed by a twenty-one gun salute.

Six days later on October 30 another Catholic graveside service was held. Father Michael Carter conducted the Catholic burial rite in the Dover cemetery before a very small assembly of twenty-two people. Quantrill's skull was 'repatriated' and re-buried in a child's casket and buried with bones previously deposited in 1889.[341]

Quantrill tombstones in Higginsville, Missouri (above) and Dover, Ohio (right)

[341] civilwarsaga.com/william-quantrills-three-graves. It is likely that the skull was returned to the Quantrill family grave plot in the Fourth Street Cemetery in Dover, Ohio. Photos courtesy William Clarke Quantrill Society Records donated by Donald R. Hale, Jackson County (Mo.) Historical Society Archives. The Portland Cemetery was later renamed St. John's Catholic Cemetery. The Kansas State Historical Society had given the leg and arm bones and hair to the Missouri Division, Sons of Confederate Veterans, for burial in the Federal Cemetery in Higginsville, Missouri.

It may never be known if other Quantrill body parts remain unburied, hidden in some museum desk or overlooked in an archival box. Regardless, there is a certain poetic irony about Quantrill's death and the plight of his skull and bones. He was hunted and killed by a wicked, unscrupulous paramilitary commander who was not officially enlisted in the Union Army. His own raid of Lawrence had resulted in the issuance of Order No. 11 that scattered to the four winds thousands of citizens. For Quantrill, his bones were likewise scattered, perhaps not into the winds but into burial spots in Kentucky, Ohio, Missouri and who knows where else. To those who detested the man—and many still do—Quantrill's death was something to be celebrated. For those who sympathized with his cause—and many continue to do so today—he had died a martyrs' death.

Major General John M. Schofield
(September 29, 1831-March 4, 1904)

Lane's failed attempt to have Schofield removed after the Lawrence disaster intensified to a hysterical pitch after Blunt's troops were annihilated at Baxter Springs on October 6, 1863. Lane was not alone in vociferously sounding his complaints to the President. As written objections to Schofield mounted on the President's desk, Lincoln eventually had no option and removed Schofield. On January 1, 1864, Schofield received joyous news from Washington: he had been transferred to Tennessee and the Army of Ohio.

Schofield's military assignment in the Army of Ohio placed him at the center of what would become part of General Sherman's overall Atlanta Campaign. After Sherman captured Atlanta, Schofield did not join Sherman and his "March to the Sea." Instead, he was assigned the task to remain behind to defend against any invasion into Tennessee by Confederate General John B. Hood.

Sherman's instincts about Hood's plans were correct. His trust in assigning the defense of Tennessee to Schofield proved to be equally sound. At both the Battle of Franklin on November 30, 1864, and two weeks later at the Battle of Nashville Schofield successfully repulsed Hood, and effectively destroyed Hood's Army. These victories resulted in Schofield's promotion to full brigadier general in the regular army. No longer faced by the threat of Confederate forces attacking Sherman's rear, Schofield was sent to fight alongside Sherman during the final months of the Civil War. Schofield's war career came to a glorious end. He officially received the surrender of General Joseph P. Johnson Confederate Army at Durham Station, North Carolina in May 1865.

Vice-President Andrew Johnson became President after Lincoln's assassination. The new President held serious doubts about the reconstruction platform and plan of the Radical Republicans. Johnson wanted to pursue the same policies of the slain President—a policy of leniency towards the South. Schofield shared Johnson and Lincoln's sentiments. Neither Johnson nor Schofield approved the sanctions that the Radicals sought to impose upon the South. They likewise held strong reservations of the post-war actions Secretary of War Stanton desired to implement. Johnson not desiring a direct confrontation with Stanton elected to retain the Secretary hoping to internally rectify their differences. Schofield's continued support of Johnson's post-bellum reconstruction plans was rewarded when the President appointed him as special diplomat to France and military governor of Virginia.

As time went by, Johnson continued to battle with Stanton and the Radical Republicans in Congress. He continued to fundamentally disagree with Stanton on almost every issue. His struggles with a Congress who desired nothing short of inflicting serve punishment upon the South became intense. When Johnson attempted to remove Stanton from his cabinet position, Stanton vehemently protested and locked himself in his office and refused to vacate his fiefdom as Secretary of War. Johnson's attempt to vanquish Stanton further infuriated the Radical Republicans. They were not going to take the President's soft reconstruction policies or efforts to eliminate Stanton. This resulted in impeachment proceedings being filed against Johnson in the U.S. House. The impeachment was a foregone conclusion as the Radicals held dominion in the House, and Johnson was overwhelmingly impeached.

The eyes of the nation turned to the Senate to see if there were the necessary two-thirds votes to convict Johnson and thereby remove him from office. The debate in the Senate was furious, contentious and divisive. The Radicals were convinced they had the required thirty-six votes to convict and rid themselves of Johnson. Friday May 15, 1868, was the date in which the vote for or against Johnson's conviction came before the Senate. The galleries and halls were filled to capacity as everyone had gathered to "eagerly and anxiously" await "the calling of the roll."[342]

Around noon the names of each Senator was called in alphabetical order. The senior Senator from Kansas, Samuel C. Pomeroy, who had been at the center of the controversy during the days of Bleeding Kansas, and a pronounced, rabid radical voted promptly and positively to convict Johnson. The author of the draconian 1865 Drake Constitution in Missouri, Senator Charles D. Drake of St. Louis joined Pomeroy's and voted for conviction. Missouri's other Senator John B. Henderson broke party ranks and voted against conviction. As the vote count continued the frenzied attention of all assembled turned to the junior Republican Senator from Kansas.

Following Lane's suicide, the Kansas Legislature in 1866 had appointed Edmund G. Ross, a new comer to Kansas, to fill the dead man's vacancy. Ross had been a thorn in Lane's side and many considered Ross' vocal opposition to Lane as being one of the factors that had led to Lane's demise. Radical Republicans were confident Ross would remain loyal and vote with them. As Senator Ross stood "a thousand pair of eyes shot into the very heart of the modest, quiet little man." Ross "nervously and quickly" responded "not guilty." There was a "suppressed condemnation" heard by the agitated proponents of conviction, but by Ross' singular vote "the fate of impeachment" was "sealed." By a vote of thirty-five to nineteen, and Ross' one vote, Johnson was not removed from the Oval Office.[343]

The irony could not be ignored. The two warring states of Missouri and Kansas continued to be as equally divided as they had been during the late War. Ross' one vote saved the last ten months of Johnson's besieged presidency. The President needed a new Secretary of War he could trust and someone the hostile Republican Congress would respect. Two decorated war veterans fit this description—General Ulysses S. Grant and

---

[342] *The New York Times*, 16 May 1868.

[343] *The New York Times*, 16 May 1868; Linder, Douglas, University of Missouri-Kansas City, *The Impeachment of Andrew Johnson*, January 2001, at www.jurist.law.pitt.edu/trials13.htm; and www.law2.umkc.edu/faculty/projects

General John McAllister Schofield. Grant had initially agreed to assume the position to mollify Stanton, but withdrew from consideration when he began considering his own presidential campaign later in the year. This resulted in the naming of Schofield as War Secretary.

Schofield's tenure as Secretary of War ended in March 1869. Although limited in duration, it proved to be in the long-term best interest of America's foreign policy. Secretary Schofield recommended establishing a naval base in the Pacific Ocean in the Sandwich Islands, at a place known as Pearl Harbor. Johnson agreed with his Secretary of War and the base was eventually established.[344]

Like Ewing, Schofield spent years addressing, explaining or defending Order No. 11. In 1877, Schofield was in charge of the military academy at West Point. As previously covered, Schofield dispatched a letter to Ewing reconfirming his endorsement of Order No. 11. Schofield's defense of Ewing received a blistering condemnation from the Missouri Adjutant General Bingham. Nonetheless, Schofield firmly held his ground to Bingham and steadfastly supported Ewing.

Schofield's military accomplishments received recognition when he was promoted to the rank of lieutenant general on February 25, 1895. His career did not last long after the promotion. He chose to retire from the armed forces later that year on September 29. When war with Spain erupted in 1898, President William McKinley turned to Schofield, as had Johnson. McKinley asked the retired General to return to active duty. At age sixty-seven, Schofield respectfully declined the President's request.

Following Schofield's retirement from active duty, he became energetically involved in the Grand Army of the Republic ("GAR"), a membership of former Union Civil War veterans. He participated in several GAR events, and he enthusiastically played a leading role in the Military Order of the Loyal Legion of the United States. He was elected to membership as a Commander of the First Class on April 12, 1871, and assigned Insignia No. 1274. He was a founding charter member of the chapter in California and served from May 3, 1871, to May 1, 1876, as Commander of the Commandery. In later years, Schofield served as Commander-in-Chief of the Loyal Legion from October 18, 1899 to October 21, 1903.

General Schofield died in St. Augustine, Florida on March 4, 1906, at the age of seventy-two. He was twice married, first to Harriet Bartlett by whom he had two sons and a daughter. After her death, he remarried Georgia Kilbourne by whom he had a daughter named Georgiana. Schofield was buried at Arlington National Cemetery with full military honors. Schofield wrote and published his memoirs entitled, "Forty-Six Years in the Army." His book was controversial but became a best seller. When Schofield died,

[344] http://www.civilwar.org/education/history/biographies/john-schofield.html. The Sandwich Islands are today known as the Hawaiian Islands.

General T. H. Ruger summed up Schofield's career: "The country has lost a great soldier, a great statesman, a great negotiator, a great patriot."

Schofield's mark upon the United States military remains enshrined today. Pearl Harbor was one of his finest military recommendations, as it has become one of the most significant naval bases in the Pacific. However, Schofield is best remembered by a commencement speech he gave to the graduates of West Point in 1879. Today it is mandatory for every officer candidate in the military institutions in the country to memorize a portion of the remarks Schofield gave that day:

> *The discipline which makes the soldiers of a free country reliable in battle is not to be gained by harsh or tyrannical treatment. On the contrary, such treatment is far more likely to destroy than to make an army. It is possible to impart instruction and give commands in such a manner and such a tone of voice as to inspire in the soldier no feeling, but an intense desire to obey, while the opposite manner and tone of voice cannot fail to excite strong resentment and a desire to disobey. The one mode or other of dealing with subordinates springs from a corresponding spirit in the breast of the commander. He who feels the respect which is due to others cannot fail to inspire in them respect for himself. While he who feels, and hence manifests, disrespect towards others, especially his subordinates, cannot fail to inspire hatred against himself.*

John M. Schofield tombstone

Abram Comingo
(January 9, 1820-November 10, 1889)

Only those who had experienced Provost Captain Comingo reign during the War would be able to state whether he had been a "Little God" or a sympathetic soul. The scant documented sources on this issue seem to support the latter opinion. Toward the end of the Civil War, complaints were lodged in Washington that challenged Comingo's Union loyalties. A federal commission was appointed to investigate the allegations. The committee's initial findings were scathing. Comingo was considered "decidedly more favorable to rebels than to loyal men" and "he always interested himself in the worst class of persons who had been arrested" in his district. Concluding negative findings accused Comingo of not only having "aided bushwhackers, but to have harbored them."[345] Despite their negative commentary, members of the investigative commission declared that Comingo possessed "good character and loyalty."[346]

The investigation ultimately resulted in the revocation of Comingo's appointment as federal Provost Marshall on April 20, 1865. Whether the claims against Comingo were true or not made little difference. Comingo and his wife were natives of Kentucky; he had known many of those engaged on both sides of the conflict long before hostilities broke-out; he and his law firm had been politically involved in Independence; he had opposed secession; and his home was only some four to five blocks south of the county courthouse and jail on Main Street in Independence. Comingo was like most other residents of Jackson County: torn between his loyalties to his friends and family and his commitment to the Union.

It is unquestioned that the revocation of Comingo's appointment as Provost Marshal of the 6th Congressional District did not negatively impact his political clout in Jackson County—in fact, it may have enhanced his reputation in the eyes of those evicted under Order No. 11. In 1868, he became an elder in the Presbyterian Church in Independence and was elected the first Recorder of Deeds of Jackson County. In 1870, and again in 1872, the people of the county sent Comingo, by then a Democrat, to Congress.

During Comingo's two Congressional terms in Washington he was afforded the rare opportunity to seek revenge against Jennison, the leader of the Kansas Red Legs. Comingo had implicitly and vehemently protested against Jennison in his memorial address to Abraham Lincoln in that fateful month of August 1863. Comingo's opportunity to retaliate came after Jennison had filed suit in Federal Court seeking damages for property he claimed was destroyed by rebel guerrillas during the War. Comingo knew the truth—Jennison and his Red Legs during their sake of Independence

345 O.R. 48:2:250.
346 O.R. 48:2:251.

in November 1861 had stolen property; it was this seized chattel that Jennison had sought compensation. Comingo returned from the Capitol to testify against Jennison and his illegitimate claim. There are no records that Jennison recovered restitution on his spurious claim.

Comingo did not seek re-election to Congress in 1874 (his second term ended on March 3, 1875). However, his local and national political influence remained. In 1876, Republican President Grant appointed him to the Federal Commission to arbitrate with the Sioux Nation to acquire their land in the Dakota Territory bordering the Black Hills.

After the conclusion of Comingo's political career, he resumed the practice of law. He formed a partnership with John Phillips and James Slover. Phillips had served as a colonel in the Union Calvary during the late War. He had received local and national military distinction at the Battle of Westport in Kansas City in October 1864, where he earned the title, "Hero of Byrum's Ford" for his gallantry at that battle. Phillips went on to serve as a Commissioner on the Missouri Supreme Court and a federal judge in the United States District Court Judge in Kansas City. James Slover had been born in Pennsylvania and moved to Independence after the War. He had been too young to have taken part in the defense of the Union. He had a remarkable career as an attorney where he gained his knowledge of the law from Phillips and Comingo, the elder members of the Firm. Slover would eventually become a Circuit Court Judge in Jackson County.

Among the firm's notable clients included Frank James. The irony of Comingo's firm's representation of the noted guerilla and post-war criminal cannot be ignored. After Frank's younger brother Jesse James was assassinated by Bob Ford on April 3, 1882, in St. Joseph, Missouri, an assemblage of leading citizens met at the Merchants Exchange Hall in Kansas City. The purpose of their gathering was to pass resolutions endorsing the role in which Governor Thomas Crittendon, Sr., Police Commissioner Henry Craig of Kansas City and Sheriff James Timberlake of Clay County had played in bringing the bandit to justice. The attendees including Comingo had grown weary and tired over the outpouring of anger and fury surrounding the death of Jesse James. Comingo and two other members of the conference were selected to draft resolutions ratifying the actions of these governmental officials. The delegates wanted to express their heartfelt appreciation for the way in which the reign of terror of Jesse James had ended, and Comingo played an integral part of ensuring their message was publically proclaimed.[347]

Within four months, the law firm of Phillips, Comingo and Slover was representing Frank James. Frank had surrendered to Governor Crittendon at the Capitol in Jefferson City on October 3, 1882, and was then transferred by train to the Jackson County Jail in Independence. Jackson County Prosecutor William Wallace filed murder charges against Frank in Independence, but when he could not successfully prosecute the elder James for this crime, Frank was transferred to Daviees County, Missouri to stand

[347] Monaco, "Son of a Bandit," 24-25.

trial for the Winston Train robbery and murder of the train's conductor that had occurred in July 1881. Throughout these criminal proceedings Comingo and his firm defended Frank James and eventually secured his acquittal of the crimes in September 1883. Wallace who had been a victim of Order No. 11 paid tribute to Comingo. He said Comingo was a true Roman who was always willing to shepherd a new attorney in the community (see, Appendix H under WALLACE for more of Wallace's personal account of Order No. 11).

Around 1883, Comingo and his wife moved to Kansas City where he died on November 10, 1889. On the day of his funeral, the county and state government offices in Jackson County closed in his honor. His wife followed him to the grave within seven months. They were buried at Elmwood Cemetery in Kansas City. Today they are buried with their infant son John, their daughter Sally and her husband Dr. G. W. Griffith in the Comingo-Griffith mausoleum at Elmwood.

Abram Comingo mausoleum

Brigadier General Thomas Ewing
(August 7, 1829-January 21, 1896)

Provost Marshal Comingo's efforts to have Lincoln remove Ewing from the District of the Border in the short-term proved unsuccessful. However, in the long-term, Comingo and his memorialists succeeded when they received support from an unlikely ally—none other than the Grim Chieftain, Senator Jim Lane. The memorialists and Lane had found common ground, or as the maximum goes: "the enemy of my enemy is my friend." Both warring factions agreed that there had been no wisdom in combining districts within Missouri and Kansas under the same command. Lincoln in the end acknowledged this fact and Ewing was relieved of his duties in the District of the Border. The President also decided to remove the Missouri counties within the District and enlarge the District of Kansas. The new Kansas district would have a new commander and the new one would consist of all of Kansas, the Indian Territory and portions of two southern states under federal control.[348] The western Missouri counties formerly within Kansas were restored to the District of Missouri—and not under the authority of Ewing or any other Kansan.

In the fall of 1864, General Sterling Price invaded southeast Missouri. Standing in the way of the rebel forces, they encountered an old foe—General Thomas Ewing, Jr. After the debacle in the District of the Border, Ewing had been transferred and by September 1864, he had assumed command of Ft. Davidson at Pilot Knob, south of St. Louis. Price's initial military strategy was to capture the river city of St. Louis. However, he made a tactical error and chose to attack Ewing; he should have gone around Ft. Davidson. Ewing defended the fort against several frontal assaults made by Price. As evening settled over beleagured Ft. Davidson, Ewing knew his position within the earthworks was precarious. He made a sound decision. He ordered his troops to abandon the fort under the protection of the night and he had the arsenal's magazine destroyed. The explosion could be heard and seen for miles as Ewing and his mean slowly retreated to safety. Ewing's action forced Price to alter his plans to seize St. Louis. Price soon began lumbering from east to west across the state where the following month he met a disastrous defeat at the Battle of Westport in Kansas City. Ewing's handling of the events at Ft. Davidson garnered him great recognition and accommodation.

When hostilities ended, Ewing's services to his country did not. On April 14, John Wilkes Booth assassinated Abraham Lincoln at Ford's Theatre in Washington. Immediately after Booth shot Lincoln, he leaped from the President's box and onto the stage breaking his leg. Booth and his co-conspirator David Herold escaped into Maryland. They went to the farmhouse owned by Dr. Samuel A. Mudd. Booth identified himself as Tsyer or Tyson and Herold said his name was Henson. Dr. Mudd proceeded to

---

[348] Smith, 210.

set Booth's broken leg, and the two fugitives spent the night at Mudd's house before departing the following day.

After an extensive manhunt for Lincoln's murderer and conspirators, John Wilkes Booth was hunted down and killed. Several men and one woman Mary Surratt would stand trial for conspiracy and murder of the President of the United States—the first assassinated President in American History. The trial would serve as the final chapter to Ewing's military career.

Secretary of War Edwin Stanton had decided not to trust the trial's outcome to a Virginia Federal jury. He insisted the crime of the conspirators was one of military terrorism and governed by a military and not civil tribunal laws. In other words, they had to be tried in a court-martial proceeding before a military court. This meant that Union field officers would determine the fate of the criminals.

Among the many defendants were Dr. Mudd, Samuel Arnold and Edward Spangler. The latter two were employees of Ford's Theatre on the night of Lincoln's murder. They had been charged with aiding Booth to gain entry into the theatre. John Ford, the theatre's owner, agreed to pay the legal fees for his employees. A Catholic priest and friend of the Mudd family approached Ewing's brother Hugh to assist in Dr. Mudd's defense. Hugh did not wish to take the lead and ultimately all three men were defended by Ewing—the author of Order No. 11.[349] In total, eight people were indicted for the murder of the late President. In the defense of his three clients, Ewing would receive the assistance from his learned father, his brother Hugh, and Orville Browning.

Ewing's legal defense of his clients was compromised from the outset. His legal talents would be taxed to the most extreme during the trial's ordeal. Ewing's defense limitations were pronounced: afforded one day to prepare for the defense of all three clients; denied a copy of the indictment or specific charges before trial; prohibited from interviewing witnesses; barred from disclosure of the identity of the federal witnesses; and bound by the strict rules of military evidence. The harshness of the evidentiary rules was self-evident: none of the witnesses could testify on their own behalf and the chief prosecutor, Joseph Holt, would be allowed to sit with the military judges during their deliberation.

Ewing's first defense strategy was to object to the case going forward on procedural grounds. He insisted that the military court was without jurisdiction to hear the case. He reminded the judges that not one of his clients had served in the military during the War. He insisted that the proper jurisdiction for their trial was in federal court, not a military court. This motion was denied. He next argued for a dismissal of the conspiracy counts against his clients. As for Mudd, Ewing told the court that the only allegation was he had provided medical care to Booth and failed to report it to authorities. These charges did not remotely comprise evidence of Mudd's involvement in conspiracy;

---

[349] Smith, 260-261.

it did not connect Mudd with the shooting of the late president or knowledge that it had occurred when he gave medical care to Booth. Once again the court sided with the prosecution.

The trial's outcome was pre-determined as had been the denial of Ewing's pre-trial motions. Ewing's client Spangler was convicted of aiding and abetting Booth "in making his escape" after the murder of Lincoln and was sentenced to six years in prison. Arnold received a life sentence for having conspired with Spangler. Mudd was also found guilty and was sent to prison for life. The only other defendant not sentenced to death was Michael O'Laughlin. David H. Herold, George A. Atzrodt, Lewis Payne and Mary E. Surratt were ordered to be executed by hanging (and they were).[350] Given the fact that Ewing's clients, with the exception of O'Laughlin, were spared their lives, the trial's outcome had been somewhat favorable to him and his co-defense attorneys. Ewing would remain in Washington for a few years following the trial where he became closely aligned with President Johnson and against Stanton and the Radical Republicans.[351]

After Ewing's senatorial defeat in 1879, he completed the final year of his term in Congress. Whether Bingham had been part of Ewing's decision to retire from politics when his Congressional term ended may be questioned. It is unquestioned that Ewing left a long political career and chose to leave Ohio as well. In 1882, Ewing moved his family to New York where he resumed a lucrative law practice on Wall Street.

Ten years later his foster brother and brother-in-law General William Tecumseh Sherman died. Ewing became the family spokesman, and handled the funeral and arrangements.[352] Some ten thousand people came to pay their final respects to a man, who, much like Ewing, had utilized total war against a civilian population during his famous or infamous "March to the Sea."

Ewing had established a successful law practice by the time of Sherman's death. The family lived in the upscale portion of Yonkers. Two of his sons, Thomas and Hampton, followed the family tradition and became lawyers. In 1893, they joined their father in business, and Ewing established the law firm of Ewing, Whitman and Ewing. Their practice included patent, investment and banking law and among their clients was the City of New York—a city in which Ewing would meet a tragic death.

Ewing visited Kansas City in January 1890, while "on a tour of the West" where he "delivered a lecture before the bar of Topeka on the jury system." The news article from the Kansas City Star on January 9, said, "He at an early day in the history of Kansas

---

[350] O.R. 2:8:698-699.

[351] Smith, 279. Before Johnson left the Oval Office he pardoned Mudd at Ewing's request. Ewing's father died in 1871 following a stroke he suffered on October 22, 1869, while he had been arguing a case before the United States Supreme Court. His brother Hugh would survive Ewing and died in 1905 at the age of 79, and was buried in the Catholic faith.

[352] Smith, 341.

City purchased about twenty acres in the West Bottoms for $20 an acre, which he sold afterward for $1,000 an acre."

On the morning of January 20, 1896, Ewing was on his way to his Wall Street law office. "As he reached 3rd avenue a cable car passed and he stepped directly behind it, not noticing that one from the opposite direction was right upon him." The corner of the approaching car struck Ewing throwing him several yards, and he landed on his head fracturing his skull. Ewing was carried to his home and despite surgical efforts he never regained consciousness and died the following day with his wife, two sons and an unmarried daughter at his side.[353] When word reached City Hall of Ewing's death, the mayor had the "flag on the building lowered to half-mast." The police determined that John Kelly, the driver of the streetcar that struck Ewing "would be arraigned on the charge of murder" for having failed to make any "apparent attempt to stop."[354] Ewing's body was taken back to his home state of Ohio where he was buried at Saint Mary Cemetery in Lancaster, Fairfield County, Ohio.

The story of Ewing's death was national news. It also covered the front page of the *Kansas City Daily Journal* on January 22. Its bold front-page headline simply read:

GENERAL EWING DEAD

Ewing's military command of the District of the Border and his Order No. 11 consumed much of the paper's ink. The paper covered the full account of the Order and its impact upon Jackson County. Colonel Theodore S. Case, a member of Ewing's staff at the time of Order No. 11 gave a *Journal* reporter the "data" relating to Ewing's "military life in and about Kansas City." Case journeyed back to that fateful summer of 1863 and

Thomas Ewing, Jr., tombstone

[353] *New-York Tribune*, 22 Jan. 1896; *The Advocate*, Topeka, Kansas, 29 Jan 1896; *Evening Star*, Washington, D. C., 21 Jan. 1896; *The Goodland Republic*, Goodland, Kansas, 24 Jan. 1896. Ewing's wife and five children survived him.

[354] *New-York Tribune*, 22 Jan. 1896.

provided an account of having toured the county with Ewing and Schofield while Order No. 11 was being enforced. He explained what they had witnessed: "Everywhere people were on the move and the roads were lined with people." In a final commentary of the war between Ewing and Bingham (a war that has never ended) the *Journal* provided the following summation:

> *Colonel Case has a steel engraving of George C. Bingham's famous picture based on this order and the scenes it caused. It is a wonderfully like engraving and portrays General Ewing astride a war horse, giving orders. Colonel Case says the issuance of order No. 11 defeated General Ewing in his race for governor of Ohio in after years, as it was used as campaign material against him.*[355]

The story of Ewing and his issuance of Order No. 11 continues to remain fodder in newspapers across the land even to this day. Nearly 50 years after the issuance of Order No. 11 a story appeared on page 2 of the *Rome* (New York) *Daily Sentinel.* On February 27, 1911, Hannahs' hometown paper revisited Order No. 11 following the death of Ewing's adjutant, Major Harrison Hannahs, and acting adjutant general at the time of Order No. 11. At the time of his death Major Hannahs, age 79, was living in Denver, Colorado (which, during the days of Bleeding Kansas, was actually located within the Kansas Territory). A large portion of the article covered Hannahs' involvement in the issuance of Order No. 11 and, in fact, the story-line attributes Hannahs with having, "issued the famous Order No. 11 which took away the support of disloyal citizens of Missouri and thus in subduing them." The story-line simply failed to mention that the Order also evicted loyal Union folks from the four counties. There is no question that Order No. 11 was issued under the under name of Major Hannahs; but, he certainly acted directly under the authority of his commander, General Ewing.[356]

It is unquestioned that in life and death Bingham's war against Ewing long survived both men. It still remains a constant issue of debate one hundred and fifty years later. The summer of 1863 in Jackson County had been and it seems to have remained engrained to this day as a year filled with blood, smoke and ashes in which everyone had been "Scattered to the Four Winds."[357]

---

[355] *Kansas City Daily Journal*, 22 Jan. 1896.

[356] Hannahs was born on 18 Feb. 1832, in Marcy, New York. In 1856, he went to Kansas with a party of six free state men to help make Kansas a free state. In 1862, Hannahs enlisted as a volunteer in Kansas, and the next year went on to General Ewing's staff as a Major and Acting Assistant Adjutant General.

[357] In the entrance way to the County Courthouse in Cass County, Missouri, there stands today a lonely rock chimney and hearth as a constant reminder of the ravaging affect of Ewing's General Order No. 11. It symbolically reads "Burnt District."

APPENDIX A

By the President of the United States of America:

A Proclamation.

Whereas, on the twenty-second day of September, in the year of our Lord one thousand eight hundred and sixty-two, a proclamation was issued by the President of the United States, containing, among other things the following, to wit:

"That on the first day of January, in the year of our Lord one thousand eight hundred and sixty-three, all persons held as slaves within any State or designated part of a State, the people whereof shall then be in rebellion against the United States, shall be then, thenceforward, and forever free; and the Executive Government of the United States, including the military and naval authority thereof, will recognize and maintain the freedom of such persons, and will do no act or acts to repress such persons, or any of them, in any efforts they may make for their actual freedom.

"That the Executive will, on the first day of January aforesaid, by proclamation, designate the States and parts of States, if any, in which the people thereof, respectively, shall then be in rebellion against the United States; and the fact that any State, or the people thereof, shall on that day be, in good faith, represented in the Congress of the United States by members chosen thereto at elections wherein a majority of the qualified voters of such State shall have participated, shall, in the absence of strong countervailing testimony, be deemed conclusive evidence that such State, and the people thereof, are not then in rebellion against the United States."

Now, therefore I, Abraham Lincoln, President of the United States, by virtue of the power in me vested as Commander-in-Chief, of the Army and Navy of the United States in time of actual armed rebellion against the authority and government of the United States, and as a fit and necessary war measure for suppressing said rebellion, do, on this first day of January, in the year of our Lord one thousand eight hundred and sixty-three, and in accordance with my purpose so to do publicly proclaimed for the full period of one hundred days, from the day first above mentioned, order and designate as the States and parts of States wherein the people thereof respectively, are this day in rebellion against the United States, the following, to wit:

Arkansas, Texas, Louisiana, (except the Parishes of St. Bernard, Plaquemines, Jefferson, St. John, St. Charles, St. James Ascension, Assumption, Terrebonne, Lafourche, St. Mary, St. Martin, and Orleans, including the City of New Orleans) Mississippi, Alabama, Florida, Georgia, South Carolina, North Carolina, and Virginia, (except the forty-eight counties designated as West Virginia, and also the counties of Berkley, Accomac, Northampton, Elizabeth City, York, Princess Ann, and Norfolk, including the cities of Norfolk and Portsmouth], and which excepted parts, are for the present, left precisely as if this proclamation were not issued.

And by virtue of the power, and for the purpose aforesaid, I do order and declare that all persons held as slaves within said designated States, and parts of States, are, and henceforward shall be free; and that the Executive government of the United States, including the military and naval authorities thereof, will recognize and maintain the freedom of said persons.

And I hereby enjoin upon the people so declared to be free to abstain from all violence, unless in necessary self-defence; and I recommend to them that, in all cases when allowed, they labor faithfully for reasonable wages.

And I further declare and make known, that such persons of suitable condition, will be received into the armed service of the United States to garrison forts, positions, stations, and other places, and to man vessels of all sorts in said service.

And upon this act, sincerely believed to be an act of justice, warranted by the Constitution, upon military necessity, I invoke the considerate judgment of mankind, and the gracious favor of Almighty God.

In witness whereof, I have hereunto set my hand and caused the seal of the United States to be affixed.

Done at the City of Washington, this first day of January, in the year of our Lord one thousand eight hundred and sixty three, and of the Independence of the United States of America the eighty-seventh.

By the President: ABRAHAM LINCOLN

WILLIAM H. SEWARD, Secretary of State.[358]

[358] O.R. 3.3.3-4.

APPENDIX B

List of Known Killed at Lawrence

Identified Union Troops

Anderson, Charles
Allen, Charles R.
Cooper, James F.
Green, John R.
Griswold, Walter B. S.
Haldermann, Aaron
Markle, David
Markle, Lewis Cass
Markle, Samuel
Parker, Ashbury
Parker, Isaac J.
Riggs, Charles F.
Speer, Robert
Watson, John
Waugh, William A.
Wilson, James
Woos, Andrew J.

Identified Civilians

Albach, George J.
Albrecht, George,
Allen, Clay
Allison, Duncan C.
Alwes, George
Anderson, John
Argel, James
Brant, R.
Brechtlesbauer, James
Brechtlesbauer, Joseph
Brown, "One Armed"
Burnes, Dennis
Burns, Michael
Burt, George
Carpenter, Judge Louis
Clona, Charles
Cloud, Charles
Coal (Coat or Coates), George W.
Coleman, Dwight L.
Collamore, Mayor George W.
Cooper, James
Cornel, John A.
Cornell, P.
Crane, John Louis
Dagle, John
Dix, Ralph C.
Dix, Stephen H.
Dulinsky, Sylvester
Dulinsky, William
Dyre, "Uncle" Frank
Early, George
Eckman, Carl

Edwards, John
Ehles (Ethels), August
Eldridge, James
Ellis, Frank
Engler, Carl
Engler, John
Englesman, Phillip
Evans, John
Fillmore, Lemuel
Finley, James B.
Finley, Joe
Fitch, Edward P.
Frank, Joseph
Fran k, "Old Uncle"
Frawley (Fromley), John
Gates, Levi
Gentry, Levi
Gerrard, George
Giebal (Gribal), Anthony
Gifler (Gufler), A.
Grill, John B.
Goldman, Mr.
Green, John B.
Gregg, Fortune
Griswold, Abner W.
Griswold, Dr. Jerome F.
Griswold, Watt
Hanson, Mr.
Hay, Chester D.
Hendrix, Mr.
Henry, "Uncle" Charle
Hoge, Calvin
Holmes, Nathan
Johnson, Ben
Johnson, Mr.
Jones (unnamed baby of Samuel Jones)
Kallmer, George
Keefe, Patrick
Kimball, Frederick
Klares, William
Klaus, Frederick
Klaus, William, Jr.
Kleffer (Kliffer), W. M. R.
Lambert, Harwick
Lambert, O. O.
Laurie (Lawrie), John W.
Laurie (Lawried) William
Leener (Laner), Christian
Leonard, Christopher
Limbach, Henry
Little, John
Longley. Otis
Loomis, Richard Rensellor
Lowe, Captain Joseph G.
McClellan, Amos
McClellan, Daniel
McFadden, Jacob
Mackin (Maclaine), Michael
McFarland, Mr.
Markle, Daniel
Martha (Murtha), Thomas
Martin, Michael
Martin, Robert "Bobby"
Meeky (McKee), Michael
Meeky (McKee), R.
Murphy, Dennis
Murphy, James
Murphy, Thomas
Martha, W.
Nichols, Mr.
Oehrle, George
O'Neil, James
Oldham, Anthony
Page, George
Palmer, Charles A.
Palmer, Daniel W.
Perine (Perrine), James
Pollock (Pollok), Jacob
Pope, George
Purington, David N.

Range, George
Range, Samuel
Reed, Peter Sr.
Reedmiller, A.
Reel, Samuel Jeremiah
Reynolds, Samuel
Roach, James
Sanger, George
Sargeant, George H.
Schmidt, Charles
Schwab, John
Smith, Charles
Snyder, Rev. Samuel S.
Speer, John M., Jr.
Stewart, Henry
Stone, Joseph
Stone, Captain Nathan
Stonestreet, Benjamin
Swan, Louis H.
Thorp, Simeon M.
Trask, Josiah C.
Tritch, E. P.
Turk, David
Turk, Addison
Watson, James E.
Watson, John,
Waugh, Mr.
West, Mr.
White, Mr.
Whitsitt, J. B.
Williams, Mr.
Williamson, William T.
Wise, Louis
Wood, James
Zimmerman, John K.

APPENDIX C

GENERAL ORDERS, ~
HDQRS. DISTRICT OF THE BORDER,

No. 9: Kansas City, Mo., August 18, 1863.

I. Lieut. Col. Walter King, Fourth Regiment Missouri State Militia, will, as often as may be necessary, visit the several military stations in that part of Missouri included in this district, and ascertain what negroes are there who desire escort out of Missouri, and were they slaves of persons who, since the 17th day of July, 1862, have been engaged in the rebellion, or have in any way given aid or comfort thereto. He will make and certify a list of all such negroes at each of such stations, and of the persons by whom the disloyalty of their master can be shown, and will deliver one copy of such list to the commander of such station and forward one to these headquarters. Before preparing such lists he will give due and public notice of the time at which he will be engaged in such duty at each station. He will be governed in the discharge of his duties by special written instructions received from or through these headquarters.

II. Commanders of such stations will furnish from time to time, as they may be called for by commanders of escorts, copies of the lists so prepared and filed with them, and will issue rations, where necessary, to negroes named in such lists who are unable to move from such station or to earn a living there, until escort can be furnished them to a place of safety where they can support themselves.

III. Commanders of companies or detachments serving in that part of Missouri included in this district will give escort and subsistence, where practicable, to all negroes named in such certified lists to Independence, Kansas City, Westport, or the State of Kansas, sending direct to these headquarters all such negroes fit for military duty and willing to enlist.

By order of Brigadier-General Ewing:
P. B. PLUMB,
Major and Chief of Staff.

APPENDIX D

GENERAL ORDERS
HDQRS. DISTRICT OF THE BORDER

No. 10: Kansas City, Mo., August 18, 1863.

I. Officers commanding companies and detachments will give escort and subsistence, as far as practicable, through that part of Missouri included in this district, to all loyal free persons desiring to remove to the State of Kansas or to permanent military stations in Missouri, including all persons who have been ascertained in the manner provided in General Orders, No. 9, of this district to have been the slaves of persons engaged in aiding the rebellion since July 17, 1862. Where necessary, the teams of persons who have aided the rebellion since September 25, 1862, will be taken to help such removal, and, after being need for that purpose, will be turned over to the officer commanding the nearest military station, who will at once report them to an assistant provost-marshal or to the district provost-marshal, and hold them subject this orders.

II. Such officers will arrest, and send to the district provost-marshal for punishment, all men (and all women not heads of families) who will-fully aid and encourage guerrillas, with a written statement of the names and residences of such persons and of the proof against them. They will discriminate as carefully as possible between those who are compelled, by threats or fears, to aid the rebels and those who aid them from disloyal motives. The wives and children of known guerrillas, and also women who are heads of families and are willfully engaged in aiding guerrillas, will be notified by such officers to remove out of this district and out of the State of Missouri forthwith. They will be permitted to take, unmolested, their stock, provisions, and household goods. If they fail to remove promptly, they will be sent by such officers, under escort, to Kansas City for shipment south with their clothes and such necessary household furniture and provision as may be worth removing.

III. Persons who have borne arms against the Government, and voluntarily lay them down and surrender themselves at a military station, will be sent, under escort, to the district provost-marshal at these headquarters. Such persons will be banished, with their families, to such State or district out of this department as the general commanding the department may direct, and will there remain exempt from other military punishment on account of their past disloyalty, but not exempt from civil trial for treason.

IV. No officer or enlisted man, without special instructions from these headquarters, will burn or destroy any buildings, fences, crops, or other property, but all furnaces and fixtures of blacksmiths shops in that part of Missouri included in this district not at military stations will be destroyed, and the tools either removed to such stations or destroyed.

V. Commanders of companies and detachments serving in Missouri will not allow persons not in the military service of the United States to accompany them on duty, except when employed as guides, and will be held responsible for the good conduct of such men employed as guides, and for their obedience to orders.

VI. Officers and enlisted men belonging to regiments or companies, organized or unorganized, are prohibited going from Kansas to the District of Northern Missouri without written permission or order from these headquarters or from the assistant provost-marshal at Leavenworth City, or the commanding officer at Fort Leavenworth, or some officer commanding a military station in the District of Northern Missouri.

By order of Brigadier-General Ewing

B. PLUMB,

Major and Chief of Staff.

APPENDIX E

General Order № 11.
Headquarters District of the Border

Kansas City, August 25, 1863

I. All persons living in Jackson, Cass, and Bates counties, Missouri, and in that part of Vernon included in this district, except those living within one mile of the limits of Independence, Hickman's Mills, Pleasant Hill, and Harrisonville, and except those in that part of Kaw Township, Jackson County, north of Brush Creek and west of Big Blue, are hereby ordered to remove from their present places of residence within fifteen days from the date hereof.

Those who within that time establish their loyalty to the satisfaction of the commanding officer of the military station near their present place of residence will receive from him a certificate stating the fact of their loyalty, and the names of the witnesses by whom it can be shown. All who receive such certificates will be permitted to remove to any military station in this district, or to any part of the State of Kansas, except the counties of the eastern border of the State. All others shall remove out of the district. Officers commanding companies and detachments serving in the counties named will see that this paragraph is promptly obeyed.

II. All grain and hay in the field or under shelter, in the district from which inhabitants are required to remove, within reach of military stations after the 9th day of September next, will be taken to such stations and turned over to the proper officers there and report of the amount so turned over made to district headquarters, specifying the names of all loyal owners and amount of such product taken from them. All grain and hay found in such district after the 9th day of September next, not convenient to such stations, will be destroyed.

III. The provisions of General Order No. 10 from these headquarters will be at once vigorously executed by officers commanding in the parts of the district and at the station not subject to the operations of paragraph 1 of this order, and especially the towns of Independence, Westport and Kansas City.

IV. Paragraph 3, General Order No. 10 is revoked as to all who have borne arms against the Government in the district since the 20th day of August, 1863.

By order of Brigadier General Ewing.
H. Hannahs, A. A. A. G.[359]

---

[359] O.R., 22:1:473-474. The reference to Hannahs alludes to his life after the War: "Major Harrison Hannahs is dangerously ill at his home in Denver from a complication of pneumonia. Major Hannahs is more than 75 years old. His death is expected soon. He wrote the famous "Order No. 11" which practically

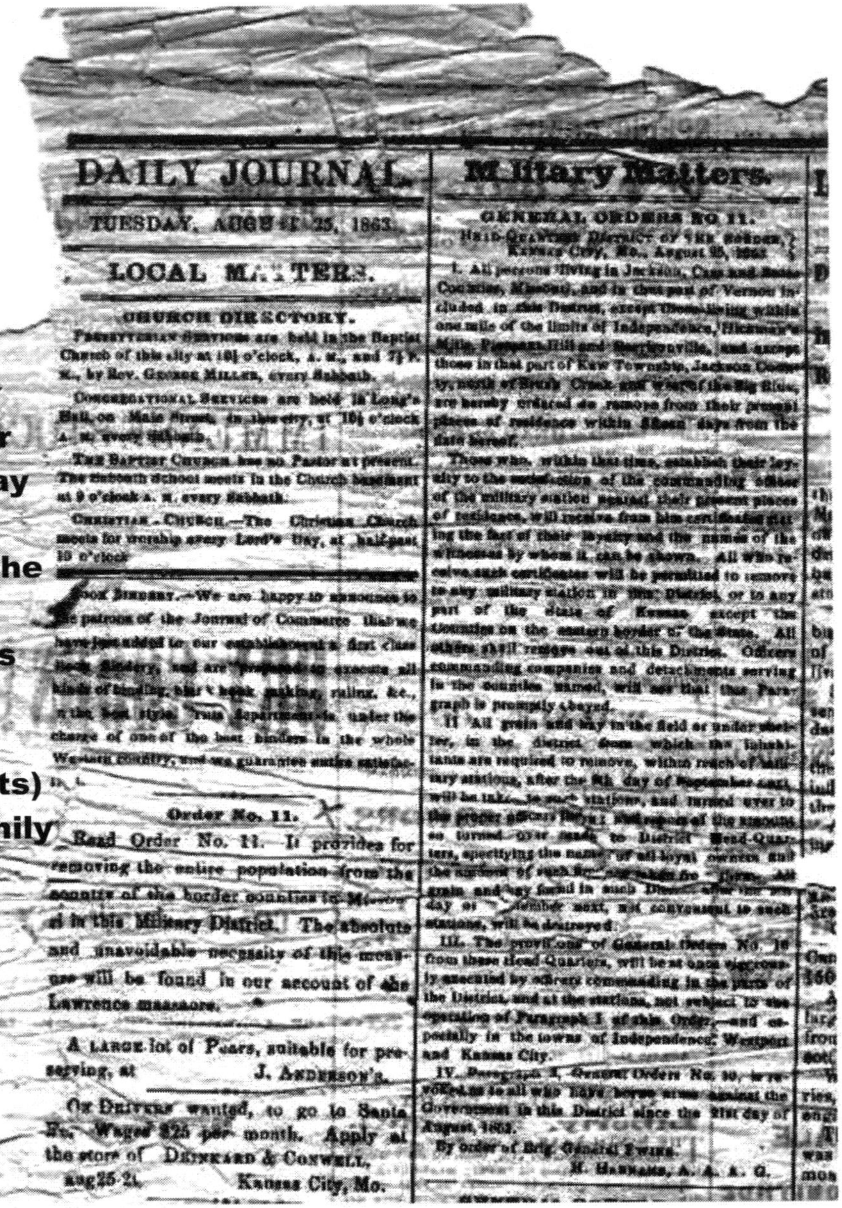

DAILY JOURNAL.

TUESDAY, AUGUST 25, 1863.

LOCAL MATTERS.

CHURCH DIRECTORY.

PRESBYTERIAN SERVICES are held in the Baptist Church of this city at 10½ o'clock, A. M., and 7½ P. M., by Rev. GEORGE MILLER, every Sabbath.

CONGREGATIONAL SERVICES are held in Long's Hall, on Main Street, in this city, at 10½ o'clock A. M. every Sabbath.

THE BAPTIST CHURCH has no Pastor at present. The Sabbath School meets in the Church basement at 9 o'clock A. M. every Sabbath.

CHRISTIAN CHURCH.—The Christian Church meets for worship every Lord's Day, at half-past 10 o'clock.

BOOK BINDERY.—We are happy to announce to the patrons of the Journal of Commerce that we have just added to our establishment a first class Book Bindery, and are prepared to execute all kinds of binding, blank book making, ruling, &c., in the best style. This department is under the charge of one of the best binders in the whole Western country, and we guarantee entire satisfaction.

Order No. 11.

Read Order No. 11. It provides for removing the entire population from the country of the border counties in Missouri in this Military District. The absolute and unavoidable necessity of this measure will be found in our account of the Lawrence massacre.

A LARGE lot of Pears, suitable for preserving, at J. ANDERSON'S.

OX DRIVERS wanted, to go to Santa Fe. Wages $25 per month. Apply at the store of DRINKARD & CONWELL, Kansas City, Mo. aug25-2t.

Military Matters.

GENERAL ORDERS NO 11.

HEAD-QUARTERS DISTRICT OF THE BORDER, Kansas City, Mo., August 25, 1863.

I. All persons living in Jackson, Cass and Bates Counties, Missouri, and in that part of Vernon included in this District, except those living within one mile of the limits of Independence, Hickman's Mills, Pleasant Hill and Harrisonville, and except those in that part of Kaw Township, Jackson County, north of Brush Creek and west of the Big Blue, are hereby ordered to remove from their present places of residence within fifteen days from the date hereof.

Those who, within that time, establish their loyalty to the satisfaction of the commanding officer of the military station nearest their present places of residence, will receive from him certificates stating the fact of their loyalty and the names of the witnesses by whom it can be shown. All who receive such certificates will be permitted to remove to any military station in this District, or to any part of the State of Kansas, except the Counties on the eastern border of the State. All others shall remove out of this District. Officers commanding companies and detachments serving in the counties named, will see that this Paragraph is promptly obeyed.

II. All grain and hay in the field or under shelter, in the district from which the inhabitants are required to remove, within reach of military stations, after the 9th day of September next, will be taken to such stations, and turned over to the proper officers there; and report of the amount so turned over made to District Head-Quarters, specifying the names of all loyal owners and the amount of such [illegible]. All grain and hay found in such District after the [illegible] day of September next, not convenient to such stations, will be destroyed.

III. The provisions of General Orders No. 10 from these Head-Quarters, will be at once vigorously executed by officers commanding in the parts of the District, and at the stations, not subject to the operation of Paragraph I of this Order,—and especially in the towns of Independence, Westport and Kansas City.

IV. Paragraph 3, General Orders No. 10, is revoked as to all who have borne arms against the Government in this District since the 21st day of August, 1863.

By order of Brig. General EWING.

H. HANNAHS, A. A. A. G.

***Kansas City Daily Journal* newspaper clipping from the day Order No. 11 was issued, archived in the Jackson County Historical Society's Archives, Colonel Upton Hays and Margaret Jane (Watts) Hays Overstreet Family Collection**

---

depopulated western Missouri. Hundreds of Jackson, Cass and Bates countians were driven into exile by the famous Civil War measure. At that time he was Adjutant General on the staff of General Thomas Ewing, commander of the district for the Union force." "Wrote General Order No. 11," Kansas City (Mo.) Star, 26 Feb. 1911, as cited in Eakin and Hale, 179. See also Chapter *XIV* at the end of Ewing's obituary.

APPENDIX F

SAINT LOUIS, *MO., August* 27, 1863.

His Excellency ABRAHAM LINCOLN,

President of the United States:

Your memorialists would respectfully state that they are loyal citizens of the United States and of the State of Missouri, and, having been such at all times, they regard it as their right and duty to represent to Your Excellency the unhappy condition of affairs now existing along the western border of their State, and to pray Your Excellency's interposition in behalf of a suffering people. Your memorialists feel that justice and humanity demand at least this much at their hands. They therefore beg Your Excellency's attention to the facts hereinafter appearing. For more than two years past our western border has been the theater of strife and bloodshed, and has been overrun by lawless bands of desperadoes, who, with a reckless and unrestrained soldiery, have rioted upon the substance of the people and have wantonly destroyed their property and trampled upon their most sacred rights. Theft, robbery, house-burning, and other crimes have been perpetrated with impunity, and to such an extent has this system of plunder and vandalism prevailed that it has impoverished and almost depopulated one of the fairest and most wealthy and prosperous parts of our State, and, unless arrested, it will certainly involve in similar ruin many other sections of the State that have hitherto, in a measure, escaped its ravages.

During the past month theft, robbery, arson, and murder have been of almost daily occurrence, and the fearful threat that the border shall be made a desolation, it appears, is about to be executed. During the past fortnight these evils have existed in a most fearful and intensified form, and but little has been done to arrest them. Why they should be allowed your memorialists cannot perceive. They had their origin as far back as the fall of 1861, in the burning of Osceola and other small villages along the border, and from that time to the present they have gradually increased, and the horrible barbarities that have uniformly attended them have at last become as appalling as those which characterized savage warfare in the early history of this country. The lives of the people and the material wealth of the country have been wantonly and wickedly destroyed in a manner and to an extent that have hitherto been unknown and unheard of among a civilized people. That which cannot be carried away is committed to the flames, and thus helpless and defenseless women and children are left destitute of food, raiment, or shelter, and without the means of escape from suffering and ruin.

These evils have produced a degree of consternation that language cannot describe, and which none can comprehend save those who have witnessed it; yet it is the natural result of the retaliatory warfare and of the unrestrained lawlessness that have existed in Western Missouri for the last two years, which, if not speedily checked, will involve in ruin by far the greater parts of this State and Kansas, and will be productive of other evils the magnitude of which no one can now estimate. Your memorialists greatly fear that the recent outrages perpetrated in both Missouri and Kansas but faintly foreshadow the future history of these States if some means cannot be adopted to allay the excitement and arrest the lawless violence now prevailing along the border. Whatever may have been the errors of many of our citizens in the beginning of this terrible rebellion, your memorialists entertain no kind of doubt that an overwhelming majority of the masses are now sincerely determined to support the Government of the United States and the provisional government of Missouri, nor the least doubt that they, in good faith, accept the ordinance of emancipation adopted by the late convention as a final and complete settlement of the question of slavery in this State. There can be no question of these facts, nor have your memorialists a shadow of doubt that a firm and just policy in the future conduct of the military affairs of this State will prove more conducive to her peace and to the interests of the Federal Government than any other that can possibly be adopted. It will do more in thirty days, if honestly carried out and rigidly enforced, to restore our State to her wonted condition of peace and prosperity than the system of pillage and burning, now enforced, will accomplish in as many years.

Your memorialists further beg leave to say that one of the most fruitful sources of trouble in Western Missouri is the attachment of a part of her territory to the District of the Border. This arrangement, however well intended, your memorialists fear will, while it is continued, occasion incessant trouble, and will greatly hinder the restoration of law and order, no matter what policy may be adopted or who may be placed in command. Old animosities existing between the people of Missouri and Kansas, embittered and intensified by the recent barbarous acts of a guerrilla band perpetrated upon the citizens of Lawrence, in the latter State, will develop themselves, and will seek gratification in retaliatory acts upon the citizens of the former, although they are, with rare exceptions, as sincerely opposed to those infamous outlaws as the people of Kansas ever have been. But this late and atrocious outrage has furnished a pretext for future and greater and infinitely more unjust acts of retaliation upon our people than any from which they have hitherto suffered. The following telegram, published in the Missouri Democrat, of this city, speaks volumes on this point. The statement that there were citizens of Missouri engaged in the raid, except such as have for nearly two years been regarded as outlaws, is not worthy of credit. It is made for effect and to palliate acts of retaliation.

[Special dispatch to the Missouri Democrat.]

LEAVENWORTH, *August* 26.

General Lane has returned to Lawrence. A meeting was held on his return. Lane said the citizens had killed 41 of Quantrill's men. Majors Clark and Plumb were denounced. The people of Baldwin disputed Quantrill in passing a ford, and say if Plumb had done his duty they could have whipped the rebels.

Lane is organizing forces, and says he will go into Missouri on the 9th of September. He left General Ewing only on a pledge that Ewing would issue an order directing all the citizens of Jackson, Cass, Bates, and part of Vernon Counties, except those in Kansas City, Westport, Harrisonville, and Independence, to leave the county within fifteen days. Ewing has issued the order, and the people of Kansas are going into Missouri to see the order executed. The people have demanded the order issued by the general commanding, and the people will see it executed. They say they will have no more of the Schofield-Ewing orders. Ewing is frightened, and in the chase after Quantrill was in a complete quandary. He is looked upon as being a general without heart and brains. About 50 of the most noted secesh of Platte County have subscribed from $1 to $10 each for the Lawrence fund. By so doing they expect to escape the anticipated devastation of Western Missouri. General Ewing has returned to Kansas City. Quantrill had with him Sam. Hays, brother of Up. Hays, Dick Yeager, Holt, George Todd, and Younger, with 150 men, on whom they could depend in a fight, with about 150 more of the citizens of Platte, Clay, La Fayette, Jackson, Cass, and Bates Counties, not over 300 in all. Quantrill's men are to-night reported scattered in Missouri.

Martial law is published in Leavenworth, but is practically null, as there is no provost-marshal or soldiers here to enforce it, and nothing to do if they were here. Martial law opened as a farce and ended in a fearful tragedy. One thousand Kansas men will be in Missouri this week. Up to this morning 183 bodies were buried in Lawrence. The remains of 7 more bodies are found. One hundred and eighty-two buildings were burned; 80 of them were brick; 65 of them were on Massachusetts street. There are 85 widows and 240 orphans made by Quantrill's raid. Lane has commenced rebuilding his house. Three men have subscribed $100,000 to rebuild the Free State Hotel, known as the Eldridge Hotel. Several merchants have commenced rebuilding. All the laboring men in town will be set to work to-morrow to clear off the ruins. In spite of the terrible calamity, the people are in good spirits. All the towns in the State have sent in large sums of money. Even the men burned out on Quantrill's retreat have sent in loads of vegetables and provisions.

A man was to-day tried in Lawrence, and found guilty of being a spy for Quantrill, and was hung. The chiefs of the civilized Indians of the Delawares and Sacs and Foxes offered their services to Lane.

Reports just in say the buildings in Cass County, Missouri, are on fire, and over 100 of the sympathizers are killed. A fearful retribution no doubt awaits Missouri. In view of these facts, your memorialists respectfully, but most earnestly, pray Your Excellency to rescind the order by which a part of Missouri is attached to the District of the Border, and to order that it be reattached to the Central District of Missouri, or to any other district in our State. All that your memorialists desire in the premises, aside from the change above indicated, is that some tried and faithful officer may be placed in command over the soldiers and people in the counties of the border--some officer whose sense of duty and of love to his country rises far above his political aspirations and party ties and prejudices, and whose sole desire and efforts will be to guard and foster the interests of the Government in that region, and to bring law and order out of the chaos that now reviles.

This is all that the masses of the people desire, and for this your memorialists will ever pray, &c.[360]

A. A. KING. R. C.
VAUGHAN.
A. COMINGO

[360] O.R., 22: 1:485-488.

APPENDIX G

Citizens of Jackson, Cass, Bates and Northern Vernon Counties Affected by General Order No. 11, 1863

(For brief biographies of those enumerated here, see the following appendix.)

Jackson Countians

Adcock, John Alfred Bills
Akers, Henry Joseph
Akers, Henry, Jr.
Akers, James "Jim" (child)
Akers, Joseph
Akers, Mary "Belle"
Alley, (Major)
Armstrong, Katherine H. A. (age 9)
(later married J. D. Kreeger)
Aspling, Thomas and Family
Austin Eliza (child)
Austin, Eliza Jane
Austin, George
Austin, Jessie G.
Austin, John H.
Austin, Robert
Austin, Sarah Ann
Austin, Thomas T.
Austin, William F.
Baker, Maggie
Barkley, W.G. and Family
Barrett, Jack (child)
Barrett, John
Barrett, Mollie (child)
Barrett, Nellie
Barrett, Robert "Bob"
Beets children
Beets, Elizabeth (Tuggle)
Beets, James
Bennett, Elizabeth (Cave)
Bennett, George (age 5)
Bennett, Jacob
Bennett, Price (age 2)
Best, John
Boggs, John O.
Boggs, Mrs. (formerly Mrs. Peace McGuire)
Bradin, Martha A.
Bradley, (Mrs.)
Bradley, Levin W.
Bradley, Margaret M. (Parish)
Bradley, Thomas
Browning, O. N.
Burns, Kate (Mrs.)
Burrus Eva (child)
Burrus, A. (Kimberlin)
Burrus, James M., Sr.
Burrus, Kate
Burrus, T. K. (child)
Bynum, Charlotte (Cave) (age 44)

Bynum, Galen C. (age 14)
Bynum, James P. (age 13)
Bynum, John (age 41)
Bynum, Robert (age 18)
Cabness, J. B.
Cabness, James
Cabness, Rachel
Cagin, John and Family
Campbell children
Campbell, Courtney
Campbell, Harriet (Buckner)
Campbell, John B.
Campbell, Rachel A. (Grubb)
Campbell, Washington and Family
Canine, Cornelius
Canine, Cornelius (Mrs.)
Cappell, Bennetter A. (child)
Cappell, Britton M.
Cappell, John
Cappell, Martha (child)
Cappell, Mary (child)
Cappell, Sarah (child)
Cappell, Susan (child)
Carlton, Ezekiel
Carlton, Mary Jane (Smith)
Carter, Al
Carter Family
Cave, Allguis (age 20)
Cave, Amanda (Potter) (age 20s)
Cave, Berilda T. (age 7)
Cave, Henry C. (age 6)
Cave, Jane
Cave, John (age 20s)
Cave, John B. (age 4)
Cave, John S. (age 45)
Cave, Lillian L. (child)
Cave, Marion
Cave, Martha "Matilda" (age 12)
Cave, Nancy (age 28)
Cave, Nancy (Hunter) (age 44)
Cave, Pleasant F. (age 4)
Cave, Susan (age 56)
Cave, Susan F. (age 17)
Cave, William (age 33)
Chiles, Elijah Young (age 3)
Chiles, H. C. (child)
Chiles, Anna Azubah (child)
Chiles, Azubah (Skinner)
Chiles, Joel Franklin (age 15)
Chiles, Mary Jane (child)
Chiles, Sarah Anne (Young)
Chiles, Sarah Margaret (age 17)
Chouteau, Cyperian
Chouteau. Nancy (Francis)
Clark, David M. and Family
Clarkson, Virginia (Saint Clair)
Clements, Hannah H. (Scrivner)
(Mrs. John Ambrose)
Clements, Milton (Mrs.)
Cogswell, Nancy Brawner "Nannie" (Rogers)
Cogswell, Oscar H.
Collins, "Buck"
Compton, George
Compton, James
Compton, John (child)
Cooper, Miss Rose
Cooper, Z. G. (Mr.)
Cooper, Z. G. (Mrs.)
Corn, George (child)
Corn, John S. (child)
Corn, Louisa (child)
Corn, Martha (Cummings) (age 40)
Corn, Martin (age 50)
Corn, Nancy E. (child)
Corn, Samuel (child)
Corn, Susan F. (child)
Cowherd, C. J. (age 36)
Cowherd, Edmond (age 47)
Cowherd, Francis K.
Cowherd, Francis K. (Mrs.)

Crawford, (Mr.)
Crowther, Benjamin
Crowther, children
Crowther, Minerva Ann Lilly
Cusenberry, (Mr.) and Family
Dalton, Lewis (age 64)
Dalton, Matilda (age 50)
Daniels, John and Family
Darby, John
Darby, Mary Jane (Austin)
Dehoney, Howard (child)
Dehoney, Leander M.
Dehoney, Lina (child)
Dehoney, Mary (Chism)
DeWitt, Daniel B. (Mrs.)
Dillingham, C. J. (age 15)
Dillingham, Family
Dillingham, Joshua (age 47)
Dillingham, Susan Jane (Walker) (age 36)
Donahue, (Mr.) and Family
Doneghy, Abraham J. (age 6)
Doneghy, Alexander (age 8)
Doneghy, James C. (age 13)
Doneghy, John T. (age 11)
Doneghy, Katherine (widow)
Doneghy, H. Weitman (age 2)
Doneghy, Herbert (age 4)
Dougal, Joseph M.
Douglass, Milda J. (Sanders) (age 29)
Douglass, Thaddeus T. (child)
Douglass, William I. (age 38)
Duncan, Charles Thomas
Duncan, Louisa Lucretia (Alley) (Mrs. Rev. Pryor)
Duvall, Carrie (O'Bryan)
Duvall, Crit (child)
Duvall, Flora (child)
Duvall, Mattie (child)
Duvall, Thomas
Duvall, W. A. (child)
Easley, Miller W.
Easley, Stephen
Ellez, A. C.
Ellez, A. C. (Mrs.)
Ferguson, Felix
Ferguson, Richard and Family
Fields, Martha Frances (Chrisman)
Fields, Amanda (Chiles)
Fields, William
Flanery, George Z.
Flanery, Horace W.
Flanery, James S. (child)
Flanery, John W.
Flanery, Laura Elizabeth (Harris) (age 25) (later Bridges)
Flanery, Laura Elizabeth (Harris) (son, about 3)
Flanery, Lucinda (age 2)
Flanery, Mac P.
Flanery, Melvina W.
Flanery, Rebecca J.
Flanery, William V. (age 3)
Flanery, Zion (age 49)
Ford Family
Ford, Emily (baby)
Ford, Lewis
Ford, Martha (baby)
Ford, Patsy (Holmes)
Franklin, J. L. H. (age 21)
Franklin, John R.
Franklin, Lucy Ann (Walker) (age 35)
Franklin, Sarah
Frick, Dr. W. E.
Frick, Mrs.
Fristoe, John
Fristoe, Mary L. (Sullivan)
Frost, Mary (Mrs.)
Gammon (baby)
Gammon, Hattie

Gammon, James
Gammon, Sally (Stuart)
Gammon, William
Garrett, Reuben and Family
Geam, Rile
Gentry, Mary (Henley) (Mrs. Joseph)
George, Amanda J.
(later Mrs. John W. Tyler)
George, N. E. (Mrs.) and Family of 5
George, Nathaniel "Nathan" Bass
George S .E (Farmer)
Gibbons, Hank
Gilky, (Mrs.)
Gilliss, William and Family
Gilpin, William and Family
Graham, Ambers (age 26)
Graham, Sarah J. (Stultz) (age 24)
Green Family
Gregg, Cornelia (child)
Gregg, D. (age 31)
Gregg, George H. (child)
Gregg, Larius (child)
Gregg, Lera H. (child)
Gregg, Maria (Bryant)
Gregg, Martha (child)
Gregg, Mary (Harper) (age 36)
Gregg, Robert L. (child)
Gregg, Samuel
Gregg, Wallace (child)
Gregg, William (Mrs)
Griffith, George W. (age 34)
Griffith, Mary A. (about 60)
Groves, (Mrs.) and Family
Hall, Charles B. (age 10)
Hall, Ella Virginia (age 13)
Hall, Franklin (age 9)
Hall, James D.
Hall, Lemuel Edward (age 33)
Hall, Mary (about 6)
Hall, Mary E. (Thompson) (age 34)
Hall, Richard (about 5)
Haller Family
Hambright, John
Hamilton, Bird
Hamilton Family
Hamilton, Sam
Hanley, Sam (Mrs.)
Hanley, Sam
Harbaugh, Amanda (age 5)
Harbaugh, George (age 7)
Harbaugh, Mary Elizabeth (Pence) (Mrs.
Franklin E.)
Harbaugh, Minnie (age 1)
Hardesty, Richard and Family
Harris, Fanny (Mrs.)
Harris, Jane (Hall) (Mrs. Samuel)
Harris, Nancy "Nannie" Elizabeth Ellen
Hatton, J. W.
Hays, Elfleda (child)
Hays, Margaret Jane (Watts)
Hays, children
Henley, Elizabeth (Belt) (Mrs. Alonzo)
Hereford, H. F. (doctor)
Hereford children
Hickman, Amelia
Hickman, J. T.
Hightower, John (age 47)
Hightower, Sarah A. (Van Ardol)
Hightower, W. H. (about 5)
Hinkel, Charley (child)
Hockensmith, Thomas Lee
Holmes, Hezekiah and Family
Holmes, Nancy (Tyler)
Hord, William
Houston, (Mrs.) and Family
Hudson, Geneva Elisabeth (Hamblen)
Hudson, James Russell
Hudspeth, Cyntha/Cynthia
Hudspeth, Elizabeth (Jesse) (age 33)
Hudspeth, George (age 43)

Hudspeth, Paralee (child)
Hudspeth, Thomas B. (child)
Hudspeth, Tybithia (age 5)
Hughes, Merritt
Hughes, Will
Hunt, E. F. (age 11)
Hunt, J. B. (age 3)
Hunt, J. M. (age 5)
Hunt, Mary A. (age 7)
Hunt, N. E. (Cave) (age 30)
Hunt, Noah (age 32)
Hunt, Noah (Mrs.)
Hunter, Berrilla (age 22)
Hunter, David (about 33)
Hunter, Elizabeth Jeans (age 71)
Hunter, John (age 75)
Hunter, Juda (Potter) Hunter (about 30)
Hunter, William (age 47)
Ingrum, Hanna (Mrs. E. A.)
Irwin, Julia (Harris) (age 20)
Irwin, Mary Jane (Miss)
Jefferys/Jeffreys/Jeffries, Charles Peter Beauchamp
Johnson, Coger W. (age 6)
Johnson, Elizabeth (Hall) (age 31)
Johnson, Greenville
Johnson, James R. (age 11)
Jones, Jemima B. (Mrs. William Sanders Jones)
Jones, Billy
Judson, Ben and Family
Keeney, Elvira (Campbell)
Keeney, Isaac
Kemper, Berilda (Bryant)
Kemper, James Harvey
Kendal, Charles (Mrs.) and Family
Keshlear, Alice (age 20)
Keshlear, George B. (child)
Keshlear, James C. (child)
Keshlear, Joseph B. (age 3)
Keshlear, Mattie A. (child)
Keshlear, Sarah (Riegs) (widow age 45)
Keshlear, William G. (child)
Kreeger, Araminta (Daniels)
Kreeger, Fannie (age 10) (later Haller)
Kreeger, George W.
Kreeger, J. D. (age 8)
Kreeger, R. R. (under age 7)
Lee, (Mr.) and Family
Livesay, (Mrs.)
Luttrell, Armilda (age 31)
Luttrell, Granvile (age 33)
Luttrell, Samuel (age 48)
Luttrell, Sarah (daughter of Samuel)
Lykins, Martha A. (Livingston) (age 39)
Lynch, Miranda (Snodgrass)
Maddox, Asa and Family
Mann, Christopher
Masters, M. C. (age 30)
Masters, Mary C. (Leader) (age 21)
Maxwell, Martha A.
Maxwell, Tom (Mrs.)
Mayes, Sarah Louisa (Wilson) (Mrs. Henry C., when a child)
McCready, (Mrs.)
Meador, Sara Angeline, a sister & 3 children
Mims, F. (professor)
Montgomery, Elizabeth (Holloway)
Montgomery, Levi
Moore, Frances Fitzhugh (George) (later Kabrick)
Moore, Lawson and 6 children
Moore, Will
Moore, Zirilda
Morrow, Jesse (age 53)
Morrow, Nathan (child)
Morrow, Rufus F. B. (child)
Morrow, Sylvia (Hudspeth) (age 56)
Muir, Bettie (child)
Muir, Buck

Muir, Buckner (age 84)
Muir, John S.
Muir, Nancy (Parr)
Muir, Sam (age 56)
Muir, Samuel (age 16)
Murray, Mary F. (Smith)
Murray, William F. (age 7)
Neat Family
Noel, James W. and wife
Nolan, R. H. and Family
Noland, Harriett (Newton)
O'Donnell, (Mrs.)
Oldham, Albert G.
Oldham, Althea (child)
Oldham, Amelia
Oldham, Mary Julia
Oldham, John R.
Oldham, John R. (Mrs.)
Oldham, Thomas (child)
Ousley, Andrew (age 17)
Ousley, Miss (daughter)
Ousley, Matilda (Hunter)
Ousley, Dana (Mrs.)
Owings, Martha E. (George)
(Mrs. Joshua) and Family of 7
Page, Granville Dyson
Page, Venitia Mary "Nish" (Colcord)
Palmer, Jacob Teaford
Park Family
Park, Annie (age 2)
Parker, Zachariah
Parker, Zachariah (Mrs.)
Parsons, Barbara (Ely)
Parsons, Daniel White
Parsons, Melvina A.
Parsons, Samuel
Patterson, Jesse (child)
Patterson, Joel (age 47)
Patterson, Joel B. (child)
Patterson, Mollie (child)
Patterson, Sarah (age 43)
Pierce, A. J. (doctor)
Pillard,Phillip D.
Potter, Amanda (child)
Potter, Benjamin (age 75)
Potter, Benjamin (son)
Potter, David
Potter, Marian (child)
Potter, Martha
Potter, Thomas (age 12)
Potts, Peter J. (Mrs. Fannie E.) (Parsons)
Powell, Alvis
Powell, John Beeler
Powell, Parthen (Meador)
Powell, six children
Ralston, Adam (age 8)
Ralston, Anna (age 10)
(later wife of Frank James)
Ralston, Sam
Ralston, Fedora (age 15)
Ralston, Harry M. (age 1)
Ralston, Katherine (age 3)
Ralston, Mary Catherine (Hill)
Ralston, Samuel (age 13)
Ralston, William H. (age 5)
Reid, John W. (age 40)
Rice, Hollow
Rice, Isaac (age 17)
Rice, Martin (age 49)
Rice, Martha (Martin's daughter)
Rice, Mary (age 70)
Rice, Mary (Martin's daughter)
Ridlesberger, Jesse and Family
Ritter, Henry T. (age 40)
Ritter, Taylor (Mrs.) (age 36)
Ross, (Mrs.)
Roupe, John
Roupe, John Henry (son of John)
Rout, Henry
Runyan, A. O. and Family

Russel, Elizabeth Southworth (Williams)
Russel Family
Ryley, Archiey (age 15)
Ryley, Emma (child)
Saint Clair, Joseph
Saunders, Drucilla (Stayton)
Saunders, John B.
Saunders, Susan (age 11)
Scrivner, Joseph
Scrivner, Pameila
Scruggs, P. T.
Sea, (Mr.) and Family
Sechrest, Charles and Family
Sevier, Henry Clay
Slaughter, children
Slaughter, Elizabeth
Slaughter, O. V.
Smart, Susan (Mrs.)
Smith, B. F. and Family
Smith, Eliza (Wallace)
Smith, Franklin (Sr.)
Smith, Franklin (Jr.)
Smith, Minor T. and Family
Staples, Abner Ford
Staples children
Staples, Elsie
Stayton, Lucy Ann (Hite)
Stayton, Moses
Stayton, Rebecca
Steele, Albert Gallatin
Steele, Emily (Hampton)
Steele, Mr.
Stegall Family
Stonestreet, James
Stonestreet, James (Mrs.)
Summers, John W. and Family
Sweeney, James and Family
Talley, David C.
Talley, Mary (Fristoe)
Tate, John (age 5)
Tate, Martha (Rice) (age 27)
Tate, Mary (age 3)
Tate, Nancy Rosalie (baby)
Tate, William Calvin (age 32)
Tatum, Elizabeth (Clark) and Family
Tatum, John W. (youth)
Tatum, Thomas Jefferson
Thatcher, J. M. (Mrs.), her father
and her 4 children
Thompson, David (age 14)
Thompson, George M. (age 47)
Thompson, James (age 16)
Thompson, Lizzie H. (age 13)
Thompson, Lucinda (Webb)
Thompson, Susan A. (age 41)
Thompson, William
Thruston, W. Creighton
Triplett, Eli and Family
Triplett, Dan (age 4)
Triplett, Margaret (Hart)
Truman, Anderson Shipp "Andy"
Truman, John Anderson (child)
Truman, Mary Jane
Trusten, Cappell (child)
Tucker, Tom
Twyman, Frances C. (Fristoe)
Twyman, George T. (age 11)
Twyman, Julia (age 13)
Twyman, Lyddall W. (doctor)
Van Ness, Alice
Vaughn, Rebecca (Harper)
Vaughn, Reuben
Vergin, William G.
Walker, Andrew J. (age 27)
Walker, Collins (age 39)
Walker, Elizabeth (Braden)
Walker, James
Walker, Jefferson (age 35)
Walker, John Riley (age 33)
Walker, Nancy C. (Montgomery)

Walker, Susan Jane
Walker, Thomas J. (age 5)
Walker, Zachaiah (age 25)
Wallace Charles H. (child)
Wallace, Addison A. (infant)
Wallace, Addison H.
Wallace, Jefferson Davis
Wallace, Jessamine (Ryley) (age 40)
Wallace, John (child)
Wallace, John D.
Wallace, Joseph (Reverend)
Wallace, Nellie (Duncan) (Mrs. John D.)
Wallace, William H. (age 14)
Walton, (Mrs.)
Walton, Thomas (age 3)
Warren, Lucinda Jane (Robertson)
Warren, Nelson
West, George Cornelius (age 7)
West, George Cornelius (Mrs.)
Whitsitt, J. B.
Wiley, William D.
Wilhite, Jefferson
Williams, J. W.
Wilson, (Mr.)
Wilson, Louisa (McMurray)
Woodruff Family
Woodruff, Julia
Woodson, M. A. (Mrs.)
(formerly Mrs. Ervin; originally Miss Bowers)
Woodson, Samuel H. (Mrs.)
Wright, Emory (Mrs.)
(her unnamed grandfather)
Wyatt, Loraine (Mrs. A. M.)
Wyatt, Wiley
Wyatt, Wiley (Mrs.)
Yankee, Jacob
Yankee, Jacob (Mrs.)
Yankee, Orlena (Beeler) (Mrs. David)
Young, Ada (child)
Young, Harriet Louisa (Gregg)
Young, Harrison (age 17)
Young, Laura Jane (age 13)
Young, Martha "Mattie" Ellen (age 11)
Young, Rebecca Jane (Smith)
Young, Sarah Ann (age 20)
Young, Solomon
Young, Susan

Cass Countians

Adams, James
Aldridge, Ben
Ament, John
Armstrong, Mr.
Barnett, Stephen
Beagles, Edward
Boren, M. F.
Bouse, Susan
Bradbury, Leonard
Brannock, A. B.
Brannock, George W.
Briscoe, Ruth
Britt, Jefferson Waller
Bronaugh, William
Brookhart, Mr.
Brown, Mary
Brown, Robert A.
Bullock, (Mr.)
Burdette, (Mr.)
Burns, Rice
Burris, F. M.
Carter, James
Cary, Edmon
Casey, Philip
Chilcote, John
Clements, (Mr.)
Cordell, Henry
Crenshaw, (Mr.)
Davenport, William
Davidson, J. N.
Davis, James
Davis, John
Dehoney, Simeon
Delevan, Alfred
Dolan, James
Dunham, (Mr.)
Dunn, Absolom
Dye, Isaac
Easley, Achilles
Estes, William
Farmer, Henry
Farnsworth, C. L.
Flood, Stephen
Franse, Celia
Freeman, George
Garrison, (Mr.)
Gilmore, Joseph
Glass, James
Graham, Mary
Harrelson, (Mr.)
Hatton, Henry
Haynes, Elizabeth
Hinshaw, Nathan
Hockensmith, G. W. R.
Holcomb, A.
Holloway, T.
Hon, Jerry V.
Hon, Richard
Hoover, S. T.
Horn, Richard
Hunt, Mrs. Noah
Ingram, Charles
Jackson, (Mr.)
Jerrard, (Mr.)
Johnston, George
Jones, James
Keeney, Thomas
Knight, Jim
Knight, Mary (France)
Laffoon, Drury
Lamar, John
Lewis, Robert
Lewis, Sarah
Lundy, Creed
Majors, Isham
Mattin, S. E.

McFarland, T. J.
McFerrin, Elizabeth
McGill, (Mr.)
Mers, Andrew
Mers, W. T.
Milam, Thomas
Miller, George (Rev.)
Moore, (Mr.)
Morris, John Calvin
Myers, Harry
Myers, William
Nailor, Fountain
Nelson, Hector
Norman, C. P.
O'Bannon, E.
Palmer, (Dr.)
Parish, (Mr.)
Parker, William H.
Payne, E. S.
Phillips, James M.
Pleasant, Edward
Poyntz, John
Powell, Jonathan R.
Quick, Cornelius
Reed, Farlow
Rice, E. N.
Richardson, Joseph
Richardson, Mary I.
Robinson, John
Rowland, Jesse
Rowland, William
Sanders, (Mr.)
Scott, (Mr.)
Settle, J. F.
Shaler, N. T.
Sharp, Matt
Shelton, Minerva
Shipley, Amanda
Shipley, Preston (youth)
Shipley, William
Sloan, F. X.
Smith, Dicey
Smith, Isaac
Smith, John
Steen, Elizabeth
Stephens, John L. L.
Storms, (Mr.)
Symington, Robert Smith
Talbot, Joshua
Taylor, Elizabeth
Thomas, Griffin
Thornton, C. T.
Tuggle, E. L.
Underwood, John
Van Kirk, (Mr.)
Wade, W.
Ward, Cyrus
Warren, Sarah
West, Lysander
West, (Mr.)
White, (Mr.)
White, William
Willett, J. N.
William(S,) David
Williams, J. R.
Willis, Alpheus
Woolery, Caroline
Woolery, J. J.
Worrell, (Mr.)
Yancey, George
Yankee, Wesley
Yoder, Solomon
Youngs, (Mr.)
Zion, Nancy

Bates Countians

Blankenbaker, Abraham
Elder, Thomas "Bud"
McCool, James
McCool, Peter V.
Ludwick Family
Shelton, William H. and father's family
Warren, Lucinda (Robertson)
Warren, Nelson

Vernon Countians

Boyd, William
Connor, John
Duncan, John
Newman, J. N.

APPENDIX H

Personal, Documented Stories of Order No. 11 Victims

The stories below are a compilation of historical accounts of those citizens affected by Order No. 11. Several committed individuals have spent decades compiling the sketches covered in this section of the book. Each of them is respected for their dedication and hard work and personally recognized in the acknowledgements that follow.

What the following chronicles attempt to accomplish:

There were thousands upon thousands of people adversely impacted by Order No. 11. They had only fifteen days to evacuate Jackson, Cass, Bates and the northern part of Vernon Counties. They were forced to leave their homes and report to the nearest military post. Failure to sign the loyalty oath would cause banishment or imprisonment. *"Imagine the untold numbers of stories and memories that 'walked out of our area' and never returned,"* said author Tom Rafiner. Unfortunately, only a fraction of the haunting tales of those victims were recorded and saved for posterity.

This study is an attempt to compile the most complete *documented and archived* chronicles of the individuals and families forced off their property under Order No. 11.

This material is provided for the novice, genealogist or scholar in identifying *sourced accounts* of those who personally experienced Order No. 11. Primary and secondary sources have been collected through previous historical works on this subject, or through donations available for research at the Jackson County Historical Society's Archives. If one's family is not included in this untoward chapter in history, the Historical Society welcomes contributions of those who wish to share and add their family's *documented* story about the Border War, Civil War, or Order No. 11. Verified and documented source materials donated *may* be incorporated into future editions of this publication.

What this list is not intended to do:

The abstracts below are abbreviated synopses that do not necessarily represent a *complete* biography of an individual or family. Readers desiring additional information are invited and encouraged to consult the cited source materials.

Furthermore, this enumeration does not include the significant number of citizens of the affected counties who were either killed, banished or voluntarily left *prior* to the issuance of Order No. 11 in August 1863. Many citizens endured the border terrorism *long before* Order No. 11; their plight was not the focus of this precise study of a specific timeperiod when U.S. citizens were under martial law.

Knowing that 40,000 residents lived in the counties affected by Order No. 11, it is clear that only a small fraction of individuals are enumerated here. The author and his colleagues have scoured all known sources and references to Order No. 11, and presented findings herein. As before stated, this study presents *documented accounts* that have been located. Any errors or omissions are unintentional by the author.

When a surname is identified more than once with the notation of an "Additional Account," it signifies that the author determined that either the families might not be directly related, or that the sources were so diverse, they required a separate entry. All entries refer to residents of Jackson County, unless another county of residence is specifically denoted.

Individuals named separately throughout this extensive appendix are *generally* indexed simply by their surname (i.e., Adams Family). Some, however, may also be indexed by full name (i.e., "Adams, James"). Readers are invited to refer to all available options presented.

ADAMS

James Adams of Cass County went to Johnson County, Missouri, because of Order No. 11.[361]

ADCOCK

John Alfred B. Adcock left Jackson County under Order No. 11. When he returned to his home in September 1865 he found "the country a desolate waste, with now and then a lone chimney to tell the story of a fire."[362] John later married Rebecca Stayton. His future wife was a native of Independence and the daughter of Moses Stayton and Lucy (Hite) Stayton. The Stayton's went to Miller County, Missouri during Order No. 11 (See Stayton below).

AKERS

Joseph Harold Akers spoke with Gary Steven Akers on August 27, 1986, and gave an oral history interview regarding his family's association with the pre-Civil War and Missouri-Kansas border conflicts. He said that his Uncle Henry and Aunt Elizabeth Jane (Miller) Akers, had two sons, James "Jim" Akers and Henry Akers, Jr. "Aunt Jane would hide one day with her two children and Grandmother," Melvina A. Parsons, "would hide the next day with her two children," Aunt Mary "Belle" Akers and his father, Joseph Henry Akers.

> *"And they did that because if they could of caught both of them they would have taken them and put them in that house in Kansas City [the prison that collapsed]. But they were afraid to take either one of them without the other because they would get to Quantrill, or get word to Grandfather Sylvester Akers with Quantrill and that would be the end of it.*

---

[361] Another man only identified by his surname, went to Lafayette County. See, National Historical Company. "History of Cass and Bates Counties, Missouri." (St. Joseph, Mo., National Historical Company, 1883), 659; and, Glenn, Allen. "History of Cass County, Missouri." (Topeka, Ks.: Historical Publishing Co, 1917). Most references to documented Cass Countians affected by Order No. 11 are courtesy of the Tom A. Rafiner Papers, Cass County Historical Society, containing research for his book, "Caught Between Three Fires: Cass County, Mo., Chaos, & Order No. 11, 1860-1865." (Xlibris, Corp., 2010). A significant amount of research data was not selected for incorporation in Rafiner's book; therefore, Rafiner's research papers should be consulted.

[362] Rice, Martin. 90-93; Eakin, "Tears and Turmoil," 1996, 33-35, citing United Daughters of the Confederacy, 124.

"Well, finally, in one of these raids of the Jayhawkers from Kansas, coming over into Missouri and playing havoc, Quantrill's bunch took after them and caught them on the prairie in Kansas and killed them all and set the prairie on fire and burnt them up. Now that's how they stopped them, stopped them from raiding. So you see, they were home guards. The Home Guards were worse than the Union Army or the Confederate either one."[363]

ALDRIDGE

Ben Aldridge of Cass County reported to the military post at Pleasant Hill during Order No. 11.[364]

AMENT

John Ament of Cass County traveled down the Missouri River to the Mississippi River and then went overland to Chicago, Cook County, Illinois after Order No. 11.[365]

ARMSTRONG

Mr. Armstrong left his home in rural Cass County and went to the military post at Pleasant Hill during Order No. 11.[366]

ARMSTRONG
see KREEGER

ASHLEY

A newspaper article published in *The Columbia* (Mo.) *Herald*, and reprinted from the *Kansas City Journal*, August 29, 1863 titled, "An Interesting Memento: List of Those Banished by Order No. 11," contained two documents identifying citizens living within Independence and Kansas City who were banished. The first is an extract of an evacuation order, "Special Orders No. 6," from the Headquarters District of the Border, Kansas City, Missouri, August 29 1863, and the other is a complete alphabetical listing of

---

[363] Jackson County (Mo.) Historical Society Archives, Document ID 10F15.
[364] Rafiner, citing National Historical Company, 463.
[365] Rafiner, Family histories compiled by Cass County Historical Society, Harrisonville, Missouri; and, The John Ament Diary, Cass County Historical Society. Harrisonville, Missouri. John kept a diary of adventures while he traveled down river to Chicago.
[366] Rafiner.

those vanquished from the cities (and from certain Kansas counties within the District of the Border).

> *A. O. Runyan and family, residents of Independence, Missouri, are ordered to remove from this district within ten days from the date hereof. They will not go to the counties of Platte, Clay, Ray, or Carroll, Missouri, to reside nor return to this district, during the rebellion, without previous express permission from competent military authority."*

Those evicted are set forth below:

Ashley, Mrs., Independence
Aspling, Thomas (and family), Kansas City
Barkley, W. C. (and family), Kansas City
Burns, Oliver P., Kansas City
Cagin, John (and family), Independence
Campbell, Wash. (and family), Independence
Chick, P. W. (and wife), Kansas City
Chipman, Ruth (Mrs.), Kansas City
Cusenberry, Mr. (and wife), Independence
Donahue, Mr. (and family), Kansas City
Dougal, Joseph M., Kansas City
Ferguson, Felix, Kansas City
Ferguson, Richard (and family), Kansas City
Filley, Mrs., Independence
Frost, Mary (Mrs.), Kansas City
Garrett, Reuben (and family), Kansas City
Gilky, Mrs., Independence
Gilliss [sic.], William, Kansas City
Gilpin, William (and family), Kansas City
Groves, Mrs. (and family), Independence
Hall, Lemuel (and wife and family), Kansas City
Hall, Mrs. (and family), Independence
Hanley, Sam (and wife), Independence
Hardesty, Richard (and family), Kansas City
Harris family, Independence
Hatton, J. W., Kansas City
Fanny (Mrs.), Independence
Holmes, Hezekiah (and family), Kansas City
Houston, Mrs. (and family), Independence

Irwin, Mary Jane (Miss), Independence
Judson, Ben (and family), Kansas City
Kendall, Charles (Mrs.), Kansas City
Lee, Mr. (and family), Independence
Lykins, Mrs. Dr., Kansas City
Maddox, Asa (and family), Kansas City
Maxwell, Tom (Mrs.), Independence
McCready, Mrs., Independence
Neat, Mr. (and family), Independence
Nicholson, George, Kansas City
Nolan, R. H. (and family) Kansas City
O'Donnell, Mrs., Independence
Pierce, A. J. (doctor), Kansas City
Pollard, Philip D., Kansas City
Rice, Hollow (and family), Osage County, Kansas
Rice, Montague, Osage County, Kansas
Ridlesberger, Jesse (and family), Kansas City
Ross, Mrs., Independence
Runyan, A. O. (and family), Independence
Scruggs, P. T., Kansas City
Smart, Mrs., Independence
Smith, B. F. (and family), Kansas City
Summers, John W. (and family), Kansas City
Sweeney, James (and family), Kansas City
Van Ness, Alice (Miss), Olathe, Johnson County, Kansas
Wiley, William D., Kansas City
Woodson, Sam H. (Mrs.), Independence.[367]

ASPLING
See ASHLEY

AUSTIN
See DARBY

[367] Jackson County (Mo.) Historical Society Archives, Document ID 68F24. *Kansas City Journal*, 29 Aug. 1863. Also, Eakin, "Tears and Turmoil," 61-63; and, Jackson County Genealogical Society, *Pioneer Wagon*: 12: 3.

AUSTIN
(Additional Account)

Robert Austin was born in St. Louis, Missouri on August 28, 1811. At about age twenty, he settled around Lone Jack with his brothers Stephen and Joseph. Later he moved to Oak Grove and purchased two hundred acres of land. In 1834, Robert married Eliza Jane Watkins in the Sni-A-Bar Township of Jackson County near Oak Grove. The couple was members of the Methodist Episcopal Church and they had seven children: Jessie G., Thomas T., Mary Jane, William F., George, John H. and Sarah Ann. In 1858, their daughter, Mary Jane married John Darby.

Austin was a strong Union man and despite his loyalty, he and his family relocated to Platte County, Missouri (their son-in-law John Darby served in the Union Army for six months during the first year of the war). They lost their property during Order No. 11 as their buildings were burned and property destroyed leaving Austin and his family nearly financially ruined. In 1864 Austin sold the Oak Grove farm and bought one hundred and sixty acres in Johnson County, Kansas. Austin was described as having done "much for churches, schools and benevolent institutions of different kinds" and being "one of those men whose advice and counsel" was "always sought."[368]

All their children moved to various states after the War, except for their daughter, Mary Jane. By 1881, Mary Jane and her husband John had nine children and lived in the Sni-A-Bar area of Jackson County.[369]

BAKER

Miss Maggie Baker was among the twenty-three families and sixty-one women and children who were piloted out of Jackson County by eighty four year old, Buckner Muir, and his 56-year-old son, Sam, under Order No. 11. Maggie Baker joined Laura during the "death-watch" over the body of three-year old Thomas Walton (for more information see Flanery in *Chapter XII*).[370]

BARKLEY
See ASHLEY

---

[368] Union, 911 and 915.
[369] Union, 326, 911 and 915.
[370] Eakin, Joanne, Chiles. *Blue and Grey Chronicle.* (August 2000); 3: 6: 3.

BARNETT

Stephen Barnett left Cass County after Order No. 11 and went to Johnson County, Missouri.[371]

BARRETT

S. M. Barrett shared a story about his parents, Nellie Barrett and Robert "Bob" Barrett, who during the War was a young wagon boss for Russell, Waddell and Majors. The story related by S. M. Barrett regarding the affects of Order No. 11 on the Barrett family follows:

> *One morning she [Mrs. Nellie Barrett] saw an immense herd of cattle being driven past her fence. It was a herd collected from the farms around, being driven over into Kansas. Order No. 11 was being put into action. Frantically she tried to sell her corn crop, and her stock, but nobody would buy because the Red Legs were sure to come and take it away.*
>
> *On the fifteenth morning, troops surrounded her log home. The officer in charge accosted Nellie Barrett, 'Madam, take the children and leave. We're ordered to burn this house.' That day the little mother felt more like fighting than praying, but with a few scraps of clothing she and the children began to walk away from their home. As they entered the woods they saw the cloud of smoke which told them their home was in flames.... Her two oldest children, Mollie and Jack, brought the extra clothing, made her a bed to lie down and rest. With a stick in his hand little Jack stood guard while Mollie bathed her mother's brow with cold water from a brook. It was there that the old grandfather, [her father-in-law, John Barrett] found her.*
>
> *She was carried to a wagon, and the family finally reached Kansas City, whence they were able to go on to Nebraska City and Bob, her husband. But in her humble way, she typified the anguish and sorrow of the Missouri border in the grip of internecine war.*[372]

---

[371] Rafiner's best guess judging from the 1860 and 1870 U. S. Census returns.

[372] Jackson County (Mo.) Historical Society Archives, Document ID Vertical File Newspaper Subject Files: Military/Wars: Civil War: 1863-"Order No. 11": "A Brave Mother's Story of Terror in War Days on Missouri Border," *Kansas City* (Mo.) *Times*, 21 Jan. 1941, by S. M. Barrett, who was for many years public relations chief of the Kansas City Board of Education. He shared this story of his parents. A variation of the family story along with other details is related in Eakin, Joanne Chiles and Donald R. Hale. "Branded as Rebels." Volume 1. Second Printing with corrections. (Independence, Mo.: Print America, 1995), 18.

BEAGLES

Edward Beagles of Cass County traveled to Johnson County, Missouri after the issuance of Order No. 11[373]

BEETS

James Beets was born in Hawkins County, Tennessee on March 11, 1816. He came to Missouri about age twenty-three years and settled in Boone County where he stayed one year before settling in the section of Missouri known as the Platte Purchase. He soon left the new settlement and traveled with a Methodist minister to establish a mission amongst the Pottawattamie Indians. After spending one-year performing mission work, Beets settled in Cass County where he stayed until 1855.

Elizabeth J. Tuggle was born in Patrick County, Tennessee. At an early age, she came to Jackson County with her parents and settled in Cass County. On June 7, 1840, she and James Beets were married. It did not take long before Elizabeth realized her new husband had a propensity for traveling and relocating. In 1848, he left the family and traveled to New Mexico where he spent six months. Upon his return he and his family settled in Cass County.

The Beet's family in April 1855 and during the early years of Bleeding Kansas moved to Paola, Kansas. They farmed and James operated a hotel until 1861. Their experience in Kansas was not pleasant. They were robbed and forced off their property because of their southern sentiments, which forced them to resettle in the Blue Township of Jackson County.

In the second year of the Civil War in February 1862, James joined Company G, 3rd Missouri Infantry of the Southern Army in Springfield, Missouri. He was engaged in all Missouri State Guard and Confederate battles until April 1863. He was discharged from military service because of failing health and paralysis he had suffered in the Battle of Corinth, Mississippi. It took him until July 4, 1863 to make the journey home.

By Order No. 11, the family moved to Lexington. The family crossed the Missouri River at the ferry, but James was forced to swim the four miles across the river. Shortly afterward, they moved to Fremont County, Iowa, where they remained about eighteen months before moving to Nebraska for about six months. Eventually, the family returned to Jackson County. By 1881 Beets owned two hundred and forty acres. He and his wife had thirteen children, six of whom were living in 1881, to wit: Joseph B., Henry R., George W., James D., William J. and Cordelia A.[374]

---

[373] Rafiner's best guess judging from the 1860 and 1870 U. S. Census returns.
[374] Union, 885.

Their son, Joseph B. Beets, was a Democrat, and Union sympathizer. This did not save him from the depredations from Kansas Unionists and in 1861 he suffered the loss of his livestock, which were stolen and run off by raiders. He had to leave the state and later joined Price's army and served two years while his family was in Iowa during Order No. 11. Upon returning to Jackson County after the War, they settled in Blue Township near Atherton, Jackson County.[375]

BENNETT

Jacob Bennett was born in Ohio on May 12, 1833. While growing-up in his native state, Jacob was educated and learned the carpentry trade in Cleveland. He moved to Jackson County in 1853 settling in Van Buren Township. On October 4, 1857, Jacob married Elizabeth Cave. Elizabeth Cave was born in Jackson County on August 31, 1837. She was one of the seven children of John and Nancy Cave.

At the time of the issuance of Order No. 11, Elizabeth and Jacob had two children, George born July 24, 1858 (who later attended the Warrensburg Normal School—today University of Central Missouri); and Price born July 17, 1861 (who later became a salesman In Kansas City). In all Jacob and Elizabeth had six children: Ella W. born September 4, 1864 and died November 16, 1874; Gius born February 24, 1868; Charles born March 10, 1872 and Frank, born February 10, 1875. Elizabeth's father had been born around 1818 in Virginia, and then moved to North Carolina before settling in Missouri.

The Bennett Family experienced first-hand the tragedies of Order No. 11 as Elizabeth's father was among the six men murdered by Kansas troops on September 6. As a result of Order No. 11, Jacob, his wife and two children left Lone Jack, not returning for two years. By 1881, Jacob was engaged in the businesses of farming, raising stock and carpentry.[376]

BEST

John Best lived near Lone Jack and left his home in September of 1863 and only returned in February 1864 after the restrictions of Order No. 11 were lifted. The family may have originally gone to Kansas City, but the banishment notation does not say where. His daughter was married to Duke Flanery and that may have been John's "crime."[377]

---

[375]Union, 885. Also, *Liberty* (Mo.) *Tribune*, 8 June 1894, as reprinted in Eakin and Hale, 22.

[376] Union, 885, 931. See also Chapter XII, and Hunter and Cave below.

[377] Order No. 47. Fold 3. NARA Publication M322: Compiled Service Records of Confederate Soldiers Who Served in Organizations from the State of Missouri. Record Group 109, 586957.

BOGGS

John O. Boggs was born in Lee County, Virginia on October 30, 1818. He came to Jackson County in 1839. On May 7, 1849, Boggs joined the California Gold rush, and following his return, he married Mrs. Peace McGuire in 1855. When the War started Boggs joined the Confederate Army and fought at the Battle of Lone Jack in August 1862. He went south and joined Captain Longhorn's Company, which formed a part of Colonel Upton Hays' Regiment in Shelby's command.

The Boggs home was destroyed and his wife was compelled to flee to Kansas. She was bitterly persecuted, at one time being banished from the State, but the order was later revoked. Upon returning, Mr. Boggs found things in a dilapidated condition. His house and fences burned to the ground, and crops destroyed. Boggs wife died March 27, 1876, and he remarried Mrs. Crane, widow of Dr. Crane, of Ashland, Ohio, on January 7, 1877. They are buried in Lee's Summit Cemetery, Lee's Summit, Missouri.[378]

BLANKENBAKER

Abraham Blankenbaker was killed in Bates County, Missouri, August 31, 1863.[379]

BOREN

M. F. Boren, Cass County, found safe haven in Liberty, Clay County, Missouri after Order No. 11.[380]

BOUSE

Susan Bouse of Cass County went to Johnson County, Missouri following the issuance of Order No. 11.[381]

BOYD

William Boyd, Vernon County, was arrested in McDonald County, August 12, 1863.[382]

---

378 Union, 949. Eakin and Hale, 33, referencing "Memorial and Biographical Record."

379 http://freepages.genealogy.rootsweb.ancestry.com/~jrbakerjr/provost3/bushwhackers.htm (viewed 10 June 2013).

380 Rafiner, citing Cass County Historical Society. "Cass County Missouri Families." (Harrisonville, Mo: Cass County Historical Society, Inc., 1976).

381 Rafiner, per personal e-mail.

BRADBURY

Leonard Bradbury lived in Cass County at the time of Order No. 11. Bradbury found a team of oxen and drove them to Macoupin County, Illinois.[383]

BRADIN

Martha A. Bradin was born on December 30, 1830. She "spent all her life" in Jackson County, "except when Order No. 11 was in force." She married Albert G. Hall on May 25, 1866, and she died on May 19, 1914.[384]

BRADLEY

Levin W. Bradley was born in Sumner County, Tennessee on March 27, 1814. Towards the end of 1833, Levin moved to Jackson County. On August 4, 1840, he married Margaret M. Parish. Levin had been reared as a farmer and gave considerable attention to tobacco culture. They lived in Van Buren Township in southeastern Jackson County at time of Order No. 11. The couple had nine children but by 1881 only four children were living, to wit: William P., Eldridge M. Bettie P. and Maggie. On December 30, 1871, Levin was listed among the old settlers who met at the Council Chamber in Kansas City to form the Old Settler Historical Society of Jackson County. As of 1881, the Levin Bradley family owned and lived on nearly one hundred sixty acres in eastern Jackson County.[385]

---

[382] http://freepages.genealogy.rootsweb.ancestry.com/~jrbakerjr/provost3/bushwhackers.htm (viewed 10 June 2013).

[383] Rafiner, citing National Historical Company, 692; also, *Cass County* (Mo.) *Democrat*, 5 and/or 15 October 1916.

[384] "Death of Mrs. Martha A. Hall," *Kansas City* (Mo.) *Star*, 20 May 1914.

[385] "Martin Rice Remembers: What I saw of Order Number Eleven." *Missouri Historical Review*. (April 2012); 106: 3; and Union, 131, 1004.

BRADLEY
(Additional Account)

Mrs. Bradley "was a widow with five children. Her husband died long before the war." Her house was burned and her slaves were stolen by Kansas' guerrillas. She and her children were forced off their property. "They lived in a wagon a few miles from their burned home for more than a year. Later, she tore all the pages on Civil War history from the school books of her grandchildren brought home for she said none of them told the truth."[386]

BRADLEY
(Additional Account)

Thomas Bradley was a neighbor of Martin Rice and the six victims killed by Kansas' troops on September 6, 1863. Bradley along with John Cave, David Hunter and Martin Rice hauled loads of personal effects to the house of Mr. Galloway in Odessa, Lafayette County, Missouri. Bradley and his family were able to escape the Lone Jack community.[387]

BRANNOCK

A. B. Brannock, Cass County, went to Kentucky to escape Order No. 11.[388]

BRANNOCK
(Additional Account)

George W. Brannock, Cass County, went to Jacksonville, Douglas County, Illinois after Order No. 11.[389]

BRISCOE

[386] "Order No. 11," a speech Alberta Wilson Constant delivered to the Independence Young Matrons, ca. 1960, Jackson County (Mo.) Historical Society Archives, Document ID 154.01F30.
[387] "Martin Rice Remembers;" www.historiclonejack.org/whatisaw5.
[388] Rafiner, citing Glenn, Allen. 360.
[389] Rafiner, citing National Historical Company, 534; and Hon Family Papers, Pleasant Hill Historical Society, Pleasant Hill, Missouri.

Ruth Briscoe, Cass County, found safe haven in Cooper County, Missouri when Order No. 11 was issued.[390]

BRITT

Jefferson Waller Britt, Cass County, left his family and went to Texas in 1861 where he remained until spring 1864. In February 1863, the Britt home located in Austin Township, Cass County was burned by Union militia. The family then stayed near Harrisonville until Order No. 11 was issued and they moved to Saline County, Missouri. They remained in Saline County until the War was over and then returned to their home in Cass County.[391]

BRONAUGH

William Bronaugh lived in Cass County and went to Clay County after the issuance of Order No. 11.[392]

BROOKHART

Mr. Brookhart, Cass County, went to Kentucky under Order No. 11.[393]

BROWN

Mary Brown left her farm in Cass County under Order No. 11 and remained at the military post in Harrisonville.[394]

BROWN
(Additional Account)

Robert A. Brown, Cass County, went to Lafayette County after being banished under Order No. 11[395]

---

390 Rafiner, citing National Archives and Record Administration. Washington, D.C. Union Provost Marshals' Papers Relating to Individual Citizens. Record Group 109. Entry M-345, Roll No. 35.

391 Eakin and Hale, 44.

392 Rafiner, citing Family histories compiled by Cass County Historical Society, Harrisonville, Missouri. Another source, National Archives and Records Administration. Washington, D.C. Records of the Quartermaster General. Record Group 92. War Claim: 215-450, noted that he went to Johnson County, Kansas.

393 Rafiner, Family histories compiled by Cass County Historical Society, Harrisonville, Missouri.

394 Rafiner, citing "The History of Wayside Rest." (Harrisonville, Mo.: Cass County Historical Society).

395 Rafiner, citing "The History of Wayside Rest."

BROWNING

The following is the affects of Order No. 11 upon the Browning family who lived at the time near Lee's Summit today:

> *Under Order No. 11, Mr. [O. N.] Browning took refuge in Carrollton and then went to Kentucky, taking a 12-year-old negro girl to accommodate a neighbor. On the bridge at St. Louis, he was interviewed about who, whose, what for, and a multitude of questions such as officers could ask about the negro girl, until Mr. Browning thought all Yankeydom conspired against him, and he would have been glad to emancipate one slave on the spot."*[396]

BULLOCK

Mr. Bullock of Cass County traveled east under Order No. 11 and settled in Pettis County, Missouri.[397]

BURDETTE

Mr. Burdette of Cass County escaped to Kentucky when Order No. 11 was issued.[398]

BURNS

Rice Burns of Cass County left Missouri after Order No. 11 and went to Jacksonville, Illinois.[399]

---

[396] Short Stories of Pioneer Days by Dollie Bee Breitenbaugh, 1959. 94-leave, typed manuscript. Two copies. Jackson County (Mo.) Historical Society Archives, Document ID 80F6. (A brief account of the Battle of Lone Jack is also provided). Willis W. Browning's "grandfather [probably O. N. Browning, though not confirmed at this writing] was evicted from his home here and the log cabin burned as a result of Order No. 11." In 1961, Willis Browning donated to the Jackson County Historical Society an original, signed, engraving of Bingham's "Order No. 11." The name of his grandfather did not get recorded at the time. *Jackson County Historical Society JOURNAL* (June 1961).

[397] Rafiner, citing Glenn, Allen, 452.

[398] Rafiner, obituary.

[399] Rafiner, citing Hon Family Papers, Pleasant Hill Historical Society, Pleasant Hill, Missouri.

BURNS
(Additional Account)

"Mrs. Kate Burns…most vivid memory of her past were during the Civil War when her family 'emptied' the house [503 North Delaware] three times for them to burn it. Each time they were able to talk them out of it until Order No. 11, they evacuated and moved to Westport."[400]

BURNS
(Additional Account)
See ASHLEY

BURRIS

F. M. Burris a resident of Cass County traveled to Texas to escape Order No. 11.[401]

BURRUS

James M. Burrus, Sr. was born in Henry County, Virginia on May 23, 1823. He came to Jackson County in the spring of 1834. Throughout his childhood and adulthood James was a farmer, except from 1859 to 1863 when he operated a saw and gristmill. He married Arminta Kimberlin on August 9, 1859. On account of Order No. 11, he moved his family to Boonville, Missouri and only after the war did they return to their one hundred and twenty acres farm. James and Arminta had seven children, five of whom were still living in 1881: Mattie K., Minnie, George, T. K., and Eva. N. Both T. K. and Eva were under the age of four at the time of Order No. 11.[402]

Mrs. Kate Burrus and her family lived for many years at 215 South Liberty in Independence. On at least three occasions, federal troops forced them to empty their house and threatened to burn it down. Each time they were able to talk the soldiers out of it until Order No. 11 when it was destroyed. After their return to Jackson County, they built a new house in Independence at 503 N. Delaware.[403]

---

[400] Jackson County (Mo.) Historical Society Archives, Document ID Vertical File Newspaper Subject Files: Military/Wars: Civil War: 1863-"Order No. 11": "Untitled," *Independence* (Mo.) *Examiner*, 11 Oct. 1940.
[401] Rafiner. best guess judging from the 1860 and 1870 U. S. Census returns.
[402] Union, 912.
[403] Wilcox, 361, citing *Independence* (Mo.) *Examiner*, September 30, 1927.

BURRUS
(Additional Account)

James M. Burrus, the son of William T. Burrus, was born in Jackson County. His father died before the War and James and his mother went to Lafayette County during Order No. 11. James joined the Confederate Army Missouri State Guard at the age of sixteen and served under Shelby. After his discharge on August 16, 1862, he enlisted in Company D (B?), 2nd Missouri Cavalry under Colonel Upton Hays and served until the end of the War, surrendering at Shreveport, Louisiana. He died at Kansas City at the age of 75 and was buried in Woodlawn Cemetery.[404]

BURRUS
(Additional Account)

Mr. Burrus was killed about two miles south of Drexel, Missouri, and just one-fourth mile inside the Kansas line. He was in the process of carrying out Order No. 11. He was related to the Pedro family.[405]

BYNUM

John Bynum, born in Rutherford County, North Carolina on July 17, 1822, was one of nine children born to his parents. His father Pomfrey Bynum was a native of South Carolina. In May 1836, John's parents along with seventy-five others settled in Jackson County near Lone Jack where his father acquired five hundred and thirty acres (located within Lone Jack today). On September 29, 1842, John Bynum married Charlotte Cave in Jackson County. His wife Charlotte had been born in Surry County, North Carolina on October 6, 1818. Her father was Bartlett Cave who had brought his family to Jackson County in the early years following the county's formation.

John and Charlotte Bynum were forced off their land by Order of No. 11. Six children were born of their marriage: Robert, born March 24, 1845; William, born August 22, 1847 and died at the age of 8; Galen C., born September 29, 1849 and died at the age of 24; James P., born January 8, 1852; Bettie born on October 6, 1855 died in infancy; and John W., born November 3, 1859. They had four children ages 18, 14, 11 and 3 at the time Order No. 11. By 1881, John and Charlotte lived on one hundred and fifty-six acres where he farmed and raised stock; he was said to have had an "excellent orchard." The surviving children all continued to live in the area.[406]

---

404 Eakin and Hale, 55.
405 Eakin and Hale, 55.
406 Union, 931-932.

CABNESS

Mrs. James Cabness was a little girl at the time of Order No. 11. She was part of the sixty-one women and children that went to Texas. Years later she told her recollections of the War stating that every day or two "Yankee soldiers would unload our wagons in search of something to steal. They said they were hunting firearms." She said they continued south subsisting on sweet potatoes and cornbread.[407]

CAGIN

See ASHLEY

CAMPBELL

See ASHLEY

CAMPBELL
(Additional Account)

"Except for the duration of Order No. 11, and the Civil War, when this couple (Courtney and Harriet (Buckner) Campbell), moved to Fayetteville, Arkansas, they lived in their home at 305 South Pleasant Street (in Independence)."[408]

CAMPBELL
(Additional Account)

J. B. Campbell was born January 20, 1822, in Loudoun County, Virginia. He married Rachel A. Grubb in Loudoun County on August 17, 1847. They had eight children: Joseph E., Mary E., Charity R., James R., John B., George G., Sarah R., and Ruth. When Campbell came to Jackson County in 1854 he was poor, but by strict economy he purchased one hundred and forty five acres and farmed forty acres of it. He and his family remained on their farm before being forced to move under Order No. 11. He moved his family to Kansas and settled on the Shawnee Reservation. In 1866 they returned to Jackson County and found their farm in almost as bad a state as it had been when he had first began to work it years earlier. By 1881, he had through "honest industry, and careful attention to his own business" improved his land and "accumulated quite a fortune" owning "1400 acres of fine land."[409]

---

[407] Leslie, 264.

[408] Johnston, Cammie. "The Buckner Family: Miss Johnston Proud of Illustrious Ancestors." *Jackson County Historical Society JOURNAL* (March 1973), 8.

[409] Union, 931-932.

CANINE

Cornelius Canine at the time of the Civil War was living east of Kansas City. His son-in-law Captain Edward West joined Sterling Price at the outbreak of the rebellion and spent the War's duration away from his family. The wife of Edward West and their son George lived with her father Cornelius' after West enlisted. Under Order No. 11 Cornelius moved the extended family to a farm until War's end.[410]

CAPELL

Britton M. Capell was born October 2, 1815 in Richmond County, North Carolina. His parents were natives of that state, and in the year of Britton's birth they moved to Simpson County, Kentucky. Britton was reared and educated in the subscription schools in Kentucky until age seventeen when he began a two- year apprenticeship as a cabinetmaker. Britton followed the trade for five years. He married Sarah A. Clayton on December 16, 1834 who had been born on January 20, 1814 in Logan County, Kentucky where she received a good education.

Britton in the fall of 1839 turned his attention to farming, and he and Sarah immigrated to Jackson County and settled on four hundred and sixty acres in the Sni-A-Bar Township. Britton and Sarah were actively involved in the Methodist Episcopal Church and Britton held positions of steward and class leader for several years in the church. In those early days, the Britton house was used as a church for their congregation. Eight children were born of their marriage, six of them were living in 1881: Bennetter A., Susan R. (later Mrs. Wm. Kirby), Mary M., Martha J. (later Mrs. J. T. Wright), John O. and Trusten P.

When Order No. 11 forced the Britton's off their property, they moved to Lafayette County for two years. After the War, the family returned to their farm. Sarah died on September 9, 1871 and by 1881 Britton's farm was described as "one of the finest farms in the county."[411]

CARLETON

The following story has been told about the disappearance of Ezekiel Carleton around the time of Order No. 11:

> *It so happened on the occasion of one very dark night that Ezekiel Carlton heard a noise that caused him to leave his home and this was the date that the mystery began.... There never was a clue as to what did*

[410] Thruston, 45.
[411] Union, 913 (See also, West below).

happen to...Carlton. It was the unanimous opinion that the noise was made by Kansas raiders and that Carlton was carried back over the line and executed. His widow, [Mary Jane (Smith) Carlton], often mentioned the times when some visitor would be approaching the front of the house. He would see four baby faces at a window looking for a daddy that never returned. The mother mentioned her struggle to raise and keep her small family together under living conditions here in Jackson County during and following the days of the Civil War. She said only friends and good neighbors made it possible for them to exist.[412]

CARTER

Al Carter's family along with the family of "Buck Collins" went to Howard County, Missouri during Order No. 11. They were each 17-years-old when they were killed and scalped by Union soldiers in Franklin, Missouri, on September 12, 1864. Al's father [unknown name] returned to Jackson County.[413]

CARTER
(Additional Account)

James Carter of Cass County went to Johnson County, Missouri under Order No. 11.[414]

CARY

Edmon Cary of Cass County went to Carroll County, Missouri following Order No. 11.[415]

CASEY

Philip Casey, Cass County, went to Miami County, Kansas to escape Order No. 11.[416]

CAVE

[412] George, Todd. "The Mystery of Ezekiel Carlton." *Jackson County Historical Society JOURNAL* (March 1963), 9.
[413] Eakin and Hale, 62-63.
[414] Rafiner's best guess judging from the 1860 and 1870 U. S. Census returns.
[415] Rafiner, citing National Historical Company, 465.
[416] Rafiner.

John S. Cave was born around 1818 in Virginia, and then moved to North Carolina. He married Nancy Hunter on September 9, 1836 in Jackson County. Nancy born in 1819 and was one of nine children born to John and Elizabeth (Jeans) Hunter. In 1842, John Cave was among the original members of the Baptist Church in Lone Jack, and he helped construct the church in 1846.[417] The 1860 census reports that John and Nancy had seven children. The children were Elizabeth Cave Bennett, Martha born in 1845, and Matilda. One of their son's married Amanda Potter Cave and they had a daughter, Lillian L. Cave.[418]

John and Nancy's daughter Elizabeth Cave married Jacob Bennett. Jacob had settled into Jackson County in 1853. Elizabeth and Jacob (and their two children at the time) were forced off their property near Lone Jack for a period of two years as a consequence of Order No. 11 (See, Jacob Bennett Family information above).

Nancy (Hunter) Cave and her children suffered first-hand the horror and personal tragedy of Order No. 11. During the enforcement of the Order, Nancy's husband John Cave and her two brothers John David Hunter and William J. Hunter, along with Benjamin Potter, age 75, William C. Tate, aged 32 and Andrew Ousley, aged 17 were killed by the 9th Kansas Volunteer Cavalry. By 1881, Nancy (Hunter) Cave remained living in the Lone Jack area of Jackson County.

Matilda Cave although having lost her father and uncle on September 6 married a former Union soldier J. K. Reid after the war. Reid was been born in Delaware County, Ohio on July 26, 1843. When the Civil War broke out, he enlisted in Company D, 20th Ohio Infantry. He became sick at Pittsburgh Landing, just after the battle of Shiloh, and was sent home and discharged from the army in the fall of 1862. Reid went to Illinois in the spring of 1864, and farmed one season in Starke County. Reid on October 12, 1864 returned to the Union ranks and enlisted in Company A, 31st Illinois where he served until the close of the War. He participated in the battles of Fort Donelson, Shiloh and many others. He returned to Starke County, Illinois, and engaged in farming. In the spring of 1866, he went to Colorado, and visited most of the Territories, and was engaged in trading with the Indians. Reid settled in the Sni-A-Bar Township of Jackson County in the fall of 1868, where he farmed and raised stock. It was following his settlement in Jackson County that he and Matilda were married. J. K. and Matilda had four children: Mary V., Lucy O., Ella W. and George B.[419]

---

[417] Union, 330.
[418] Eakin, "Tears &Turmoil," 3.
[419] Union, 925.

CHILCOTE

John Chilcote left Cass County and the State of Missouri and moved to Muhuska County, Iowa because of Order No. 11.[420]

CHILES

At the outbreak of hostilities, James "Jim Crow" Chiles was married to Sarah "Sallie" Anne (Young) Chiles. She was the daughter of Solomon and Louisa Young of southern Jackson County. By 1863, Jim Crow had become a bushwhacker serving under Quantrill and was not with his wife and three year old son Elijah Young Chiles at the time of Order No. 11. Sallie and her son were forced off their property under Order No. 11.

Jim Crow returned after the War and he and Sallie had three more children, Suda (Chiles) Wells (1865-1954), William D. Chiles (1868- 1885) and George Hamilton Chiles (1870-1878). Sallie's nephew (the son of her sister, Harriett Young Solomon) would become President of the United States, Harry S. Truman). Sallie died on November 7, 1907 in Jackson County and was buried at Woodlawn Cemetery in Independence. Sallie's husband Jim Crow Chiles and her son Elijah preceded her in death after having been gunned down in 1873 (for more information about the Jim Crow Chiles family see story of Solomon Young family in *Chapter XII, Impact Upon the Family of a Future President*).

Azubah Skinner Chiles lived in the area of Six Mile in northeast Jackson County. She was the mother of H. C. Chiles and Joel Chiles. They were cousins to Henry Clay Chiles (a contractor who bought, sold and hauled freight and cattle for the federal government to forts out west) and Mary Jane Chiles (who attended school in Kentucky at the time of Order No. 11). After Order No. 11, Azubah removed the family to Platte County where they temporarily lived and her son H. C. Chiles stayed with Mrs. Franklin in Plattsburg. H. C. married Mrs. Franklin's daughter Sarah Ann.

Azubah's home was the only one within the community not destroyed. Before she left her home under Order No. 11, Federal troops had come onto her property and began to take most of her personal belongings and made preparations to burn down her house. At that moment, Mrs. Chiles appeared. Many of the men were acquainted with her and new her as the "Good Samaritans." Azubah had nursed the wife of one of the soldier's who had suffered from a sever fever. Azubah helped a physician deliver a baby of the wife of another member of the troops. She approached the man and asked about the well being of his baby. She then turned to another man and inquired whether he had enjoyed the meat and vegetables she had sent to him. Azubah held similar discussions with other

[420] Rafiner's best guess judging from the 1860 and 1870 U. S. Census returns.

men in the ranks. After she had spoken to the men, they left her property with heads hanging low in shame. They removed nothing from the property and did not strike a match to the house.[421]

CHIPMAN

See ASHLEY

CHOUTEAU

Cyperian Chouteau married Nancy (Francis) at the old Shawnee Mission in 1851, there years before the land became known as the Kansas Territory. Nancy had lived at the Indian mission with her parents since she was eight years old. They had immigrated to the mission from Wapakoneta, Ohio. After their marriage "[t]he Chouteau's began their residence on a farm near what is at present 24th Street and Kensington Avenue [Kansas City, Missouri]. They lived there ten years, but soon after the Civil War began they were compelled to move into 'the city'" where they resided "on the old 'Jarboe Hill' at 5th and Charlotte Streets." Whenever Kansas' soldiers passed their farm, they never failed to disturb them in all possible ways, taking horses, mules and provisions. "Order No. 11 also disturbed their residence on the farm. Mr. and Mrs. Chouteau remained at the old home place until his death in 1879." Mrs. Chouteau died February 6, 1912, at the home of her daughter, Mrs. Carl Guinotte, 3709 East 9th Street, Kansas City, and she was buried in Mount St. Mary's Cemetery.[422]

CLARK

David M. Clark, from Richmond, Virginia, came to Missouri in 1856. "During the War the family had to move to Greenton [Odessa, Lafayette County, Missouri] and remained there during Order No. 11."[423]

CLARKSON

Virginia "Jennie" (St. Clair) Clarkson was married to Alfred Clarkson. They lived in the Pink Hill area of eastern Jackson County. During Order No. 11, she watched federal troops march through the township burning most of the homes and removing the

---

[421] Eakin, Joanne Chiles. *Blue & Grey Chronicle*. (April 1999): 2: 4: 5; and, Wilcox, 364, citing *Zion's Ensign*, 31 Oct. 1901).

[422] "Mrs. Nancy Chouteau Dead," *Kansas City* (Mo.) *Star*, 7 Feb 1912. Chouteau's Missouri death certificate.

[423] Eakin and Hale, 76, referencing Hickman and *Independence* (Mo.) *Examiner*, 4 Oct. 1905.

citizens. She wrote that as she was standing on her front porch she was able to count "13 chimneys, all that was left of 13 homes."[424]

CLEMENTS

Following the issuance of Order No. 11, Mrs. Milton Clements reported to the post headquarters located in a house in Hickman Mills. The house had a window that overlooked a front porch. It was at that window citizens were ordered to take the loyalty oath. When Mrs. Clements approached the window, she told Union officials that she was a Union woman who did not have any relatives in the Confederate army. Although her statements were untrue, she did take the loyalty oath because "she was alone and filled with fright.[425]

CLEMENTS
(Additional Account)

> *In Civil War days, Hickman Mills was Post No. 1 under General Order No. 11 issued by the Federal commander which required rural folk to move into federally controlled areas, or leave the country. There were 32 families living in two large residences of C[harles] P[eter Beauchamp] Jefferys named 'Hawthorne' and 'Blythesdale.' Hawthorne was located at what is now (1965) the southwest corner of Craig and Blue Ridge Boulevard [another source said it was where the first Ruskin High School was later erected]. Blythesdale was located about five hundred feet north of where Orchard and Longview Roads intersect [modern-day Stratford Estates subdivision]. The Federal troops occupied the church property. Pameila Scrivner who was then 59-years-old arranged to stay at this post and she served as the cook for the Federal officers. Her daughter, Mrs.* [John] *Ambrose* [Hannah H. (Scrivner)] *Clements, with her three small children on leaving the farm went into Kansas City and stayed with Mrs. Solomon Young* [Harry Truman's grandparents] *until she could get boat passage to Kentucky. She left the two oldest children with her grandfather, Joseph Scrivner, and she with her baby stayed with her other grandfather, Zachariah Clements, at Lockport, Kentucky....*
>
> *When a band of raiders came plundering and burning homes, the home across the road from [James Harvey and Berilda (Bryant) Kemper] home was burned and Henry Rout's home a half a mile south of the*

[424] Jackson, David W. "A History of Pink Hill Park and Veterans Way Memorial." (Independence, Mo.: Jackson County Historical Society, 2013), 14.
[425]Eakin, "Tears and Turmoil," 32-32.

Kemper home was also burned but the Kemper home was spared. The family thought this happened because Mrs. Kemper was sitting in the yard and was wearing a Masonic emblem. As they looked, they counted 19 houses burning. The Kemper family were moved to the Hickman Mills post where they lived during the enforcement of Order No. 11.[426]

CLEMENTS
(Additional Account)

Mr. Clements of Cass County went to Higginsville and lived with two other families.[427]

COGSWELL

Oscar H. Cogswell had been born in Vermont. His grandson gave the following account of the affect of the border war on his grandparents:

> *My grandfather father was a strong 'Union Man,' though he did own slaves. Loyalty to him and a strong desire to do nothing to increase the distrust and hatred of the times caused [his wife, Nancy Brawner "Nannie" (Rogers) Cogswell] her never to take sides verbally.... In spite of my grandfather's devotion to the Union, 'The Wah' ruined him, destroying the fortune he had made in the gold fields and hastening his death. Federal troops maddened by battle losses rode into town in 1863 and wrecked the fine store grandfather had built. The fact that he was a Vermonter and a Union man made no difference to these men. He had married a Virginia woman and that was all the excuse that they needed. They stole thousands of dollars worth of Brussels carpeting which they used for saddle blankets. After helping themselves to additional thousands of dollars worth of various kinds of stock they smashed everything that would break, scattered everything else, poured a barrel of molasses over it and tramped it into hopeless ruin.... Grandmother continued to live in what she always referred to as 'The Old House,' at*

---

[426] An unpublished article titled, "A Panoramic View of the Lost Townships," by Mrs. Greenberry (Jesse Crosby) Ragan, ca. 1965, Jackson County (Mo.) Historical Society Archives, Document ID 68F25. See also Mrs. Ragan's article in the July 1965 edition of the *Jackson County Historical Society JOURNAL*; "Civil War Ordeal for Hickman Mills," *Kansas City* (Mo.) *Times*, 24 Aug. 1969; Eakin, "Tears and Turmoil," 31. Burton, Angela M., and Hannah Clements Montgomery. "Order No. 11 in Hickman Mills." *Kansas City Genealogist;* 24: 1: 16.
[427] Rafiner, citing Glenn, Allen, 464; obituary.

311 South Main Street (Independence, Missouri) until her death in 1919.[428]

COLLINS

The families of Buck Collins and Al Carter went to Howard County, Missouri. They were only seventeen years old when they were both killed and scalped by Union soldiers in Franklin, Missouri, on September 12, 1864.[429]

COMPTON

James Compton was the father of five-year old George L. Compton. James gathered as much of the family belongings as he could onto a big wagon under Order No. 11. The children were placed on top to ride while James walked alongside the wagon. During their exodus, Union soldiers threatened the lives of James entire family, even threatening to hang them.[430] "The little son, John Compton, heard the threats and oaths against his family that always remained fresh in his memory, as a business man of Independence." In later years, George often related the harrowing experiences he and his family faced at the hands of federal troops under Order Number 11.[431]

CONNOR

John Connor, Harrison Township, Vernon County, was killed by Jim James in the winter 1863.[432]

COOPER

A short vignette of Jackson County pioneers, with emphasis on Lee's Summit families, includes the affects of Order No. 11 on the Cooper family:

---

[428] Oscar H. Cogswell died in 1881. Cogswell, Burdette. "Aunt Nannie [Nancy Brawner (Rogers)] Cogswell's Life Colorful." *Jackson County Historical Society JOURNAL* (Fall 1966), 3-4, 8. Eakin, "Tears and Turmoil," 78.

[429] Eakin and Hale, 87.

[430] Wilcox, 363-364, citing *Independence (Mo.) Examiner*, January 4, 1951.

[431] Wilcox, 363-364, citing *Independence (Mo.) Examiner*, January 4, 1951. Also, an undated, typewritten recollection (which may have come from the *Examiner* article), Jackson County (Mo.) Historical Society Archives, Document ID Vertical File Newspaper Subject Files: Military/Wars: Civil War: 1863-"Order No. 11."

[432] http://freepages.genealogy.rootsweb.ancestry.com/~jrbakerjr/provost3/bushwhackers.htm (viewed 10 June 2013).

I don't like to tell war stories," said Miss Roe Cooper [daughter of Z. G. Cooper]," the war is over; it's one country." Lincoln was a great man and would have paid for every slave if he had lived. Many things were due to the Kansas and Missouri feud. It was thought that the women acted as spies, for, when our fathers came home, soldiers would come and rob. The abolitionists came, apparently to buy apples, and stole slaves, either to free or sell again.

"In October 1863, we were robbed of four negroes and 21 horses. Dr. [Pleasant Graves] Lea recovered one mule for us, and we had only one gray horse. We had a chore boy named Joe who helped Aunt bury the bed clothes, the china and the silver in the hen house, the garden and orchard. Aunt was wise and moved the stump marking the silver to another row of trees, so, when the lad came back with the Kansas Red Legs, to rob us and all the neighbors for whom he had worked, he dug in vain for our silver. He took our bed clothes, dresses, the baby's clothes, and father's spurs before our eyes.

"One stormy night in February, 1863, the first night mother had undressed for weeks, Penick [sic.] sent soldiers from Independence to burn houses. They said it was a shame to turn us out, but such were the orders. Then they helped mother to dress the children and set out the furniture in the carriage house and fixed a shelter. Billy Jones saw the flames and came and took us to his home. Fourteen houses were burned that night.

"Mother had fixed up the carriage house for us when an order came to leave the country. We departed in a wagon with ox team taking all we could with us, for Waverly. Going down a steep hill the pin holding the yoke fell out, the wagon raced down a steep hill through the woods across a sapling, while the ox enjoyed a drink....

"Three months we spent in Waverly, then went to Kentucky as refugees, returning to Missouri in 1865, August, by train as far as Pleasant Hill, as no more track had been laid then. The section house was at the Parry farm.[433]

---

[433] Short Stories of Pioneer Days by Dollie Bee Breitenbaugh, 1959. 94-leave, typed manuscript. Two copies. Jackson County (Mo.) Historical Society Archives, Document ID 80F6. If the incident relating to the Dr. Pleasant Lea occurred in October 1863, the author is either mistaken about the month and year or the identity of Dr. Lea as the benefactor. Kansas Jayhawkers' murdered Dr. Lea on September 12, 1862, near what is today Lee's Summit.

CORDELL

"Mr. Henry Cordell, an elder in the Presbyterian Church at Pleasant Hill, walked with his family to Jefferson City. He had only a wagon containing household effects pulled by two yearling calves, the only animals missed by marauding forces under Jennison and Lane." [434]

CORN

Martin Corn was born on a farm on May 27, 1813 in Patrick County, Virginia. At age twenty, Martin and his parents moved to Jackson County and settled in Van Buren Township. In 1840 Martin married Martha Cummings who had been born in Missouri in 1823. Martin and Martha were members of the Baptist Church and they had eleven children, six of whom were still living in 1881: Samuel, George, W., Susan F. (later Susan Brown); Nancy E. (later Nancy Temple); John S.; and Louisa (later Louisa Bale).

Following Order No. 11, Martin moved his family to Lafayette County where they lived for two years. They returned to their farm near Lone Jack where Martin resumed his profession as a farmer in Section 25. Martin was described as "pleasant and agreeable" and respected by "a large circle of friends."[435]

COWHERD

Francis K. Cowherd, wife and children lived in eastern Jackson County. They were strong southern supporters and slave owners. When Red Legs came to their property during Order No. 11, they burned the granary, which held three thousand bushels of wheat. The soldiers told his wife to take the children and "go." The troops ordered the slaves to take from the house what they wanted. The loyal Cowherd slaves took only some children clothing and the family Bible. Later the slaves secretly returned the items to the Cowherd's. The Red Legs loaded the family furniture in wagons, except for the sewing machine, which was destroyed. They concealed their silver in a bag and hung it

---

[434] 1961 article about the Jackson County Historical Society-sponsored exhibit of Bingham paintings at the Nelson Atkins Art Museum. Also, *Kansas City* (Mo.) *Times*, 24 Aug. 1963, "Order No. 11, Crux of Civil War Bitterness." And, a third article, "Order No. 11 Drove Families from Border Homes in 1863," Kansas City (Mo.) Star, 9 Apr 1961, lists three families affected by Order No. 11. Jackson County (Mo.) Historical Society Archives, Document ID 144.01F33. Eakin, Joanne Chiles. "General Order Number Eleven." (First of the Series on Order No. 11*). Kansas City Genealogist;* 23: 1&2: 11-12. See also MILLER and POYNTZ in this enumeration. Rafiner, citing Miller, George. "Missouri's Memorable Decade: 1860-1870." (Columbia, Mo: Press of E. W. Stephens. 1898). 105; and, Cass County Historical Society. "Cass County Missouri Families." (Harrisonville, Mo: Cass County Historical Society, Inc., 1976).

[435] Union, 933.

into a tree among the leaves. Unfortunately, a piece of metal caught the sun and the silver went into the hands of the plunderers'.[436]

> *When the dreaded Kansans arrived in war times, they burned the granary with 3,000 bushels of wheat and told [the wife of] Mr. Francis K. Cowherd to take the children and go. The Negroes were told to take what they wanted from the house. But, the Negroes, being loyal, selected the children's clothes and the family Bible, which they secretly returned to Mrs. Cowherd. Then the Kansans loaded the furniture into wagons all except the sewing machine, expensive in those days, was too large and cumbersome, so it was chopped up in the yard.*
>
> *The family had put their silver in a bag and hung it in a tree among the leaves for safety, but a piece of metal in the bag caught the sunlight, and the silver went into the plunderers' hands. Though Mr. Cowherd was not in war service, three times he had a rope around his neck to be hanged, and three times he talked himself to freedom. In 1849, Mr. Cowherd went to California and returned with a small amount of gold from which he had a ring of unusual design made. This was lost with everything else, and Mr. Cowherd did not dare speak about it when he saw a man in Independence wearing the ring, for he was afraid of being shot on the street.*
>
> *Under Order No. 11, the Cowherd family went to Colorado....*[437]

CRAWFORD

A Mr. Crawford was affected by Order No. 11.[438]

CRENSHAW

Mr. Crenshaw of Cass County went to Kentucky after General Ewing issued Order No. 11.[439]

---

[436] Wilcox, 360, citing *Jackson* (Independence, Mo.) *Examiner*, 28 July 1905, and *Independence* (Mo.) *Examiner*, November 6, 1922.

[437] Short Stories of Pioneer Days by Dollie Bee Breitenbaugh, 1959. 94-leave, typed manuscript. Two copies. Jackson County (Mo.) Historical Society Archives, Document ID 80F6. This incident was also published in the Lee's Summit Journal, 18 Apr. 1929, publishing an interview given by Mrs. E. H. Graves, granddaughter to Dolly Breitenbaugh.

[438] Bartels, 151.

[439] Rafiner, citing National Historical Company, 683.

CROWTHER

Benjamin and Minerva Ann Lilly Crowther settled in the Sibley area of Jackson County sometime between 1853 and 1856. "His family was evicted from Jackson County by Order No. 11 and his wife and children were banished from Missouri…."[440]

CUSENBERRY

See ASHLEY

DALTON

See LUTTRELL

DANIELS

John Daniels and his family were pioneers of Jackson County having lived near Lone Jack since 1830. Before being forced off their land under Order No. 11, Kansas Red Legs came onto their farm, slaughtered chickens with their sabers, ordered the Daniels to dress and cook meals for the troops, and were compelled to feed the soldiers for two days. At the same time, they counted twenty-one buildings burning in their neighborhood.[441]

> *With Order No. 11 the [John] Daniels family went to Lafayette County. All was packed up to go and as they were closing the gate, two men rode up for meals and would not move an inch until they got them. "The family's provisions were given to the men, and, while they were eating, six guns fired some distance away. One killed the father of William Cave and others accused of having been in the Lawrence raids. Immediately, the men ran their horses and camped in the woods near Chapel Hill.*
>
> *On the trip to Lafayette County the Daniels family was not permitted to water their stock for a distance of 10 miles. Hank Gibbons led one horse to drink with Mrs. Daniels on it. Soldiers had taken the carriage, so the family had only one wagon, an ox team hitched to a buggy was also enroute. The winter was very cold; the family had but two rooms,*

---

[440] Eakin and Hale, 98-99.
[441] Wilcox, 359.

one large and one small, and one of the boys had typhoid fever. The Daniels family returned to their home in March.[442]

DARBY

John Darby was born on May 7, 1836 near the Oak Grove area of Jackson County. John's formal education was limited but he did receive some schooling at home. On April 29, 1858, he married Mary Jane Austin who was born in the same vicinity. Mary Jane was the daughter of Robert and Eliza Jane (Watkins) Austin. John learned how to farm and operate a milling business from his father. He served in the Union Army for six months and then returned home to his family. After Order No. 11, John took his wife and small children to Platte County out of fear of Quantrill and his men. After the War, John sold his interest in the mill and farm and bought a store and engaged in raising stock. He was appointed as postmaster for Oak Grove during the administration of President Grant—a post he held for well over eleven years. He and his Mary Jane had nine children: R. Belle, Samuel T., Robert Winfield, Eliza J., William T., Alice, Mittie, Bessie and Bertha.[443]

DAVENPORT

William Davenport of Cass County reported to the military post at Harrisonville and went to Cole County, Missouri to escape Order No. 11.[444]

DAVIDSON

J. N. Davidson, Cass County, described drawing lots to eat. His wife, Eliza, tended a "squalling baby." They went to Boone County, Missouri to avoid conditions under Order No. 11.[445]

---

[442] Short Stories of Pioneer Days by Dollie Bee Breitenbaugh, 1959. 94-leave, typed manuscript. Two copies. Jackson County (Mo.) Historical Society Archives, Document ID 80F6. A brief description of the Battle of Lone Jack is provided in the materials.

[443] Union, 915.

[444] Rafiner, citing National Archives and Records Administration. Washington, D.C. Records of the Quartermaster General. Record Group 92. War Claim: M-1213.

[445] Rafiner, citing National Archives and Records Administration. Washington, D.C. Records of the Quartermaster General. Record Group 92. War Claim: L-3292; See National Historical Company, 684; also, Harrisonville Heritage Association. The History of Harrisonville. (Harrisonville, Mo.: Harrisonville Heritage Association. 1988). 125.

DAVIS

James Davis of Cass County traveled to Kentucky after Ewing decreed Order No. 11.[446]

DAVIS
(Additional Account)

John Davis who resided in Cass County went to Johnson County, Missouri after Order No. 11.[447]

DEHONEY

Leander M. DeHoney was born on October 28, 1826 in Scott County, Kentucky. His father Thomas brought the family to Jackson County in 1849 and purchased one hundred and twenty acres homestead near Raytown for five hundred and twenty dollars. When Leander's father died in 1853, he bought from his siblings their shares of their late father's farm. Leander erected a "substantial home" to replace the log cabin built by his father. Their house was within two miles east of Raytown. Leander built their home out of hewed logs from walnut trees that were said to have been one hundred years old.

In 1854 Leander married Miss Mary Chism. At the time of Order No. 11, they had two children, Howard and Lisa (Lisa lived until 1922). They would have three more sons and two more daughters after the issuance of the Order: Leander, Jr., Robert, Buford born in 1874, Carrie and Cora (his daughter Carrie married Will Robinson, a nephew of Parson J. J. Robinson of Raytown).

From Leander's front yard, he saw about twenty-four houses burning near the Blue Springs during Order No. 11. Leander's family also experienced the wrath of the arsonists. Some twenty-five to thirty mounted, guerilla Red Legs came to the house demanding to be fed. Leander insisted the Red Legs had come from the direction of the burning houses with wagons loaded with spoils, driving before them horses and cattle. The captain of the Red Legs told Leander they had burned every house east of their property and had orders to burn down his house as well. Leander offered food and rest to the captain and his men. It is believed that Leander's treatment of the Red Legs and his membership in the Masons saved his property from ruin.

The family left Raytown by Order No. 11 taking with them two horses, Sam and Jim. Leander placed his two children into an old farm wagon. The DeHoney's traveled with their neighbors, W. Creighton Thruston family, to Clay County. They took only

---

446 Rafiner, citing National Historical Company. "History of Cass and Bates Counties, Missouri." (St. Joseph, Mo.: National Historical Company, 1883), 643.

447 Rafiner's best assessment, judging from the 1860 and 1870 U. S. Census returns.

what crop and livestock they could gather together in such extreme haste. The hogs they were unable to capture were shot for food. They returned to their Raytown home after the War.

In 1874, Leander was elected county assessor. He also held for about forty years the office of justice of the peace for Brooking Township. Leander received the honorary title of "Squire" in recognition to his public service. He died on August 20, 1910 in the same house in which he had built after his father's death in 1853. All of the DeHoney children remained life-long residents in Raytown and became leading citizens with the community.[448]

DEHONEY
(Additonal Account)

Simeon Dehoney, Cass County, went to Clay County, Missouri under Order No. 11.[449]

DELEVAN

Alfred Delevan left Cass County following the issuance of Order No. 11 and chose to take refuge in Iowa.[450]

DEWITT

Past president of the Jackson County Historical Society, Mary Mildred Zick DeWitt, mentioned, "Forty some years ago (ca. 1924), I had the privilege of knowing Mrs. Daniel B. DeWitt, to me, Grandmother DeWitt. She lived on the R. D. Mize Road just east of Little Blue. This was only a part of the farm that was her home before Order No. 11."

> *"When General Ewing gave that Order her home was burned. She left in a dilapidated wagon drawn by a horse that was old and slow, taking with her family and one or two keepsakes.... The DeWitt family went to Illinois, returning to Jackson County to rebuild their home and raise their*

---

[448] Wilcox, 359, citing *Kansas City Star*, 7 Aug. 1932; Thruston, 21-25; and *Jackson* (Independence, Mo.) *Examiner* 5 May 1905.

[449] Rafiner, citing Belton Star-Herald. "The First Hundred Years: Belton, Missouri 1872-1972." (Belton, Mo.: Belton Star-Herald, 1972).

[450] Rafiner, citing National Archives and Records Administration. Washington, D.C. Records of the Quartermaster General. Record Group 393: Entry 3107.

family.... Grandmother DeWitt lived to be 94 and is buried in Lobb Cemetery near her home. I never knew her husband, Daniel DeWitt."[451]

DILLINGHAM

C. J. Dillingham was born in Jackson County on November 2, 1847. He received his education in the common schools of Jackson County and became a farmer and owned ninety acres in the Sni-A-Bar Township (near Blue Springs today). Following Order No. 11, he stayed in Clay County through 1864 when he returned to his farm. On November 19, 1868, he married Miss Sarah Luttrell, also a native of Jackson County whose father Samuel Luttrell and family had also been forced off their property. They had three children: Lulu L., Willis J. and Annie M. [452]

DILLINGHAM
(Additonal Account)
See LUTTRELL

DOLAN

James Dolan was forced to leave Cass County after Order No. 11 was issued and sought safe haven in Calloway County, Missouri.[453]

DONAHUE[454]
See ASHLEY

DONEGHY

Between the birth of their sons between 1855 and 1857, James and Catherine Sophia "Kate" (Campbell) Doneghy moved from Kentucky to Missouri. Kate, born February 12, 1830, was the daughter of James and Kate (Bradshaw) Campbell. After establishing their home in Jackson County, they experienced numerous tragedies during the Civil War. Their first encounter with Federal troops occurred in February 1862.

During the Civil War, in the month of Feb. 1862, between the 12 and 18 days, and about 8 o'clock at night, no one at home but myself, and children (6

---

[451] "Story Hour Manuscripts," by the Independence Young Matrons from 1964. "Uprooting of Families of Order No. 11," by Jackson County (Mo.) Historical Society Archives, Document ID 66F40.
[452] Union at 915, (See also, Luttrell Story in *Chapter XII* and below).
[453] Rafiner, citing Glenn, Allen, 381.
[454] Possibly James Donohue. *Pioneer Wagon;* 12: 3, as referenced in Eakin and Hale, 115.

> *little boys) 50 Federal soldiers entered the house saying, "they were determined to burn the house, they came for that purpose," "as we were Rebels." They were in full uniform and on the guns and pistols. My husband was then in the South as no Southern man was safe in his home at that time in Jackson Co. Mo. This Bible lay on a small table, its usual place, and in evenings I read to my children from its pages. It was in its customary place, when the soldiers entered, wild with vengeance. The little boys standing back, not knowing what to do, only to be still. At last the Capt roared, "take out, we are determined to burn," with that my son Alexander Doneghy just 7 years old, exclaimed at the top of his child voice, "Commander I will take out my mother's family Bible, I know she don't want to lose it," and running to the table, he clasped this book in his little arms ready to take it out. The scene was more than the enraged soldier could stand, the child having clasped the book. The cocked pistols fell to their sides, and heads bowed with humility. Not a word was spoken for some minutes, when all at once the Capt said, 'Come boys lets go.' There was no burning done that night. They were conquered by the powers of the Bible: 'A little child shall lead them.'*[455]

Later in August that same year, Kate's husband James was killed in the Battle of Lone Jack. He left a widow and six children (James C., born ca. 1850; John T., 1852; Alexander, 1855; Abraham J. 1857; Herbert, 1859; and, H. Weitman, 1861). Under the terms of James' Last Will and Testament he bequeathed the five hundred acre farm to his widow Kate (that would later become part of Grandview, Jackson County). The "widow and family, with their slaves, were driven from the home along with other settlers as a result of the famous Order No. 11."

In 1865, Mrs. Doneghy sold the entire tract of land she had inherited under the terms of her husband's Will. She sold the land for the sum of $8,000. Mrs. Kate Doneghy died on June 24, 1914, and was buried in Oakwood Cemetery, Macon, Macon County, Missouri.[456]

Controversy over the land sold by Mrs. Doneghy erupted fifty years later in 1915. Her son, James Doneghy, an attorney in Kirksville, Missouri, sued the new owner of the property for $250,000. The lawsuit asserted that his mother, "the widow had no right to sell the property, but only to hold it as a dower (or lifetime title) right." In other words, it was the family position that the only interest their late mother could have sold in the farm

---

[455] Genealogical material relating to Miss Cammie Johnston, focusing on the Doneghy family. Included are copies from the Holy Bible of Mrs. James (Kate S.) Doneghy, where she related the story, Jackson County (Mo.) Historical Society Archives Document ID L62F3. Also interpreted in "Kansas Citian Kate Doneghy Performed Her Own Version of 'Damn Yankees' More Than a Century Before it Hit Broadway," *Presidential Pie Eating Contests and Lesser-Known Facts from the Jackson County Historical Society*, organizational promotional brochure, ca. 2007.

[456] Findagrave.com and Doneghy's Missouri death certificate.

was her life interest in the property and that upon her death the land and all improvements thereon reverted to the children of his late father—such was the law of dower. The most interesting witness in the suit was John P. St. John, former governor of Kansas, who had been a resident of Independence during the Civil War, and who had actually executed the deed by which Mrs. Doneghy disposed of the tract.

DOUGAL

See ASHLEY

DOUGLASS

William I. Douglass was born in 1825 in Madison County, Kentucky. He came to Jackson County with his parents in October 1835. In 1846 Douglass traveled to Mexico, and by 1849 he was in California where he spent about four years. When he returned to Jackson County, he married Miss Milda J. Sanders in 1855. Milda was born in Kentucky on December 5, 1833. They had one child, Thaddeus T. and they lived near Lake City, Ft. Osage Township, in Jackson County. Under Order No. 11, Douglass removed his family to Ray County where they remained for eighteen months before returning to their property. By 1881, Douglass owned two hundred acres of land, part of which was improved. The Douglass family was members of the Baptist Church. Douglass was described as being one of the oldest pioneers in the county, and had the respect of a large circle of acquaintances.[457]

DUNCAN

"Mrs. Rev. Pryor (Louisa Lucretia Alley) Duncan…who was born and reared on the site of Grain Valley, and who under Order No. 11 during the Civil War made a long trek to Oklahoma on foot, died this morning [April 15, 1940] at her home, five miles south of Grain Valley.... Her father, Major Alley, took his family to Oklahoma during the war and Mrs. Duncan, being strong and young, walked while some of the older persons in the caravan rode."[458]

---

[457] Union, 901.

[458] Jackson County (Mo.) Historical Society Archives, Document ID Vertical File Newspaper Subject Files: Military/Wars: Civil War: 1863-"Order No. 11": "Her Life Span Would Fill Book of County History," *Independence* (Mo.) *Examiner*, 15 Apr. 1940.

DUNCAN
(Additional Account)

Charles Thomas Duncan surrendered at Schreveport, Louisiana, in 1865 and went to Texas to reunite with his family, which had been banished as a result of Order No. 11. Two years later, he returned to Jackson County where he lived for 10 years south of Oak Grove. He is buried in the Duncan Cemetery.[459]

DUNCAN
(Additional Account)

John Duncan was arrested in Vernon County by the 3rd Wisconsin Cavalry, August 19, 1863, and sent to Fort Scott to stand trial.[460]

DUNHAM

Mr. Dunham a resident of Cass County went to Johnson County during Order No. 11.[461]

DUNN

Absolom Dunn who was living in Cass County at the time of Order No. 11 went to Illinois.[462]

DUVALL

W. A. Duvall who was five-years old at the time of Order No. 11 related his memories and family saga:

> *At the time General Ewing issued Order No. 11, I [W. A. Duvall] was about five years old. We were living on the family farm consisting of 640-acres in Cass County, two and a half miles south east of Gunn City. The family consisted of my father, Thomas Duvall, mother Carrie (O'Bryan) Duvall and brothers and sisters, Mattie, Crit, Flora and myself. We had a wagon with a cow and an ox hitched to it. We had no horses as I*

---

459 Eakin and Hale, 121.

460 http://freepages.genealogy.rootsweb.ancestry.com/~jrbakerjr/provost3/bushwhackers.htm (viewed 10 June 2013).

461 Rafiner, citing Glenn, Allen, 825.

462 Rafiner, citing Glenn, Allen, 561.

recall, as every one worked cattle at that time. We packed our bed clothes and cooking outfit in the wagon and left our house and belongings in the care of an old neighbor named Tackett.

They had my father in jail two or three days trying to get him to sign the Oath of Allegiance, but he refused to do so and was ordered to leave the country.

We left 80 head of hogs, hay in the stacks, and grain in the fields. The grain was afterwards burned in the field and not a rail left on the fences. We stayed in what was called a Kentucky lynch-pin wagon.

After about a week traveling, most of the time we children walking behind the wagon, we reached Cooper County. We rented a farm from a man named Goley, two or three miles south of Rocheport on the Missouri River. Ten miles east of Booneville. We lived close to the O'Bryans, my mother's parents. We raised tobacco the first year and I can remember hauling it across the River on the ice to Rocheport in the winter.

After being away three years, we returned to Cass County. My father had three negro slaves, an old woman and two children (boys) by the name of Walker. My Grandmother O'Bryan had about 30 slaves. She lived 10 miles east of Boonville. My father was a Southern sympathizer, but on account of disability, never served in the Army.[463]

DYE

Isaac Dye from his home in Cass County drove chickens and geese north to Clay County to escape Order No. 11.[464]

EASLEY

Achilles Easley of Cass County first went to Lafayette County to escape Order No. 11 before settling in Linn County, Missouri.[465]

[463] Duvall genealogy, and a 22 Aug 1927 affidavit by W. A. Duvall recollecting Order No. 11. Jackson County (Mo.) Historical Society Archives, Document ID 121F38.

[464] Rafiner, citing Belton Star-Herald. "The First Hundred Years: Belton, Missouri 1872-1972." (Belton, Mo.: Belton Star-Herald, 1972).

[465] Rafiner, citing "History of Peculiar, Missouri."

EASLEY
(Additional Account)

The following is an account of the impact of Order No. 11 upon the Easley family of eastern Jackson County:

> *With Order No. 11, Mr. [Miller W.] Easley turned his hogs into a fine crop of corn, and the family went to Ray County. With a permit he came back and drove some of the hogs over for their winter's meat. In the spring, the Easley family returned to find fences burned and no corn in the country for hogs. Then, Stephen Easley drove an ox team and the hogs to Lexington, fed them and brought them back. He returned with the hogs, sold some and peddled the rest in Pleasant Hill.*[466]

ELDER

Thomas "Bud" Elder, Bates County, was in Vernon County with Marchbank' when killed in the Stockton raid, late 1863.[467]

ELLEZ
See RUSSEL

Mr. and Mrs. A. C. Ellez resided near the Russel family in eastern Jackson County. Both families left their properties during Order No. 11. The Ellez family ultimately settled permanently in the area of West Line, Cass County.[468]

ESTES

William Estes of Cass County went to Clay County after Order No. 11.[469]

---

[466] Short Stories of Pioneer Days by Dollie Bee Breitenbaugh, 1959. 94-leave, typed manuscript. Two copies. Jackson County (Mo.) Historical Society Archives, Document ID 80F6.

[467] http://freepages.genealogy.rootsweb.ancestry.com/~jrbakerjr/provost3/bushwhackers.htm (viewed 10 June 2013).

[468] "As to the Battle of Lone Jack," *Kansas City* (Mo.) *Star*, 26 April 1911. And, Russel's Missouri death certificate.

[469] Rafiner, citing National Archives and Records Administration. Washington, D.C., Records of the Quartermaster General. Record Group 92. War Claim: N-1422.

FARMER

Henry Farmer traveled from his home in Cass County and reported to the military post at Pleasant Hill during Order No. 11.[470]

FARNSWORTH

C. L. Farnsworth of Cass County escaped Order No. 11 by relocating to Johnson County, Missouri.[471]

FERGUSON

See ASHLEY

FIELDS

Martha Frances (Chrisman) Fields was married to Christopher Columbus Fields. Field's was the son of William and Amanda (Chiles) Fields. He enlisted in the Confederate Army under General Sterling Price. Because of Order No. 11, Martha went to Cooper County, Missouri where she stayed until returning home in 1864. She lived with her mother, Elizabeth (Yeary) Chrisman until her husband returned home after the War.[472] Martha described their experience under Order No. 11.

> *[We} put some of her smaller things in a box and buried them out in the orchard.... They went to Cooper County. Upon their return, they found their house, barn, all the outbuildings and rail fence burned. Some of the fruit trees had been cut down, but the tree near where the box was buried still stood. The box was dug up [and its contents recovered]. The Fields farm was located on Holke Road across the road from the Fairmount Methodist Church.*[473]

FILLEY

See ASHLEY

[470] Rafiner, citing Glenn, Allen, 346.
[471] Rafiner, citing "History of Johnson County, Missouri," 844.
[472] Eakin, "Tears and Turmoil," 49.
[473] "Story Hour Manuscripts," by the Independence Young Matrons from 1964. "Uprooting of Families of Order No. 11." Jackson County (Mo.) Historical Society Archives, Document ID 66F40.

FLANERY

Zion Flanery and his wife Lucinda had eight children born of their marriage: Martin Van Buren Flanery, James S., Rebecca J., Melvina W., John W., Mac P., George Z. and Horace W. Zion was about forty nine at the time of Order No. 11. Their son Martin Van Buren Flanery had married Laura Harris before the rebellion and they had a three-year old son William V. Zion. The entire Zion Flanery and extended family frequently suffered at the hands of Kansas Jayhawkers. (See *Chapter XII* for story of Flanery and Harris families).

Years later, Laura Harris Flanery related to her son William V. the story of the destruction of their property on October 27, 1861.

> *Before the church was fired, the soldiers took the Bible out and placed it securely on a stump, away from the flames. After the battle eleven of us girls gave first aid to the wounded and dying soldiers. Southern people after the battle of White Oak drove the federal troops into Kansas City, killing all but 23. Captain Washington Wells and Captain John Williams were the only two rebels killed in this engagement.*

William had only limited memories of his father, as he was a young boy when his father was killed in 1864. After Order No. 11 was issued Laura Harris Flanery did not go with her father-in-law Zion and his family to Saline County. Instead she took Little William along with her sister Mrs. Julia Irwin and twenty-three families whose fathers, husbands and sons were in the rebel army and "set out eight days after the order was issued" under the guidance of the Muirs. As they left their home, Laura loaded into one wagon one hoarded feather bed, two pair of pillows and one small trunk that she described as "a prized object brought from Virginia" by her parents. Inside the trunk was just "a few bare necessities in the way of cooking utensils, etc., for myself and small son" (for more on this story refer to Chapter XII, *The Story of Laura Harris Flanery*).

After her husband was killed in the war, Laura and William V. resettled in Jackson County. She remarried Joseph P. Bridges, a widower on January 6, 1867. They had two children, Lillie B. Bridges (who was the mother of future US Congressman William J. Randall of Independence) and Mary J. Bridges who married Adron J. Randall. Laura died in Independence on September 24, 1930 at the age of 92.[474]

---

[474] Eakin, Joanne Chiles. *Blue & Grey Chronicle*. (August 2000): 3: 6: 4. "Independence Woman's Heroic Walk to Texas is Still Vivid Memory," *Kansas City* (Mo) *Journal-Post*, 16 May 1926; "Mrs. Laura Harris Bridges, 90, Believes in Present Generation," *Independence* (Mo) *Examiner*, 26 Apr. 1929; and, "Mrs. J. P. Bridges's Life a Story of Hardy Pioneers," *Kansas City* (Mo.) *Star*, September 1930; as located in the Jackson County (Mo.) Historical Society Archives, Document ID Vertical File Newspaper Subject Files: Biography: "Bridges." Also, Laura Elizabeth Bridges Missouri Death Certificate, 24 Sept. 1930.

FLOOD

Stephen Flood was living in Cass County at the time of Order No. 11 and relocated to Calloway County, Missouri to escape the Order.[475]

FORD

"Another memory dear to me is that of the 'Ford' house, one of two left standing after the ravaging of homes as the result of Order No. 11. The other house was the 'Wallace' house. They were left for use as headquarters. At one time one of my Uncles lived there.... I have been told that the house was built with wooden pegs instead of nails. It had slaves' quarters in the upstairs back of the house which were completely closed off, so there was no approach except from the outside."[476]

FORD
(Additional Account)

See HOLMES

FRANKLIN

J. L. H. Franklin was born March 30, 1842 in Fort Osage Township. He received a common school education and was reared on a farm. He became a farmer and stock raiser near Section 25, with a post-office Lake City, in Jackson County. At the age of twenty-one, Franklin left his parents farm to start his own farm. He was not involved in the politics of the War preferring to tend to his farm and cattle business. When Order No. 11 went into effect he moved to Lexington where he resided for about six months. By 1881, he owned six hundred and thirty four acres of improved land, and ninety acres of timber in eastern Jackson County. Franklin's farm was described as being "well stocked" with "a fine orchard and a comfortable house." He was said to have had "the respect of a large circle of friends."

Franklin was married twice. He married his wife Miss Marietta Tar in 1871 but she died two years later in 1873. Franklin remarried Miss Laura C. Turner on September 16, 1874. His second wife had been born in South Carolina on November 27, 1846. They were members of the Baptist Church near Lake City.[477]

---

[475] Rafiner, National Archives and Records Administration. Washington, D.C. Records of the Quartermaster General. Record Group 92. War Claim: M-485.

[476] Svobada, Mary Frances. "Stories My Mother Told Me of Order No. 11." *Kansas City Genealogist*; 23: 1&2: 14.

[477] Union, 902.

FRANSE
(see also ZION)

Celia Franse of Cass County was driven off her farm and found safe haven in Boone County, Missouri under Order No. 11.[478]

FREEMAN

George Freeman a resident of Cass County traveled to the military post at Pleasant Hill during Order No. 11.[479]

FRICK

Dr. Frick, his wife and children lived in the eastern part of Jackson County. In the early years of the War, "Dr. Frick" took his slaves to Arkansas and freed them. "His wife had ridden off on horseback to safety, but he would not leave Oak Grove. He was needed there, he said, to care for the remaining sick. He stood on Frick Hill, where Frick Park is now (1976), and watched the homes of his friends fired by Union soldiers. Pink Hill was also burned. It is reported that as many as 20 houses and buildings were burning at the same time."[480]

Eventually Dr. Frick's home would suffer the fate of those of his neighbors. "Oak Grove was raided by soldiers and the Frick home was burned. Not only that but the entire Frick family was forced to leave Oak Grove through the provisions of Order No. 11, and Mr. Frick settled anew at Montgomery City," Montgomery County, Missouri. "Once back in Oak Grove the log cabin on the hill was rebuilt," which "gave way to a more modern home, the home of Dr. W. E. Frick until his death in 1888." Around 1921, the home, after it had been rented for over thirty years, came in to possession of Dr. W. J. Frick who remodeled the home on "Frick Hill."[481]

---

[478] Rafiner, citing Cass County Historical Society. "Cass County Missouri Families." (Harrisonville, Mo: Cass County Historical Society, Inc., 1976), 105.

[479] Rafiner, National Archives and Records Administration. Washington, D.C. Records of the Quartermaster General. Record Group 92. War Claim: L-3288.

[480] Jackson County (Mo.) Historical Society Archives, Document ID Vertical File Newspaper Subject Files: Military/Wars: Civil War: 1863-"Order No. 11": "Saddest History is Recalled," *Pictorial News*, 1 Dec. 1976.

[481] "Made Old Home His New," *Kansas City* (Mo.) *Star*, 4 Sept. 1921.

FRISTOE

Richard Marshall Fristoe's grave located in the Six Mile Baptist Church Cemetery was destroyed during the Order No. 11 occupation.[482]

FRISTOE
(Additional Accounts)
See TALLEY and TWYMAN

FROST
See ASHLEY

GAMMON

"[William] Gammon took his wife [Sally (Stuart) Gammon, and children], Hattie, James and the baby to Independence where they stayed until close of the war."[483]

GARRETT
See ASHLEY

GARRISON

Mr. Garrison, Cass County, went to Lafayette County as a consequence of Order No. 11.[484]

GEAM

Rile Geam lived near Lone Jack. He was a friend to the Roupe family. He had attempted to assist the Roupe's in relocating to the military station at Pleasant Hill in compliance with Order No. 11.[485]

---

482 Page 6 of a letter from Mrs. G. T. Twyman to a Mrs. Green, 19 Apr. 1926, relating history of Richard Marshall Fristoe. Jackson County (Mo.) Historical Society Archives, Document ID 87.01F2. It also contains information about Fristoe's War of 1812 military service; his role as one of the earliest Jackson County Judges on the Jackson County Court; as well as anecdotal family history, such as the claim that Fristoe named Jackson County in honor of General Andrew Jackson who he served under during the War of 1812; and genealogical information concerning Fristoe's descendants.
483 Short Stories of Pioneer Days by Dollie Bee Breitenbaugh, 1959. 94-leave, typed manuscript. Two copies. Jackson County (Mo.) Historical Society Archives, Document ID 80F6.
484 Rafiner, citing National Historical Company, 565.
485 Eakin, "Tears and Turmoil," 11.

GENTRY

At the time of Order No. 11, Mary Henley Gentry lived with her children and her mother Elizabeth (Belt) Henley. They did not have any men to help them. The women and children lived on South Chrysler near Independence and were known for their southern sympathies. Within days before traveling to Kentucky under Order No. 11, Mary had put her wash out on the clothesline. When she discovered her clothing missing, she went to the provost office in Independence and complained to the commanding officer. She was able to positively identify the men who had taken the clothing; her clothes were returned to her that evening. The night before they left their residence under Order No. 11, Mary had put her children to bed and blown out the lamps when she heard a banging at her front door. Mary gave the following account of what happened, and which story has been passed down through the Gentry family.[486]

> *My forbearers were Southern by birth and by loyalty, and to live under Yankee rule was too unthinkable for words. When the order was issued it was withheld until three days remained, before it was to be in effect. The two courageous women of my family [grandmother, Mrs. Joseph [Mary (Henley) Gentry, and great-grandmother, Mrs. Alonzo [Elizabeth (Belt) Henley], and were alone at the time, except for the children. All arrangements to leave their home [on South Crysler, owned by the Williamson family in 1965), their furniture, their stock, their crops, etc. were made and transportation to St. Louis by steamboat had to be secured. From there they would go to Kentucky.*
>
> *They were known to be Southerners and for weeks they had been disturbed at night by roving marauders. On the night before their departure, they had put the children to bed, had blown out the lamps, and were quietly and fearfully laying plans for their departure the next day. There was a banging at the door. The two women lit a candle and went to the door. The intruders demanded money. "We know you are leaving these parts and we know you have money and we are going to have it." Grandmother had hastily hidden most of her money in the flour bin. She carefully left some in her purse which she handed to the men. They knocked the candle out of her hand, took the purse out into the moonlight, cursed, and came back demanding more and trying to trap her by questions. 'How much money do you have in the bank?' 'I do not know, as the bank has recently been taken over by the Chase Bank in St. Louis,' she replied. The men knew that was true. 'How much money did you have*

[486] Wilcox, 361-363, citing *Jackson* (Independence, Mo.) *Examiner*, 26 June 1914.

in your purse?' 'I do not know. I have been paying bills all day and do not know how much I had left.' Two women against four ruffians...[who] left cursing. A few days before in broad daylight, a group of Union soldiers rode into their yard, stripped the clothes line of all clothing and rode away.... Grandmother...mounted her horse, rode to the Union headquarters and voiced her complaint to the commanding officer. She identified the soldiers and her clothes were returned to her that evening.

For days, Grandmother and her younger sisters had carried small bundles of silver tied to their petticoats. Growing more weary and exasperated by the minute, she finally exclaimed, 'If they don't come and steal this silver tonight, let's just give it to them!' The marauders didn't come and she didn't have to give it to them. Instead she and a trusted servant buried it. She also hid her rings in a stone wall that ran in front of the house. There is a stone wall there now (1965)--whether it is the same one or not I do not know. The little rings have likely been washed away many years ago. My Grandmother was never able to find them. Later she was able to find the silver. However, she missed the burying place a little and the spoons are nicked by the spade.... On the day of their departure for Kentucky, they traveled to the Old Wayne City landing which is now (1965) Cement City. There, they were to take a steamboat to St. Louis. It grew late. The boat did not come. For three days they waited...filled with fear of a raiding party from Independence. There were small children to cope with, shelter was inadequate and food was scarce.... Old Aunt Betty, who provided an additional problem, for traveling with a colored person in those days was difficult.

Finally the boat arrived and they were on their way. A day or two later they ran aground a sand bar and there was danger of the smokestack falling. However, they did reach St. Louis...and continued to Kentucky.... My family returned from Kentucky before the war ended.... Of course, the house had been emptied of all its contents, and the hemp which was the main crop was long gone. There was no help except Aunt Betty who was as steadfast in her allegiance as ever.... [They] moved into the old brick house which later became Woodland College. [Two days before the Battle of Westport] Union soldiers [set up camp] on the grounds of Woodland College. The soldiers were half starved. They invaded the kitchen and my two grandmothers and old Aunt Betty were mustered into service. They made coffee and hot biscuits all through the night and part of the next day. The women would put the biscuits in the oven on one side and the hungry soldiers would take them out on the other and eat them half-cooked. The women were careful not to let it be known they were Southerners...."[487]

GEORGE

An extract from a letter written by Frances Fitzhugh (George) Moore Kabrick contains information about the George and Owings family's ordeal through Order No. 11. Frances married Ezra Moore on March 21, 1860, and that was her surname during this time period. Frances Moore married Henry V. P. Kabrick in November 1870 (misspellings and lack of punctuation preserved for authenticity):

> *That was the terblest drive that was ever experianced by women and children I know. Every body in the country moveing at one time the wether dry hot and dusty the dust so thick on the fences a person could gether it up by hand fulls. White women and children looked like negroes they had to moove any way they could. I started to moove the last of my things a horse back with my bed waying 35 lbs on the horse behind me a larg amount in my lap a tea kettle on the horn of my saddle a shovel under one arm carrying them that way two miles or more. Then puting them in sister Martha Owings' wagon. Not knowing when we started where we could get a house to stop in. About four or five miles from where I put the bed in the wagon the tire came off of the wagon my mother was in and they sent a runner a head for me so I had to go back and put the tire on which I did but worked and worked and tride every way before I got it on at last. I got mad and got stout so it went on and I fastened it on with telegraph wire and the lastime I saw the old wheel it still had that wire and tire on it at J. White's south of Hebron Church in Lafayette Co., Mo. Somthing like five miles east of Odessa. That was my first wagon. We drove till dark over took us stoped the wagons at the side of the road and the dust bunked upon the grown till morning. Sister Amanda and I rode in the night horse back to Elder Josih [Josiah] Warders near Wellington, Lafayette Co., Mo. He (Mr. Warder) told us he thought we could get to moove in the school house in his district Sister and I went the next morning to see the trustees of the school house. So my mother, Mrs. N. E. George and family of five and sister Martha Owings and family of seven and myself all lived there a while which was 13 including the negro girl. I and sister manda rode around and hunted till we found a house about nine*

[487] Shaw, Mary Gentry. "My Most Unforgettable Character." *Jackson County Historical Society JOURNAL* (July 1965), 5, 15. Also, Wilcox, 361-363, citing *Jackson* (Independence, Mo.) *Examiner*, 26 June 1914.

> *miles southeast from the above named schoolhouse. mother and others her family mooved up there in the fall or first of winter 1863. That fall and winter my sisters two three[--?--] nephews and I stoped at Dink Campbells with two wagons loads of hogs when we got there our clothes was frozen on us as it had been raining and sleeting the most of the evenign turning colder all the time that night I hated to get in a clean bed so bad I told Mrs. Campbell to throw some old thing in the corner and let me sleep there but that great and noble hearted woman put us in her clean bed with our mudy clothes on. Next morning when we got up it was a snowing and snowed verry hard all day I got on my horse riding all day to sell the hogs as they had to bee turn loose as they were tyed the night before when we got to Mr. C. I came back and staid all night a Mr. C the next night. The second morning I started to try to sell them I was directed to a man by the name of Wm. Morrison geting his description that evening starting to Greenton to see him on my way I met him, I knew in resan it was him before I got in fifty yard of him. When we ment I asked him he was the man (I said I supposed so by the description I got of you.) So he went back with me to where the hogs were I sold them to him for just what I could get, him making a big proffit. Now this was one of my hardships with many others likewise before the war closed.*[488]

GEORGE
(Additional Account)

Nathaniel "Nathan" Bass George and his wife S. E. (Farmer) George fled their home in Jackson County. While they were exiled in Lafayette County "the former S. E. Farmer, taught school to the refugees of Order No. 11."[489]

GIBBONS
See KREEGER

GILKY
See ASHLEY

---

[488] Jackson County (Mo.) Historical Society Archives, Document ID 144.01F4. Amanda J. George married John W. Tyler in Jackson County on 27 Feb. 1868. Bartels, 180-190, citing *Oak Grove* (Mo.) *Banner*, 1 March 1897.
[489] Eakin and Hale, 162.

GILLISS [sic.][490]

See ASHLEY

GILMORE

Joseph Gilmore of Cass County sought refuge in Clay County, Missouri after Order No. 11.[491]

GILPIN

See ASHLEY

GLASS

James Glass of Cass County reported to the provost marshal's station in Harrisonville and then to Johnson County, Missouri during Order No. 11.[492]

GRAHAM

Ambers Graham was born in the Van Buren Township of Jackson County on March 4, 1837. His father Daniel Graham was one of the first settlers of Jackson County. He began a career in farming and raising stock. Ambers married Sarah J. Stultz on September 13, 1857. His wife was born on July 17, 1839 in Jackson County. They had three children born after the War: Mollie, born June 24, 1866; Lulu, born July 9, 1869; and Della, born June 25, 1872.

As a result of Order No. 11, Ambers and his wife lost everything except "one poor pony that was not worth taking from him." They returned to the Lone Jack area after the War where Ambers cultivated one hundred and twenty acres of property that was described as being "under a fine state of cultivation, with good buildings and orchards."[493]

GRAHAM
(Additional Account)

Mary Graham of Cass County sought safe haven in Howard County, Missouri when she was evicted under Order No. 11.[494]

---

[490] The surname is most likely Gilliss, from an image of his hotel on the Missouri River, Gilliss House.

[491] Rafiner, National Archives and Records Administration. Washington, D.C., Records of the Quartermaster General. Record Group 92. War Claim: K-2869.

[492] Rafiner, National Archives and Records Administration. Washington, D.C., Records of the Quartermaster General. Record Group 92. War Claim: M-785; also, obituary.

[493] Union, 934.

[494] Rafiner, per personal e-mail in author's collection.

GREEN

The Green family was noted in a vignette connected with the Young family, as having been affected by Order No. 11, according to Joanne Chiles Eakin in her book, "Tears and Turmoil."

GREGG

D. Gregg was a farmer in Section 16, and his post-office was in Independence. He was born in Jessamine County, Kentucky on born July 7, 1831 and reared on a farm until 1847 when he came to Missouri. He first settled in Clarke County where he lived for two years attending school. In 1849, he went to Randolph County and took a two-year course in the high school of Huntsville before coming to Jackson County where he engaged in farming until 1855.

He married May Harper of Bath County, Kentucky on July 17, 1856. May had been born on July 4, 1827. She had settled in Jackson County in the spring of 1850. They had five children: George H., Laura E., Lera H., Robert L., and Larius M. In the year Gregg and May were married, they moved to Cass County where they remained for two years before returning to Jackson County.

In November 1862, Gregg enlisted in the Confederate service, and was taken prisoner in June 1863. He was held for about six weeks at Butler, Bates County before being released and returning home. By Order No. 11 Gregg moved his family to Calloway County, where they remained until the close of the Civil War. In 1874 Gregg was appointed as overseer of the County Farm, having charge of this institution through the spring of 1881. Gregg then relocated to his sixty acres farm.[495]

GREGG
(Additional Account)

Samuel Gregg and his wife Maria lived on a farm south of Hickman Mills with their three children, Martha, Cornelia and Wallace and their eight slaves. Around 1862 Sam Gregg took their slaves to Sherman, Texas and kept them there throughout the war.

When Order No. 11 was issued, Samuel took his family east to Montgomery County, Missouri. They traveled by wagon and remained there until 1869 when they returned to their home in Jackson County. They found their home badly worn but not destroyed. Maria later learned the reason the house survived. The good fortune was because one of the Federal soldiers sent to destroy the house was Maria's cousin. As they

---

[495] Union, 888.

prepared to burn down the house, he intervened and informed the captain that the home was that of his cousin. Reluctantly, the captain agreed not to burn the house.[496]

GREGG
(Additional Account)

Mrs. William H. Gregg's husband was a well-known Quantrill guerilla who participated in the raid on Lawrence, Kansas on August 21, 1863. Mrs. Gregg in her memoirs written years after the War, "Can Forgive, But Never Forget" provided an account of Federal soldiers coming onto their farm and forcing their slaves to Kansas. Her father was imprisoned several times during the War, once for feeding families of known guerrillas. She was made a refugee by Order No. 11, which she described as "O, the misery! Old men, women and children, plodding the dusty roads barefooted, with nothing to eat save what was furnished by friendly citizens."[497]

GREGG
(Additional Account)

Martha F. Reed relayed a story about Samuel and Maria (Bryant) Gregg, who "hitched up his teams again to take his family to Montgomery County for safety. The roads where choked with refugees. Sam and Maria assisted those in trouble when they could. All around them the saw houses, great and small, sending clouds of flame and smoke into the sky." They added the following account of the journey and experience.

> *Union soldiers were blackening the earth from the Missouri River south.... The Greggs thought they would never see their house again. But a kind Providence had sent a man by the name of (shall we say) Robinson. Those were the days when families split over the question of preserving the Union, and Robinson was a Union soldier. As they were spreading destruction, they came to the home of Sam and Maria.*
>
> *As a soldier started to apply the torch, Lieutenant Robinson intervened. 'Hey, there, lieutenant, get away from that,' the captain yelled with an oath. 'Don't you know others?' But Robinson stood his ground. 'Please, Captain, just this one! It is the home of my good cousin, Maria. I won't interfere again.' After some heated cussing and argument, the captain relented. 'All right!' he exclaimed, 'We won't burn it. But we'll use it--for a stable!'*

---

496 Eakin, "Tears and Turmoil," 65.
497 Fellman, 256-257.

When the war was over, the Gregg family returned to their badly abused house, but at least they had a roof over their heads.[498]

GREGG
(Additional Account)

See YOUNG

GRIFFITH

Mary A. Griffith was born in Kentucky and her parents were of German descent. She married John Griffith also of German heritage in Mason County, Kentucky. John had been a veteran of the War of 1812, and participated in the Battles of Fort McHenry and Baltimore in 1813. John and Mary's son George W. Griffith was born on January 29, 1829 in Mason County, Kentucky. The Griffith's settled in Lafayette County, Missouri in 1836 when George was seven years old. In the spring of 1837, John permanently settled the family in Van Buren Township of Jackson County (John died in November 1859).

In 1852, George went to California with only one cow. After working in California for about nine years, George returned to Jackson County in 1861 with some $12,000 in hand—all of which he had earned from his farming operations out west. George joined General Price, but by the time of Order No. 11 he was living with his mother Mary when they were forced off their property. After the War, he returned to Van Buren Township and developed a farming operation on four hundred and twenty acres. Mary died in September 1875.[499]

GROVES

See ASHLEY

HALL

See ASHLEY

HALL
(Additional Account)

Lemuel Edward Hall was born on October 8, 1819 in Virginia. He was the third of thirteen children born to William Hall, Jr. and Mary Ann (Lowther) Hall. Lemuel in his youth was a friend of Thomas "Stonewall" Jackson as they lived near each other in

---

[498] Jackson County (Mo.) Historical Society Archives, Document ID Vertical File Newspaper Subject Files: Military/Wars: Civil War: 1863-"Order No. 11": "Maria and Samuel Gregg's Home Spared During War," undated, unattributed newspaper clipping.

[499] Union, 935.

Lewis County, Virginia (which became part of West Virginia during the Civil War). Lemuel and Jackson even attended school together. Lemuel felt so fondly of General Jackson that he named two of his sons after him. Lemuel had a brother named Cyrus who was a federal judge in Virginia. Judge Cyrus was one of the last and decisive votes that resulted in separating Virginia and West Virginia. Lemuel like his brother Cyrus studied and became a lawyer, but Lemuel's poor health forced him to retire from the legal profession and devote his life to farming.

Lemuel married Mary E. Thompson on February 6, 1850 in Virginia. Mary was born on January 15, 1829 in Louden County, Virginia. The first three children, Ella Virginia, William and Charles B., were born in Virginia. In 1855, Lemuel moved his young family to Jackson County near Kansas City. They had more children, Franklin, Richard and William who died in infancy.

According to the 1860 census, the Hall household included Leumuel and Mary, their five surviving children, along with Albert Buininn, an eighteen year old laborer from Massachusetts and one fifteen Mulatto female slave (she would later become counted as the seventy fifth fugitive from Missouri). During the early years of the Civil War, the Hall's experienced theft, vandalism, arson and threats. From the steps of the courthouse in Kansas City, Lemuel voiced his objection to Order No. 11 and protested the conduct of Red Legs, Jayhawkers and Federal troops. This produced a cry for his imprisonment and Lemuel was forced to hide in the brush. Federal troops demanded that Mary disclose her husband's whereabouts, but she refused. Despite Mary's pregnancy, she was imprisoned for further interrogation. The family house was burned and their livestock confiscated by federal troops. Ella, their teenage daughter, gathered a few things together and tied them into a "log cabin" quilt and then she led her younger brothers and sister to the house of a family friend. Eventually, the Hall's reunited and bordered a steamboat and settled in Glasgow, Missouri where they lived on farm property owned by Mr. Hughes.

The loss of their property would prove not to be the only disaster experienced by the Hall family during the War. While living on the Hughes farm, their small daughter, Ida, fell into an iron kettle filled with hot water for washing clothes and died. Little Ida was buried on the farm in an unmarked grave. The Hall's never returned to Jackson County. In 1869 Lemuel purchased a two twenty-three acres farm in Fayette, Missouri. Mary Hall died on February 14, 1890 and was buried in the Mount Gilead Cemetery in Howard County. By 1998 the Hall's Fayette property was owned by the fourth generation of the Hall family.[500]

---

[500] Eakin, Joanne Chiles. *Blue & Grey Chronicle*. (October 1998): 2: 1: 10-11.

HALL
(Additional Account)

James D. Hall resided in Jackson County. He served for about two years in the War before returning home. He remained in his house "and stayed until Order No. 11 was issued when he went to Lafayette County. From there he went to Nebraska City for one year.... He returned to Jackson County after the War."[501]

HALLER
See KREEGER

HAMBRIGHT

Thomas Lorraine (Mrs. Paul M.) Campbell relayed the history of the Hambright family that had been passed down by family members. Included is their plight under Order No. 11:

> *For a time [during the Civil War], Missouri Hambright and her daughters and youngest son, Amos (born 1845) continued on the farm, managing as best they could." When federal troops occupied most of eastern Jackson County, they began their burning and pillaging. "A detachment of them camped at the John Hambright place. Missouri became so fearful for her daughters and herself that she appealed to the officer in charge, asking permission to load a wagon with necessary provisions and clothing, and to leave. The officer told her that as long as she was there she and her family would be safe from either Redlegs or Union Army; but, he added that he had to leave the next day, or the day following, so he would help her to go." Missouri hurriedly loaded a couple of wagons, and with her family, departed for Ray County. They crossed the Missouri River at Lexington. "Thereafter, the Federals burned the Hambright Inn and all the other buildings....*
>
> *Only two or three houses in the Six Mile survived the brutal edict of Order No. 11 and the raids of Jim Lane's scoundrels. The Chiles house on Blue Mills Road, less than a mile west of the John Hambright home, was fired, but a mail carrier happened along in time to see the fire set, and after the soldiers had gone he put out the fire." A solid walnut armoire was*

[501] Eakin and Hale, 183, citing *Independence* (Mo.) *Examiner*, 25 Aug 1924.

among the possessions rescued and taken down the river by the Hambright's.[502]

HAMILTON

See also TRIPLETT

Bird Hamilton and his brother Sam Hamilton and their respective families were removed from their properties under Order No. 11.[503]

HANLEY

See ASHLEY

HARBAUGH

Franklin and Mary Elizabeth (Pence) Harbaugh and their three children, George, age 7; Amanda, age 5; and Minnie, under the age of 2, lived about two miles north of what would become Lee's Summit. Prior to the issuance of Order No. 11, Franklin had been banished from the family property, and took refuge in Liberty, Missouri working as a wagon maker. Franklin did make secretive trips to visit his family. "Mrs. Harbaugh and her sister kept the home fires burning and provided food for other Southern sympathizers who, like Mr. Harbaugh, spent most of the time in hiding, fearful of life if they remained at home."[504]

Mary Harbaugh with assistance by her two young children George and Amanda loaded their wagon and cart with their belongings. Amanda and George were placed on the front board seat in the wagon while little Minnie was tucked into a hand-hewn cradle. Mrs. Harbaugh took the family to Westport Landing in Kansas City where they were ferried north across the Missouri River to Harlem Landing in Clay County where Franklin was awaiting the arrival.[505]

---

502 Manuscript written by Thomas Lorraine (Mrs. Paul M.) Campbell from Minden, Louisiana, in November, 1959, for the Jackson County Historical Society Contest, entitled "Some Interesting Ancestors Who Lived in Jackson County." Jackson County (Mo.) Historical Society Archives, Document ID 77.01F16. The house was still standing as late as November 1959. The solid walnut armoire was lovingly handed down through the generations, Dora (Hudspeth) Slaughter, of Buckner, Missouri, had the kindness and foresight to posthumously gift the wardrobe to the Jackson County (Mo.) Historical Society in the spring of 2013 for continued use, preservation and interpretation of the Hambright family and Order No. 11.

503 Wilcox, 365, citing *Independence* (Mo.) *Examiner*, May 1906.

504 Jackson County (Mo.) Historical Society Archives, Document ID Vertical File Newspaper Subject Files: Military/Wars: Civil War: 1863-"Order No. 11": "Lee's Summit Resident Recalls Flight Under Order No. 11 Ninety Years Ago," unattributed newspaper likely published 25 Aug 1953.

505 Wilcox, 361, citing *Kansas City* (Mo.) *Star*, 1 Mar. 1912.

Although Amanda Harbaugh was only five at the time of Order No. 11, she vividly recalled years later their "enforced exodus by oxen team and wagon to Liberty, Missouri," but she only vaguely remembered circumstances surrounding the move. Most of what she related came from her mother's first-hand account. Mary told Amanda about the "anxiety that surrounded preparation to comply" with that Order. Amanda did recall the intense "excitement as the wagon and cart were loaded with household and other belongings" as she and her brother, George were lifted to the board seat in front while her baby sister, Minnie, "was tucked in her blanket just back of the seat, in a hand hewn cradle made by her father for his first born." The Harbaugh's moved back to Jackson County in the fall of 1867.[506]

HARDESTY

See ASHLEY

HARRELSON

Mr. Harrelson left Cass County under Order No. 11 and went to Texas.[507]

HARRIS

See ASHLEY

HARRIS
(Additional Account)

Jane (Hall) (Mrs. Samuel) Harris, mother of Laura (Harris) Flanery, left her home south of Blue Springs, Missouri, moved across the county line to Lafayette County during Order No. 11.[508]

HARRIS
(Additional Account)

Nancy "Nannie" Elizabeth Ellen Harris was born April 4, 1844. She married Jabez McCorkle in March 1863 and he was killed in June. She had been among the

---

[506] Jackson County (Mo.) Historical Society Archives, Document ID Vertical File Newspaper Subject Files: Military/Wars: Civil War: 1863-"Order No. 11": "Lee's Summit Resident Recalls Flight Under Order No. 11 Ninety Years Ago," non-attributed newspaper likely published 25 Aug 1953.

[507] Rafiner, citing Belton Star-Herald. "The First Hundred Years: Belton, Missouri 1872-1972." (Belton, Mo., Belton Star-Herald, 1972).

[508] "Mrs. J. P. Bridges's Life a Story of Hardy Pioneers," *Kansas City* (Mo.) *Star*, September 1930, as located in the Jackson County (Mo.) Historical Society Archives, Document ID Vertical File Newspaper Subject Files: Biography: "Bridges." (See also, Flanery above, and Chapter XII).

victims of the Women's Prison collapse in Kansas City. Nannie suffered a broken ankle in the tragedy. After the disaster, Nannie was allowed to return to her parents' home in eastern Jackson County. The entire family was forced to leave their house when Federal troops burned their home and gristmill on September 6. They left the turmoil in Jackson County and Nannie walked, ailing ankle and all, more than one hundred miles to Howard County, Missouri. When the rebellion ended, Nannie McCorkle married James E. Lilley. Nannie's life was cut short when she died November 10, 1872, at the age of twenty-eight. Her body was buried in the Davis-Smith Cemetery in Raytown, Missouri.[509]

HATTON

See ASHLEY

HATTON
(Additional Account)

Henry Hatton of Cass County went to Iowa to escape Missouri and Order No. 11.[510]

HAYNES

Elizabeth Haynes left Cass County and found safe haven in Johnson County, Missouri following Order No. 11[511]

[509] Petersen and Jackson, 52, 53.
[510] Rafiner, obituary.
[511] Rafiner, citing Glenn, Allen, 353.

HAYS

Margaret Jane (Watts) Hays was the widow of Colonel Upton Hays. Her husband had been killed during the Battle of Newtonia in Bates County, Missouri on September 15, 1862.[512] Margaret fled the Westport area after Order No. 11 and went to Williamsburg, Calloway County, Missouri. She wrote a letter to her mother three weeks after she had arrived at her temporary residence on September 26, 1863 (her spelling, grammar and punctuation retained for authenticity):

Colonel Upton Hays

Dear Mother

It being my Birthday I thought I would spend a part of it in writing to you. I have writen one letter since I came to this place, three weeks this eavening. Since I got down to this place I have met with a great many friends hear. Three of my children has been going to School every day Since I came down, friends sends them. I have been up to Fulton since I wrote to you. I was at Uncle Sams [Samuel H. Berry (1815-?)] *family, all well. Uncle Dicks Aunt* Becky [Richard C. and Mary Rebecca Hamilton Berry (1813-1900)] *has got down. Uncle Eads* [Edward Gaither Berry and his wife Sally Ann Galbreth (1810-1867)] *family well, their son John* [John W. Berry (1835-1863)] *was Drowned a short time ago. They all ensisted on me stay with them. I am living at Ams* [Amazon Hays], *he lives in the Burg. Times is quiet hear as could be expected. We have a company of Soldiers Stations hear. I heard from Jackson today, they have everthing else but peace up their. People all left the country some was alowed to move to the posts, others Banished. I was not alowed to stay up their, left their on Fridy, my house burned on Mundy after I left home. I had left part of my furniture in the house. Mother they have done all they can do. Burned me out three times and at last drove me from my country. They*

[512] Monaco, "Blood on the Streets," 21, note 11. Upton Hays was born on March 29, 1832, in Callaway County, Missouri, but he grew-up in Westport in Jackson County. His father was Boone Hays, a grandson of Daniel Boone, the famous trailblazer and pioneer. His father Boone Hays settled his family in Westport, Jackson County, Missouri in 1837, but Boone Hays died in California in 1850. Within a few weeks after the Battle of Lone Jack, Col. Hays would be killed in a skirmish near Newtonia, Missouri when he met his death leading a charge against Federal troops from Wisconsin.

have nothing against me, they say. My Husband did his Duty, now he is gone. They will take their spite out on myself and little Children.

Mother their is a great emergration going out in the Spring to California. If I do not go between now and then I will go then. Mother John Ed [John Edward Berry] *is in Alton prison, has been their some time. I received one letter from him, he wished me to write to you and let you know where he is at and says tell you where he belongs to. He dont want you to think that he holds to the side Bunion [John Bunyan Rowland was in the Union Army] belongs to. Jemmy* [James Samuel Berry] *was a prisinor at Vicsburg. Uncle Jemmy wishes me to asqe you to write a few lines to Uncle Nealy for him concerning his going their. If he thinks he can do anything their to make a living he will go in the Spring. He is almost blind, this war has left him with but little to go on. Uncle Calup B.* [Caleb Ewing Berry] *will go out their in the Spring if he is able in the Spring. He is very feble now. Mother take good care of yourself, it is so mutch pleasure to know that I have a Mother. If fate has ordered it so that I cant be with you do not troubble yourself about my troubles. I am able to bear them all. My love to my brothers tell them I thank god that they are not hear. So fare well Dear Mother from your Daughter*

Margaret J.

ps mother let me know Uncle Neales post office. Uncle Jemmy wants to write to him. Tell Cousin Van [Van Hays] *his mother is well, she is living hear in the Burg. I stayed with her last night, Margaret*

Margaret Jane (Watts) Hays, wife of Colonel Upton Hays. They married in 1852 and had four children. Photo courtesy Colonel Upton Hays and Margaret Jane (Watts) Hays Overstreet Family Collection, 1849-1865, Jackson County Historical Society Archives, gift of Marian Franklin and family. More information is available at wattshaysletters.com

Margaret wrote two more letters to her mother from Williamsburg further detailing the tragedies of Order No. 11 and troubles on the border in Jackson County. The first letter that follows was written on November 7 and the second on December 28, 1863 (again they are translated verbatim for authentic purposes):

Dear Mother

Once more I have the pleasure of answering a kind letter received a few days ago. The letter was directed to Westport, my friends forewarded on to me. This is the fourth letter I have writen since i have been in this place. We have peaceble times hear. Have never had but little trouble hear with the exception of negros running off. They are leaving every few days, my last one has gone, Andy. I had run him down hear for two years thinking I could save him, but they are but little profit to anyone now.

Mother I had quite an agreable surprise last Saturday. Eliza Dickerson [Eliza Watts Dickerson lived in Fayette, Howard Co., Missouri] *heard that I was down hear, she came to hunt me up, one of her sons was with her. I did not think that she could get this far from home, she is one of the fleshest women I ever saw. She seldom ever leaves home but when she heard that I was down hear she could not rest without coming to see me. Our friends is all well in Ky is good union people. Aunt Nancy* [Nancy Watts Dickerson (1779-after 1863) lived in Kentucky.] *is still* [erased] *Cousin Beverly Dickerson* [Aunt Nancy's son Beverly James Dickerson, Jr. (1820-1863) was the Marshal of Franklin, Johnson Co., Indiana. He was killed while waiting in a train station in Franklin. He was a witness in a lawsuit in Louisville, Kentucky.] *was killed in Indeania, he was killed by some unknown person, he was a rebel. Cousin Eliza Kimberlain* [Aunt Nancy's daughter Eliza Dickerson Kimberlin (1814-1895) was the widow of Richard Samuel "Sam" Kimberlin] *is in howard* [Howard County, Missouri]. *Cousin Jourden L.* [Jordan R. Lowe (1819- after 1880)] *is living near Fayette, sent perticular word for me to come and stay with him. Eliza wanted me to go an live with her. I am living with Am and Cousin Pap [Amazon and Mary B. Berry Hays]. They are very kind to me and the children, they are all going to school.*

I heard from home a few days ago or the place that was once my home. Times was more quiet than they had been for some time. That country above Blue is all burned, that is all the fencing and a great many houses. My last house was burned a few days after I left their. My crop is all distroyed, if the land would sink I would be satisfied. Cousin Lis Berry [Elizabeth Ramsey Berry] *is living in Kansas.*

The Bushwhackers has all gone South. Shelby [General Shelby, CSA] *made a Raid in Mo. He had some 25 or twenty eight hundred, he came to Boonvill. When he went back he had upwards of five thousand and three fine canan. He came in after recruits, he got all he came for and went back. You know what accounts their is in the papers about the raid. Uncle Jemmy [James B. Yager] is living hear, he is almost intirely Blind, has been for a long time, has to be led about the house. Write to him Mother, he is so low spirited, it would please him so mutch. Send him word how to direct letters to Uncle Nealy* [Rev. Cornelius Yager]. *Our friends hear is all well. I have been to see all my Uncles, Uncle Dick and Aunt Becky* [Richard C. and Mary Rebecca Hamilton Berry] *is living down hear. Old man Hamilton* [Charles Hamilton (1787-1881)] *and Mary and her husband is down hear at Uncle Dicks. They have not heard from Hue* [Hugh Hamilton was in California] *for a long time. The old man wishes me to asque you to write to Hugh and let him know where they are and tell him the old man has lost all he has in the world, the old man says he has not enough of money to buy himself a plug of tobaco. He thought if Hugh was able to help him he would do it and Dick is very Needy. Fulton is their posoffice.*

mother I send you a thousand thanks for the present you sent me in my last letter, a ten dollar bill. That kind of money is worth more than any other kind hear, exceping gold. Greenback is all the kind we have hear. You said the Boys was going to send their minatures to me and yours. I am looking anxiously for them. My love to them, if you see Cousin Wade [Wade Hays] *tell him his mother is well, living hear alone with her three negros. One left her a few weeks ago. I go to see her every few days. Mother, Am and Cousin Pap sends their love to you also Uncle Jemmy and family, my love to Brother John [John S. Watts, Jr.]. I seldom ever hear from him but think a great deal about him. So fare well from your Daughter Mag J.*[513]

[513] The Colonel Upton Hays and Margaret Jane (Watts) Hays Overstreet Family Collection, 1849-1865 [early 1900s], Jackson County (Mo.) Historical Society Archives, with gratitude to Marian Franklin and family for donating the originals for preservation, for providing annotations in [brackets], and for posting transcripts and other data about this important family online at wattshaysletters.com.

The third letter written on December 28, 1863 reads:

Dear Mother:

Once more I am permited to write a few lines to you, we are all well at this time. I have been very sick with cold but feel quite well now. Since I wrote to you last I have been off on a visit, went to spend a week or so with an old Jackson Neighbor, Now a refugee, I stayed over two weeks their and at Uncle Sams [Samuel H. Berry]. We are very lonely hear altho it is Chrismas but that is the Sadest time with me. The Children has gone to spend Chrismas with their Aunt Mary Hughs [Mary Boone Hays Hughes (1829-1872)] and Aunt Elenor Chick [Elenor Hays Chick (1818-1896)]. *Three of my children attends school regular every day. Elfleda learns very fast, says she is going to learn fast so she can write to you. I have a great many friends hear. I often get embarest when out among strangers. When they find out who I am they try to show me so mutch attention. My Husbands deeds is not forgoten hear, all seems to feell so mutch for me for the manner in which I have been treated. Times is quiet hear at this time. Some excitement about the Draft, but I think it will wear off. I think they will get negros enough without taking any of the white men. They give the State credit with the negros that is sent. I do hope that they will take them all. Andy went some time ago, the last time I heard of him he was dressed in his uniform and had his gun to fight for his freedom.*

I heard from Jackson a short time ago, times was quiet up their. Some two or three of my neighbors had mooved on their homes. My house was burned a few days after I left their. I had about thirty achors of corn, as fine corn as ever I saw raised, besides my hay and oats and vegetables, all taken and distroyed. I have received my beads [beds] *and clothing at last some two weeks ago. They cost me more than I could for them to put them up at the highest Bider. I had but few things, they cost me near 60 Dollars, then when I came down it cost me between thirty five and forty dollars. I do not know what want is yet, I always have plenty to eat and wear for myself and Children. I am still living at Amberzons [Amazon Hays], he had to leave his home yet he can make enough to live on and divide with me. John Ed [John Edward Berry] is still in prison. Uncle Dick* [Richard C. Berry] *and family is going to move to California this winter. They have received a letter from Hugh* [Hugh Hamilton (1822-1883)], *he is going to send money for them to go out with Uncle Jemmie. He has got so he can see to walk about the house. Aunt Mary is very sick, Rachel* [Mary Rachel Yager] *is Married, married some three weeks ago to a yong man by the name of Henry Harper. He is very nice yong men, one of my old Jackson Co. Neighbors. He is very well off, a study industrous young man. Mother Old Mrs Graham dide a few weeks ago, she left*

Cousin Lis Berry one Hundred Barrels of corn, McSpaden, Do and made over six hundred dollars to a woman in Westport. Mother you wished to know whether I would come this winter or Spring. I do not know what to do, if I go know my children will never get anything hear, if I stay a while longer I think that their will be some arrangement made for them to have something. Take care of yourself, I still have hope that we will meet again, my love to my Brothers, farewell from your Daughter Mag J.

please send all your likenesses, I will try to make arrangements to come as soon as posible, write often.[514]

HENLEY

See also GENTRY

Elizabeth (Belt) Henley was the mother of Mary (Henley) Gentry. The women were alone, except for Mary's children, when they were forced to leave their property on South Chrysler near Independence. The family was known for their southern sympathies. Elizabeth and her daughter, Mary Gentry sought refuge in Kentucky under Order No. 11. After the rebellion ended they returned home. Of the jewelry they hid before they left, the silver was found but the rings were never located.[515]

HEREFORD

During Order No. 11, the Hereford family of Jackson County faced great hardships. Dr. H. F. Hereford escaped to St. Louis. "His wife was arrested and banished from the Border District. She was secreted in the Gillis [sic.] House until a boat, giving no whistle or announcement of its arrival, stopped quickly and picked her up.... She and her children finally got to her husband and went to California and stayed until the end of the war."[516]

HICKMAN

"The [J. T.] Hickman family returned from Johnson County, where they went during Order No. 11, and the father finished 60-years residence on his farm."[517]

---

[514] The Colonel Upton Hays and Margaret Jane (Watts) Hays Overstreet Family Collection, 1849-1865 [early 1900s], Jackson County (Mo.) Historical Society Archives.

[515] Wilcox, 361-36, citing *Jackson* (Independence, Mo.) *Examiner*, June 26, 1914.

[516] United Daughters of the Confederacy, as cited in Eakin and Hale, 201.

[517] Short Stories of Pioneer Days by Dollie Bee Breitenbaugh, 1959. 94-leave, typed manuscript. Two copies. Jackson County (Mo.) Historical Society Archives, Document ID 80F6.

HICKMAN
(Additional Account)
See OLDHAM

HIGHTOWER

John Hightower was born March 16, 1816, in Jessamine County, Kentucky, but he was principally reared in Mason County, Kentucky. He came to Jackson County in October 1855 where he settled in Fort Osage Township and engaged in farming. In November 1858, he married Miss Sarah A. Van Ardoll, a native of Kentucky who had been born in March 1822. They had eleven children, eight of whom were living in 1881: W. H., C. C., J. T., George P., Margie, Janie, Alice, and Pearly. Mr. Hightower and wife were members of the Baptist Church. They lived on their farm until Order No. 11 compelled him to retire from the county, and he then moved his family to Lafayette County. They returned to their two hundred and forty acres farm at the close of the War, and continued farming until 1881. By the first of March 1881, John went into the mercantile business in Fort Osage Township and carried on a general stock of dry goods and groceries.

HINSHAW

Nathan Hinshaw a resident of Cass County reported to the provost marshal's post at Pleasant Hill during Order No. 11.[518]

HOCKENSMITH

The Jackson County pioneers, with emphasis on Lee's Summit families, included an account of the affect of Order No. 11 on the Hockensmith family:

> *Thomas Lee Hockensmith...entered the Missouri Militia and was ordered into active service, operating on the border at Harrisonville and Pleasant Hill, where they guarded government supplies, with only a little picket fighting. After Gen. Price's last raid, Tommy Lee was ordered back to Harrisonville and stayed on the ground witnessing the distress, destruction an hardship of settlers under Order No. 11.*
>
> *At the close of the war, he was relieved of service; father dead, no home, no friends, no money, no clothes; facing the duty of taking his beloved stepmother to her home in Indiana, with his five half-sisters.*

[518] Rafiner, citing National Historical Company, 466.

Returning, the 18-year-old boy found a home with Uncle Lycurus Jones, and wife, Annie (Younger) Jones, sister of the Younger boys.[519]

HOCKENSMITH
(Additional Account)

G. W. R. Hockensmith, Cass County, went to the military post in Harrisonville during Order No. 11.[520]

HOLCOMB

A. Holcomb of Cass County traveled to Higginsville in Lafayette County following Order No. 11.[521]

HOLLOWAY

T. Holloway of Cass County went to Pettis County, Missouri because of Order No. 11.[522]

HOLMES
See ASHLEY

HOLMES
(Additional Account)

Nancy (Tyler) Holmes was a widow with ten children. After her older sons settled in Jackson County, Nancy in about 1845 joined them. She brought with her a sack of tea and her late husband's beaver hat. She traveled with several of her slaves and her two oldest daughters. One of her girls, Mary Jane Holmes, had been writing to Anderson Shipp Truman, a young man in Shelby County, Kentucky. Despite her mother's strenuous objections, Mary Jane Holmes married the Truman boy in 1846.[523]

---

[519] Short Stories of Pioneer Days by Dollie Bee Breitenbaugh, 1959. 94-leave, typed manuscript. Two copies. Jackson County (Mo.) Historical Society Archives, Document ID 80F6.

[520] Rafiner, National Archives and Records Administration. Washington, D.C., Records of the Quartermaster General. Record Group 92. War Claim: 114-563.

[521] Rafiner, citing Cass County Historical Society. "Captain Alexander Robinson's Co.: Enrolled Mo. Militia Muster Roll." (Harrisonville, Mo.: Cass County Historical Society, 2002).

[522] Rafiner, citing National Historical Company, 450.

[523] McCullough, 16-19. Anderson Truman was one of twelve children and according to McCullough Nancy's objection to the marriage was because her daughter was "marrying down" since the Truman's owned no slaves.

In an article written by Miss Mary Ethel Noland she talked about her most interesting ancestor, Nancy Tyler Holmes, the great grandmother of Harry S. Truman:

> *The peace of* [the happy couple] *was disturbed by the border warfare of the Civil War period. They were in constant fear of the Kansas Redlegs. Lewis Ford joined a little band of militia for the protection of his family and his property. For that, he and his family were banished to Alton, Illinois, under Order No. 11. Mother Nancy* [(Tyler) Holmes] *went to live with* [her daughter] *Mary Jane and* [Anderson Shipp] *"Andy" Truman, and their five children. One autumn night the alarm was spread that the Redlegs were coming. The Negroes had heard the rumor. Andy got his gun ready to stand his ground along with other men of the neighborhood. The women, children, and Negroes fled to a neighboring cornfield. Mother Nancy, now 82-years-old, went with the rest. The two babies, little Emily and Martha, were left in the care of Hannah (a slave). John Anderson Truman and a little neighbor boy, Charley Hinkel, shivered and shook in their night clothes. Mammy--for that was her grandchildren's name for Nancy--drew the two little boys to her side and covered them with her voluminous skirts.... John Anderson Truman was to become the father of a president of the United States, Harry S. Truman.*[524]

Nancy (Tyler) Holmes, Great Grandmother of U.S. President Harry S. Truman

---

[524] Noland, Mary Ethel. "Nancy Tyler Holmes' Motto Fitting: 'Difficulties No Dismay.'" *Jackson County Historical Society JOURNAL* (Fall 1967), 8-9; and, McCullough, 32-33. The incident involving the Red Legs was reason enough for Anderson Truman to move his family and slaves from Jackson County prior to the issuance of Order No. 11. They went to Platte County, Missouri and remained throughout the War. Truman took his slaves to freedom in Leavenworth, Kansas in April 1865 and then returned to Jackson County. Lewis and Patsy (Holmes) Ford were married at Shawnee Mission, probably by cousin, Isaac McCoy.

The widow Holmes and her family, like their future relatives by marriage, the Solomon Young family, was forced off their property under Order No. 11.

HON

The Hon's took different directions after Order No. 11. Jerry V. Hon, Cass County, went with two sisters to Greencastle County, Indiana.[525] Richard Hon of Cass County reported to the military post in Harrisonville and then went to Pettis County, Missouri.[526]

HOOVER

S. T. Hoover of Cass County traveled to Paola, Kansas to get away from Order No. 11.[527]

HORD

William Hord, a farmer in Blue Springs, said, "I was driven from my home and this county by it [Order No. 11]."[528]

HORN

Richard Horn of Cass County left Harrisonville and went to Pettis County, Missouri because of Order No. 11.[529]

HOUSTON

See ASHLEY

HUDSON

Oak Grove newspaper columnist, Dorothy (Pierson) Butler, described James Russell Hudson and Geneva Elisabeth (Hamblen) Hudson's land being "on both sides of

---

[525] Rafiner, citing Hon Family Papers, Pleasant Hill Historical Society, Pleasant Hill, Missouri.

[526] Rafiner, National Archives and Records Administration. Washington, D.C., Records of the Quartermaster General. Record Group 92. War Claim: M-861.

[527] Rafiner, National Archives and Records Administration. Washington, D.C., Records of the Quartermaster General. Record Group 92. War Claim: 115-224.

[528] "Juror Knew Order No. 11," *Kansas City* (Mo.) *Star*, 7 Mar. 1911, as located in the Jackson County (Mo.) Historical Society Archives, Document ID Vertical File Newspaper Subject Files: Military/Wars: Civil War: 1863 "Order No. 11".

[529] Rafiner, citing United Daughters of the Confederacy, 42.

R. D. Mize Road, as it enters the Bucker-Tarsney slab. Their woodlot was Monkey Mountain and the place where Mrs. Hudson hid in her best black silk dress with her small child while their home burned during Order No. 11."[530]

HUDSPETH

George Hudspeth was born January 27, 1820 in Simpson County, Kentucky, and came to Jackson Country with his parents in 1828. Before George settled down to marry, he took two trips to Mexico and made several journeys to California where he spent five years on the west coast. His brother Robert, two years his junior, journeyed with him. George married Miss Elizabeth Jessie in March, 1857. Elizabeth had been born in Russell County, Virginia on May 6, 1830. After she moved to Jackson County she and George were married. They lived in Fort Osage Township near the Lake City post office on a farm that consisted of four hundred and eighty seven acres.

During the Civil War, George's younger brother Robert served under William Quantrill, and a younger brother J. R. Hudspeth served under Jo Shelby. Following the issuance of Order No. 11, George and his family crossed the Missouri River where they remained for about eighteen months. They had eight children with six children living as of 1881: Tybithia P., Anna E., Lola G., Ida M., Joel E., and Silas B. The family was members of the Baptist Church, and George was a Mason and director of the school district for some twenty years.[531]

HUDSPETH
(Additional Account)

William Hudspeth, patriarch of the family, was born in North Carolina in 1778. He married Tabitha Beal, born in 1786. They had eleven children before Tabitha died; she is buried in Franklin, Kentucky. The Hudspeth's moved to Jackson County in 1828, two years after the County had been formed on December 15, 1826. Five of William's sons were "'49ers," having rushed to California Territory after the famous gold discovery at John Sutter's fort (Sutter was also formerly from Jackson County).

One son, Benoni, is remembered for finding a "cut-off" route—Hudspeth Cut-Off—used by thousands of travelers to California for years. With California riches brought back in a trunk, the Hudspeth family ended up owning extensive property and operations in Jackson County.

---

[530] Butler, Dorothy (Pierson) Butler, "In Early Days," Oak Grove (Mo.) Banner, undated: A newspaper clipping describing James Russell Hudson and his 1852 trip across the plains on the California Trail for the California Gold Rush. Jackson County (Mo.) Historical Society Archives, Document ID 167.5F32.

[531] Union, 904

Most 49ers' had hopes of striking it rich quickly and returning home swiftly. The dreams of the vast majority were unfulfilled; young wives were often left at home for months and years; some men never returned. Thomas Jefferson Hudspeth was one such man, but his failure to return was due to an untimely death upon reaching California. His faithful widow Cyntha (or Cynthia) continued to run the Hudspeth home, farm and the management of their slaves.

Thomas Lorraine (Mrs. Paul M.) Campbell relayed the history of the Hudspeth's that had been passed down by family members. Included is the following account of Order No. 11:

> *On the same day the Federals burned the Hambright home, they probably also destroyed the house of John Hambright's sister, Cynthia Hudspeth. "Cynthia had been summoned to Independence by Order No. 11, and while she was away her house was set afire by Federal troops (Kansas Redlegs). One of Cynthia's daughters, Paralee, ran into the burning building to save her new side saddle, but when she came out a soldier wrested the saddle from her, and with a jeering laugh threw it back into the flames.*
>
> *In later years, Cynthia's son, Thomas B. Hudspeth, said that on that evening he stood on a rise of ground and saw the smoke of nine windows' homes burning.*
>
> *The William Hudspeth house, which was located on a side road west from present-day Highway 24 was not burned, because, according to Thomas B. Hudspeth, of Frank Shepherd's protection. Shepherd had been the husband of Henrietta Hudspeth, oldest daughter of Cynthia and Thomas J. Hudspeth, and during their brief married life until Henrietta's death in 1855, they lived in the William Hudspeth home.... Cynthia Hudpeth and her family spent the remaining war years in Ray County.*[532]

A manuscript written about Narcissa Hudspeth includes another recollection of Order No. 11:

> *When they approached the Hudspeth home with torches aflame, the Masonic emblem called a halt. They did not know that the women in the family had gathered the jewelry and silver and hidden them in huge holes dug by the slaves in the orchard. Certainly they passed on leaving*

[532] Manuscript written by Thomas Lorraine (Mrs. Paul M.) Campbell from Minden, Louisiana, in November 1959, for the Jackson County Historical Society Contest, entitled "Some Interesting Ancestors Who Lived in Jackson County." Jackson County (Mo.) Historical Society Archives, Document ID 77.01F16.

> *the house in tact, the silver safe in its hiding place, to be used generations afterwards at wedding and family reunions and the like.*[533]

A separate account provided yet another account of the burning of the Hudspeth property during Order No. 11 "Mrs. Hudspeth's home was maliciously burned. The embers still red hot, Cynthia instructed one of her slaves, Sam, to rake through the ashes and salvage any possessions that might have survived. The family's Bible, which had been hidden in a barrel of beans, was the only item to be found. Although the cover was charred, the pages were intact and readable." The story continues:

> *When Cynthia* [Hudspeth] *returned home, near nightfall, she found only ashes where her husband had built a home 35 years before. Cynthia told* [her slave] (the slave that Thomas J. Hudspeth had purchased a few days before leaving for California, from his brother-in-law, John Hambright) *to get a hoe and dig in the still-smoldering ashes, in case something might have survived.* [The slave] *protested that it wasn't going to do any good, that nothing was left, but Cynthia told him to 'keep digging.' Just then [his] hoe struck something, and seconds later [he] lifted up the family Bible, undamaged except for a corner knocked off by the hoe. The Bible survived because of having been buried in a barrel of dried beans--hidden, as was the practice of most Jackson County folk, where prying Feds couldn't examine it for records of male members of a family who might be in service in the Confederate Army, or with Quantrill....*
>
> *Cynthia rebuilt her house, it is said, from bricks of Sibley store buildings, destroyed during the war.*[534]

---

[533] Jackson County (Mo.) Historical Society Archives, Document ID 77.01F21.

[534] Manuscript written by Thomas Lorraine (Mrs. Paul M.) Campbell from Minden, Louisiana, in November, 1959, for the Jackson County Historical Society Contest, entitled "Some Interesting Ancestors Who Lived in Jackson County." Jackson County (Mo.) Historical Society Archives, Document ID 77.01F22. The Jackson County Historical Society accepts more than 100 donations of historical documents and photographs into its collections annually. One addition in 2007 represented the first artifact directly related to Order No. 11—an 1834 *Holy Bible* from one of Jackson County's most prominent pioneer families. The Bible contains between the Old and New Testaments the handwritten recordings of births, marriages, and deaths of Jackson County's Hudspeth family. These hard-to-find statistics may be desirable to countless descendants, today and into the future. There is more to the story beyond the good book's usefulness as a permanent family record, as you will see. But, first here's a peek into the main characters.

The salvaging of the bible continued to attract attention even after the death of Sam Jackson, a former Hudspeth slave. The obituary reveals additional information:

> *Sam Jackson was owned by the Hamiltons and the Hudspeths of Fort Osage Township.... [He] was fond of telling stories of the times before the war, and one was of a miracle he saw with his own eyes. The old log house known as the Hudspeth home in Fort Osage Township was burned by the Federals under the famous Order No. 11. Mrs. Hudspeth was a very old and devout lady and she had the record of her family in an old family Bible which, at the time of the fire, was in a cupboard in the old house. When the house was smoldering the old lady called to Sam and point to a place where the fire had been the hottest and ordered him to take a long handled rake and dig into the coals. She told him he would find her Bible there unharmed. [He] dug into the coals and got out the Bible, which was scorched around the edges of the heavy leather cover, but unharmed inside. Old Sam always referred to this as a miracle.*[535]

To the right and next page: William and Cyntha/Cynthia (Hambright) Hudspeth 1834 *Holy Bible* recovered by her slave, Sam Jackson, after the family home was burned to the ground. Dora (Hudspeth) Slaughter, of Buckner, Missouri, had the foresight in 2007 to donate the artifact to the Jackson County Historical Society for continued preservation and future access through its Archives.

[535] Lovingly handed down through the generations, Dora (Hudspeth) Slaughter, of Buckner, Missouri, had the foresight to donate the 1834 Bible to the Jackson County (Mo.) Historical Society for continued preservation and future access through its Archives. The trunk referenced above was donated to the Jackson County (Mo.) Historical Society, which has placed in on long-term loan for exhibit at the the National Frontier Trails Museum in Independence, Missouri.

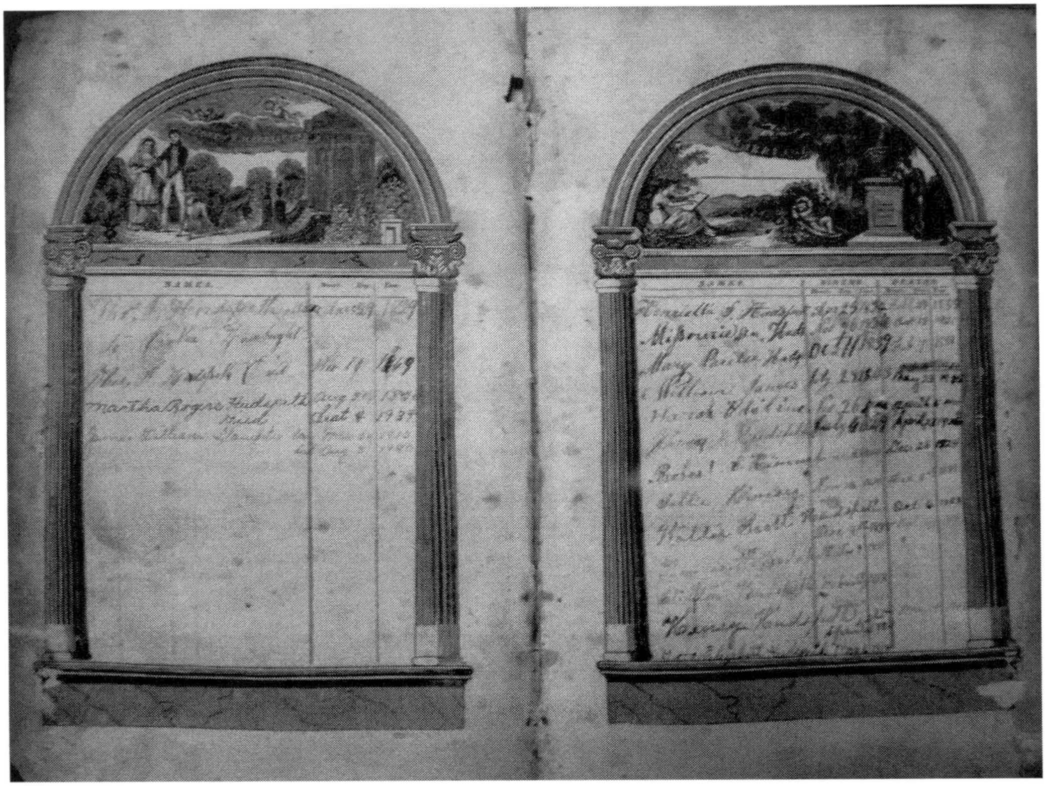

HUDSPETH
(Additional Account)

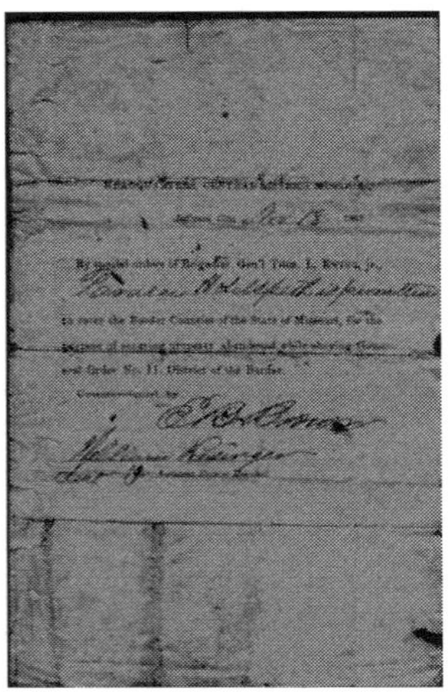

Cynthia and her daughter, Paralee Hudspeth, were issued provost passes in November 1863 that permitted travel within the border counties of Missouri.[536] Although Cynthia Hudspeth signed the Loyalty Oath and received a protection order by the United States Army on February 2, 1862, she was forced to leave her property while following Order No. 11.[537]

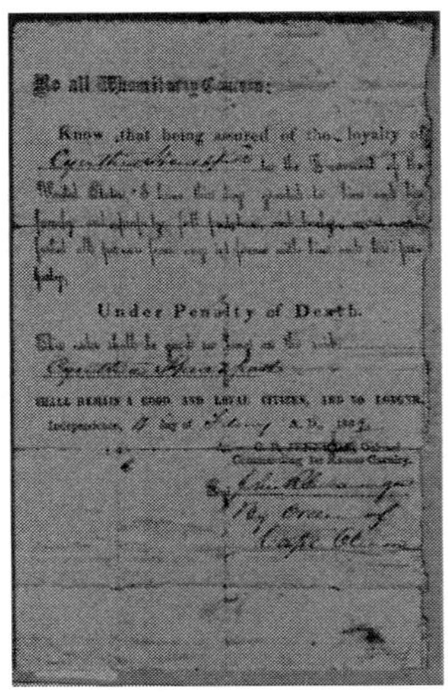

To all Whom it may Concern:

Know, that being assured of the loyalty of

Under Penalty of Death.

[536] Jackson County (Mo.) Historical Society Archives, Document ID 237F4.
[537] Jackson County (Mo.) Historical Society Archives, Document ID 237F5.

HUGHES

Will Hughes was related to the Steele Family and he went to St. Louis, Missouri following Order No. 11.[538]

HUGHES
(Additional Account)

Merritt Hughes, cousin of John T. Hughes, left their 1857 house and eighty acres farm (located today at the northwest corner of Parker and College Streets in Independence) under Order No. 11. After their departure, federal soldiers took-up residence in their home. A cannon ball hit the gabled roof and another hit the kitchen, demolishing the stove. Their property served as the main route of the march from the Battle of the Little Blue through Independence for the Second Battle of Independence in October 1864.[539]

HUNT

Mrs. Noah Hunt, a widow from Cass County, found safe haven in Johnson County after Order No. 11.[540]

HUNT
(Additional Account)

Noah Hunt was born on February 20, 1831 in Davie County, North Carolina. He immigrated with his father H. C. Hunt to Jackson County in 1842. Before the War, Noah Hunt operated a horse-powered Grist Mill six miles northeast of Lone Jack. On August 21, 1851, Noah married Miss N. E. Cave who had been born on April 1, 1833. After their marriage they moved less than a mile west of Lone Jack, where Noah began the wool-carding business. In 1856, he built another horse-powered gristmill and he continued in both businesses for years. At the time of Order No. 11, Noah and his wife had five surviving children (they had twelve children in all—see Appendix G under Jackson County for list of children living at time of Order No. 11, their son G. C. died at the age of three in 1857 and their daughter Mary born April 11, 1856 who was seven at

---

538 Eakin, "Tears and Turmoil," 27.

539 *Independence* (Mo.) *Examiner* 20 Dec. 1932, as referenced in Eakin and Hale, 220-221.

540 Rafiner, citing National Archives and Records Administration. Washington, D.C. Records of the Quartermaster General. Record Group 92. War Claim: 215-450; also, Glenn, Allen. "History of Cass County, Missouri." (Topeka, Ks.: Historical Publishing Co, 1917.), 577. It could not be determined by this author if Johnson County referred to Missouri or Kansas; but, presumably Missouri.

the time of Order No. 11 died on November 28, 1878). They like all other neighbors were forced off their property under Order No. 11.

After the Hunt's returned to the Lone Jack area their property holdings grew to two hundred and eighty three acres. By 1869 Noah had built a steam-powered gristmill and operated it on the farm. Noah's wife N. E. (Cave) Hunt died on March 7, 1878. Two years later, Noah remarried Mrs. Mary E. Travis, the widow of R. Travis, on September 13, 1880. By 1881, Noah and his new wife had a large household that included Noah's mother-in-law from his first marriage, seventy-year old Mrs. Susan Cave, and Zellah Travs, a daughter of his second wife, who had been born on November 20, 1863.[541]

HUNT
(Additional Account)
See CAVE

HUNTER

John Hunter was born in Stokes County, North Carolina on March 16, 1789. His parents, David and Dorothy Hunter were both natives of Virginia. On July 18, 1812, he married Elizabeth Jeans who was born in 1792. Hunter operated a distillery until he and his family moved to Missouri in April 1836. They had nine children born of the marriage, but only three were living by 1881. At the time of the issuance of Order No. 11, John was seventy-five and lived within one mile of the Rice property near Lone Jack.

John and Elizabeth suffered horribly during Order No. 11. Their two sons, William J. and John David Hunter, and their son-in-law, John Cave (husband of their daughter Nancy Hunter Cave) was murdered on September 6, 1863. They had to experience first-hand the sight of their family members dead in the field. "The family rolled their dead in sheets and buried them in the woods. Through the center of the grave runs the section line and around the spot, deeded to the family, is a concrete wall and a stone bearing the names."[542] John and his neighbor Martin Rice had the daunting task of burying all six victims.

Order No. 11 forced John, his wife and remaining family off their property. The Hunter's returned to their property after the War. After Elizabeth died, John continued to live on the same property he had first purchased in 1836. John passed away in 1883 at the

[541] Union, 936. The following are the list of the Hunt family children born after Order No. 11: Lucinda M., born December 24, 1864; W. W., born March 11, 1867; Virginia L, born February 7, 1870 and died November 13, 1870; Benjamin B., born September 25, 1871; R. T., born January 11, 1874; and S. A., born December 25, 1876.

[542] Short Stories of Days by Dollie Bee Breitenbaugh, 1959. 94-leave, typed manuscript. Two copies. Jackson County (Mo.) Historical Society Archives, Document ID 80F6.

age of 94. At the time of his death, his daughter Berrilla, who had been born on April 3, 1831, was caring for him.[543] (See also: *Chapter XII*, Cave, Potter, Rice and Tate.)

INGRAM

Charles Ingram who lived in Cass County went to Morgan County, Missouri under Order No. 11.[544]

INGRUM

E. A. Ingrum drowned in August 1863 while in the Confederate Army. His widow Hanna was ordered out of Jackson County under Order No. 11. The couple was married on November 11, 1847, and they had seven children. Hanna died on August 15, 1914.[545]

IRWIN

Julia Irwin was born Julia Harris. Her parents were Samuel Harris and Jane Hall Harris from the "Old Dominion State" Virginia. She was one of eleven children born of the marriage of her parents. Her parents settled in Jackson County and in 1827, her father purchased land from the federal government in 1827 for one dollar and twenty-five cents per acre. Her father was involved in the business of raising hemp and mules until he was killed in 1854 when a mule kicked him. Julia and her sister Laura Harris Flanery (who had been born on May 14, 1838 two miles southeast of Independence on land that is now Drumm Farm) were both compelled to abandon their property under Order No. 11 and Julia joined her sister Laura and the other women and children who went to Sherman, Texas (See also *Chapter XII* and Flanery.)

IRWIN
(Additional Account)

See ASHLEY

JACKSON

Mr. Jackson a resident of Cass County went to Monroe County, Missouri following the issuance of Order No. 11.[546]

---

[543] Union, 933-934.
[544] Rafiner, citing family histories compiled by Cass County Historical Society, Harrisonville, Missouri.
[545] *Lee's Summit* (Mo.) *Journal*, 20 Aug. 1914, as cited in Eakin and Hale, 227.
[546] Rafiner.

JEFFERYS/ JEFFREYS/ JEFFRIES

See CLEMENTS

JERRARD

Mr. Jerrard of Cass County went to the military station in Westport where he sought refuge during Order No. 11.[547]

JOHNSON

Greenville T. Johnson was born in Virginia on October 25, 1820. His wife Elizabeth Hall was also born in Virginia. They were married in February 1846. Four children were born of their marriage, but only sons, James and Coger W. survived. When the Civil War broke out, Greenville enlisted into the Confederate Army serving in Company D, First Missouri Calvary where he served until the War's end. This forced Elizabeth to tend to her boys and the family farm in the Sni-A-Bar Township of Jackson County.

When Order No. 11 was issued, Elizabeth was compelled to abandon the family farm and take her boys to Lafayette County. When Greenville returned home, he found his family living in Lafayette County and took them back to their farm in Jackson County. They found the property destroyed and land dilapidated. It took several years to restore the farm and property to its antebellum appearance.[548]

Coger W. Johnson, the son of Green and Elizabeth Johnson, was born in Jackson County on April 13, 1857. On March 10, 1874, Coger married Miss Mattie Knight of Kansas City and by 1881 they had one child born of their marriage, David Knight Johnson.[549]

James R. Johnson, the son of Green and Elizabeth Johnson, was born in Jackson County, Missouri on May 16, 1852. On April 25, 1878 James married Miss Louvina R. Montgomery of Jackson County, and the daughter of J. N. Montgomery. They owned one hundred and fifty six acres and James engaged in the stock trade during his lifetime.[550]

JOHNSTON

George Johnston of Cass County went to Cedar County, Missouri under Order No. 11.[551]

---

547 Rafiner, obituary; and *Oak Grove* (Mo.) *Banner*, 20 Nov 1908, as cited in Eakin and Hale, 238.
548 Union, 919.
549 Union, 919.
550 Union, 919.
551 Rafiner's best guess judging from the 1860 and 1870 U. S. Census returns.

JONES

The history of the Jones family was provided to Maude Lee (Burgess) by her grandfather, William Sanders Jones. Sanders related to her an incident he experienced under Order No. 11 that she later reported:

> *Jones went with Upton Hays at Big Blue where he had a recruiting station. About November 1861 they were attacked by the Federals under Lt. Col. Anthony at White Oak [Creek]. During the engagement John P. Williams, a friend and neighbor was fatally wounded, and while grandfather was trying to raise him up, he himself was shot through the lung with a mini ball, which came out under the shoulder blade. He was placed with Mr. Williams and another neighbor in a wagon and taken to their respective homes.*
>
> *He lay from November 11, 1861, until May 1862 and when just able to rise from his bed was taken prisoner by the Federals to Kansas City where he was kept about three months. When he was given his freedom, because he could not stay in Jackson County, he went to Fort Union, New Mexico, with a freight train owned by George Bryant.*
>
> *While on this trip, Order No. 11 was issued and his family was forced to leave Jackson County. Grandmother took what possessions she could pile in an old wagon drawn by a pair of two year old oxen and with her children went to Clinton County across the river to friends by the name of J. R. Dougherty at Plattsburg. They shared their two story log house with her. They were friends who also came from Kentucky. In December 1863, Grandfather returned and after much searching found his family and remained with them until the close of the War. They stayed near Liberty, Missouri....*
>
> *It was in the fall of 1866 when Grandfather returned to his farm to find everything in a dilapidated condition, for ruin and devastation had followed in the wake of the armies. With determination of true southerners he and Grandmother started to reestablish their home, restock the farm and cultivate the land. They again became prosperous farmers.*[552]

---

[552] Manuscript written by Maude Lee (Burgess) Immele, Kansas City, Missouri ca. 1960, for the Jackson County Historical Society Contest, entitled "Some Interesting Ancestors Who Lived in Jackson County." Jackson County (Mo.) Historical Society Archives, Document ID 77.03F13. Also, *Lee's Summit* (Mo.) *Journal* 29 May 1908, as cited in Eakin and Hale, 243-244.

JONES
(Additional Account)

Maude Lee (Burgess) Immele relayed family history of her grandmother, Jemima (Best) Jones, including an incident of Order No. 11. On August 30th, her grandmother "hitched up old 'Doll' to the buggy and with a heavy heart and tears in her eyes made the hot trip to Hickman Mills to get the slip of paper that meant giving up all they owned to those that were to take over this territory. Many were there for the same purpose. It was something they were forced to do." After questioning her she was given a signed piece of paper--the permit of removal--it read:

> *This certifies that Jemima B. Jones, lately living at Washington Township in the County of Jackson, Missouri, has established her loyalty to my satisfaction by the following witnesses: J. Collins and J. A. Moore and is permitted to remove to any Military Station in this District, or to any part of Kansas included in this District, except the counties on the Eastern border of the state. Witness my hand at Hickman Mills this 30th day of August, 1863, SIGNED L. D. Rowell (Capt. 2 Reg. Col. Vol. Commanding Station)....*

"With a heavy heart she gathered what she could take of her treasured possessions and clothing for the children and piled them into an old wagon drawn by a pair of two year old oxen.... Joining a group of neighbors the little caravan was on the way for the long trek across the hot, dusty roads in the late summer of 1863. They picked up others along the way. They could see smoke from smoldering homes as they passed deserted farms. They left a desolate picture behind...."[553]

JONES
(Additional Account)

James Jones of Cass County went to the military post at Harrisonville during Order No. 11.[554]

JUDSON
See ASHLEY

---

553 Manuscript written by Maude Lee (Burgess) Immele, Kansas City, Missouri ca. 1960, for the Jackson County Historical Society Contest, entitled "Some Interesting Ancestors Who Lived in Jackson County." Jackson County (Mo.) Historical Society Archives, Document ID 77.03F14.
554 Rafiner, obituary.

KEENEY

Isaac Keeney married Elvira Campbell of Platte County, Missouri in 1859. His wife Elvira (Campbell) Keeney was born in 1840 at Parkville, Missouri. Her father George Washington Campbell came to Missouri in 1833 but settled in land that did not become part of Missouri until the Platte Purchase. Because of the wilderness nature of the country, Elvira's father built a blockhouse for refuge against marauders and Indians.

After their marriage, Isaac moved his new bride to a farm near Martin City in southern Jackson County. The Keeney's lived in Martin City except for the years they were exiled by Order No. 11. During the War a party of raiders burned their farmhouse and drove away the stock. When they returned to the property, they rebuilt the farm buildings and house. The Keeney's had seven daughters: Mrs. Kate Estus, Mrs. Louise Hocker, Mrs. Rhodie Porter, Mrs. Sadie Pinkstown, Mrs. Temple Maxwell, Miss Fannie and Miss May Keeney. Isaac died in 1915 and Elvira died on January 31, 1919. They are buried in the Belton Cemetery in Belton, Missouri.[555]

KEENEY
(Additional Account)

Thomas Keeney, Cass County, sent his daughter to Kansas City after Order No. 11 was issued.[556]

KEMPER
See CLEMENTS

KENDALL
See ASHLEY

KERN
See WOODRUFF

KESHLEAR

Sarah (Riegs) Keshlear was a native of Montgomery County, Kentucky, where she was born on October 7, 1817. When she was about three years of age her parents moved to Boone County, Kentucky, and then relocated in 1829 to Jackson County. Sarah married Jacob Keshlear on July 25, 1837. Her husband had been a native of St. Charles

---

[555] Eakin, "Tears and Turmoil," 51.

[556] Rafiner, citing Belton Star-Herald. "The First Hundred Years: Belton, Missouri 1872-1972." (Belton, Mo. Belton Star-Herald, 1972). 30F.

City, Missouri and was born February 12, 1813. Jacob was reared on a farm, and came to Jackson County some time before 1829. Jacob died on May 16, 1858, leaving a family of eleven children, seven of whom were still living as of 1881: Mattie A. (later Mrs. Blackman) William G., James C., Joseph B., George B., Jacob B., and Alice.

After her husband's death, Mrs. Keshlear relied on the assistance of her sons, who helped care for their two hundred and twenty two acres farm. By Order No. 11, Sarah was compelled to leave the property. All the stock was taken and Sarah and her sons were imprisoned as a consequence of Order No. 11. The family returned to their destroyed property after the Order was rescinded.

Joseph. B. Keshlear was one of the eleven children born to Jacob and Sarah Keshlear. He was born in Jackson County on July 12, 1855. Joseph spent his boyhood days on a farm, receiving a common school education. After the Civil War, Joseph turned his attention to agricultural pursuits until 1875. In that year he became engaged as a mail contractor and followed the business four years; first, between Independence and Lexington, then between Buckner and Pleasant Hill. In 1879 he operated a hotel at Grain Valley until 1880 when he went into the livery business as the owner of Keshlear's Livery Stable. On March 1, 1877, Joseph married Miss Mary E. Karr, a native of Jackson County. As of 1881, the young couple had one child: Lola F.

KNIGHT

The plight of the Knight Family of Cass County was provided in the history of the Jim Knight Family:

> *When Order No. 11 was issued...Jim [and Mary (France)] Knight chose to leave his home and business located six miles north of Harrisonville, Missouri, rather than sign the Oath of Allegiance. He [his wife] and his family [of four children] moved to Henry County, Missouri. Times were hard before they moved. The unrest and feeling of distrust between neighbors in Cass County was strong. Neighbors would not discuss the War. Many friends had no idea how the other felt... Union or Confederate.... The hogs were turned loose to go wild when they left the County.... The Knights couldn't take them and didn't want the Union Army to get them.... The first winter after the War was the most difficult time in the life of Jim Knight and his family. The only things they had was the barren land they owned before the move to Henry County, a team of horses, a wagon, a small tent, and some personal items. The Union Army had burned the house, barn and destroyed anything of value on the land. A tent was erected in the timber on his land in Cass County. The winter was cold and the snow was deep. There was very little money to buy food. A*

> *rifle was used to hunt wild hog and turkey.... [The hogs they'd left behind became] the major source of food on their return to Cass County.*[557]

KREEGER

In the Jackson County pioneers there is an account of the affect of Order No. 11 on the Kreeger family:

> *With Order No. 11, the* [George W. and Araminta (Daniels)] *Kreeger family went to the vicinity of Carrollton in company with Hank Gibbons, a son-in-law, and Tom Tucker, a neighbor, in great haste. These families took only necessities with a cow and chickens, so they got along nicely together, dividing the milk, eggs, and whatever they had with the others.... The family returned to all that was left of their home--the* [1852] *log cabin--to find the fences gone and hogs wild. Mr. Kreeger built a rail pen with a gap and inveigled the hogs with food, closed the gap and got his winter supply of meat. Mrs. Kreeger was artistic and designed beautiful quilts, one of which was taken from her bed during the war. Years later she saw a copy of this quilt at [the] Independence Fair and named the design "Order No. 11."* [Quilt and appliqué designer, Ruby Short McKim, when publishing the design in the newspaper in 1929, told the story of Fannie Kreeger Haller who, as a ten-year-old girl, 'saw her mother's choice new quilt snatched from their bed by marauders back in 18[63] when Order No. 11 was the issue. She carried the treasured design in her mind and years after reproduced the quilt, christening it 'Order No. 11'.' The design, also known as Hickory Leaf or the Reel, is actually a pattern that dates back to the 1840s.] *Another design, taken from Seth Thomas grandfather clock, a wedding present from her father, she named, 'Missouri Rose.'*[558]

---

[557] Brigforth, Charles B., contributor. "How Order No 11 Affected the Jim Knight Family of Cass County, Missouri." *Kansas City Genealogist*; 23: 1&2: 13. Also, Rafiner, citing Harrisonville Heritage Association. *History of Harrisonville, Missouri*. (Harrisonville, Mo.: Harrisonville Heritage Association, 1988), 173.

[558] Short Stories of Pioneer Days by Dollie Bee Breitenbaugh, 1959. 94-leave, typed manuscript. Two copies. Jackson County (Mo.) Historical Society Archives, Document ID 80F6. See also the Jackson County (Mo.) Historical Society Archives newspaper subject files for a couple of newspaper references to the Kreeger family. Ruby Short McKim also publicized the "Order No. 11" quilt design in the *Kansas City* (Mo.) *Star* in 1929, and in her ca. 1931 book, "101 Patchwork Patterns." The design is also featured in Barbara Brackman's book, "Butternut and Blue: Threads of the Civil War," who masterfully identified the 1840s design.

Fannie Kreeger was the daughter of George W. and Araminta Kreeger. Fannie was "born August 21, 1853, on a farm two and a half miles west of Lone Jack" where the family resided until Order No. 11 was issued. They left their home driving "an oxen team to Davis, Lafayette County, Missouri." When they crossed the Tebo River, the team of oxen was "so thirsty they broke from the driver's control and plunged into the stream."

Fifty years later Fannie now Mrs. Haller crossed the stream in a large touring car. On the night before the Battle of Lone Jack, Mrs. Haller remembered that General F. M. Cockrell with his force of Confederates camped in the pastures on her father's farm and that the sounds of the guns and the roar of the battle were heard the next day.[559]

After the War, their son J. D. Kreeger received a college education at Central College in Howard County, Missouri and became a teacher. He later married Katherine H. A. Armstrong a native of Mercer County, Kentucky who had been born on May 6, 1854. J. D. was appointed census enumerator for the thirty-third district of Missouri. By 1881, J. D. and his wife owned three hundred and forty acres "of choice and well improved land" where his younger brother R. R. Kreeger worked with him "extensively in stock." The family was members of the Presbyterian Church.[560]

KREEGER [561]
(Additional Account)

Mrs. Kathryn Kreeger, widow of J. W. Kreeger, died October 17, 1919. Like many people affected by Order No. 11 in their earlier years, their obituaries mentioned the ordeal. Mrs. Kreeger "had lived in Jackson County 71 years except for a short period in the Civil War when the famous Order No. 11 caused her to go to Carrollton, Missouri."[562]

LAFFOON

Drury Laffoon left his home in Cass County after Order No. 11 and became a freighter.[563]

---

[559] Jackson County (Mo.) Historical Society Archives, Document ID Vertical File Newspaper Subject Files: Military/Wars: Civil War: 1863-"Order No. 11": "Untitled," undated, unattributed typewritten transcript. Another article appeared in the *Lee's Summit* (Mo.) *Journal*, 26 Sept. 1929. Also, Eakin, "Tears and Turmoil," 77.

[560] Union, 937.

[561] "Deaths in Kansas City--Kreeger," *Kansas City* (Mo.) *Star*, 18 Oct 1919.

[562] "Deaths in Kansas City--Kreeger," *Kansas City* (Mo.) *Star*, 18 Oct 1919.

[563] Rafiner, citing Cass County Historical Society. "Cass County Missouri Families." (Harrisonville, Mo: Cass County Historical Society, Inc., 1976).

LAMAR

John Lamar of Cass County went to Platte County, Missouri when Order No. 11 was issued.[564]

LEE

See ASHLEY

LEWIS

Robert Lewis of Cass County traveled to Henry County, Missouri where he resided in slave quarters.[565]

LEWIS
(Additional Account)

Sarah Lewis of Cass County went to Montgomery County, Missouri following the issuance of Order No. 11.[566]

LIVESAY

"When [Mrs. Livesay, mother of Mrs. A. H. Smith] looked out of her kitchen door she could see seven burnt chimneys of the homes of her friends."[567]

LUDWICK

John Lafayette Ludwick and his family lived in Butler, Bates County. John served in the Confederate Army and became a prisoner of war in the summer of 1863 following the surrender of the City of Little Rock. "He was taken to St. Louis then to Alton and Cairo, Illinois Military Prisons. He was finally taken to New Orleans for exchange…just in time to surrender with his command at Shreveport, Louisiana. On his return, he found his family in Henry County where they had gone on account of Order No. 11. He and his family finally returned to Butler in 1880."[568]

---

564 Rafiner, citing Belton Star-Herald. "The First Hundred Years: Belton, Missouri 1872-1972." (Belton, Mo.: Belton Star-Herald, 1972), 10F.

565 Rafiner, Family histories compiled by Cass County Historical Society, Harrisonville, Missouri.

566 Rafiner's best guess judging from the 1860 and 1870 U. S. Census returns.

567 "Order No. 11," a speech Alberta Wilson Constant delivered to the Independence Young Matrons, ca. 1960, Jackson County (Mo.) Historical Society Archives, Document ID 154.01F30.

568 "History of Bates County," (1883), as cited in Eakin and Hale, 275.

LUNDY

Creed Lundy of Cass County went to Boone County, Missouri to escape Order No. 11.[569]

LUTTRELL

Mrs. Ethel Milton Hughes gave the following account of the history of the Luttrell family during the War. "My older sister, Bess Milton, now Mrs. Forrest C. 'Phog' Allen of Lawrence, Kansas, was a constant playmate of Myra Carr, Mrs. Luttrell's granddaughter, at their old Luttrell home. She remembered Mrs. Luttrell's account of the hardships which the family encountered in their preparation to vacate their home under Order No. 11." The story continues on how the Luttrell's worked hard digging trenches to bury prized positions and specifically a feather mattress.

Sarah Luttrell, the daughter of Samuel Luttrell, married C. J. Dillingham on November 19, 1868. Dillingham was a farmer born on November 24, 1847 in Jackson County and was educated in the common schools. Sarah like her future husband was evicted from their land under Order No. 11. Sarah and C. J. later had three children: Lulu L., Willis J. and Annie M.[570]

LYKINS

Martha A. (Livingston) "Mattie" Lykins was born Martha A. Livingston near Frankfort, Kentucky in January 1824; she was orphaned at an early age. She was a descendant of Phillip Livingston, one of the signers of the Declaration of Independence. Mattie grandmother took her in when her parents died and ensured her granddaughter's education in the schools of Frankfort and Shelbyville, Kentucky. Mattie at the age of sixteen went to live with her sister Mrs. Thomas J. Hughes in Jefferson City, Missouri. When her sister and her husband moved to Lexington in 1847, Mattie went with them

---

[569] Rafiner, S: 12.

[570] Christensen, 6-7; and, Union, 915. (See also, *Chapter XII* for the story of Samuel Luttrell and Family.) The older sister of Mrs. Ethel Milton Hughes was Bessie E. Milton. Bessie later married Phog Allen (1885-1974) on June 25, 1907. Phog was born in Jamestown, Missouri—his parents were from Virginia. Phog grew-up on Delaware Street in Independence (just ½ block north of the future home of President Harry S. Truman), and he attended and coached at William Chrisman High School. Bessie and Phog Allen had two boys and two girls. Today Phog's name is synonymous with basketball, and especially the University of Kansas Basketball Team whose field house is named after him.

where she opened a private school for young ladies. While teaching, she met Dr. Johnston Lykins, the father of Sarah Lykins, one of Mattie's pupils. On October 12, 1851, Mattie married Dr. Lykins. They moved to Kansas City where Dr. Lykins served on the Kansas City Council and as the city's mayor.

Mattie, a southern sympathizer, was banished from Jackson County during Order No. 11 in August of 1863. Her banishment was pursuant to orders of the provost marshal. Mattie's husband was allowed to remain in Kansas City. Mattie bordered the steamship at Westport Landing in Kansas City while her husband stood-by for the sad parting. When hostilities began to calm in Jackson County, Mattie returned to her husband in Kansas City. On June 11, 1866, Mattie wrote a letter to President Andrew Johnson describing the horror she had experienced during the Civil War. Dr. Lykins died on August 15, 1876.

After the death of Dr. Lykins, Mattie on June 24, 1878, became the third wife of George Caleb Bingham. While she and the "Missouri Artist" had known each other for many years, their marriage was shortened by Bingham's death during the summer of 1879. After Bingham died, Mattie devoted the next twenty years serving the needs of Confederate War Veterans. She established the Widow's and Orphans Home in Kansas City and the Veterans Home in Higginsville. When Mattie died in 1890 her Last Will and Testament required the bulk of Bingham's art to be sold for the benefit of Confederate veteran projects and programs, including the orphanage, the Lykins Institute (later known as the "Little Sisters of the Poor"). Mattie is buried between her two husbands Dr. Lykins and George Caleb Bingham at Union Cemetery in Kansas City.[571]

LYKINS
(Additional Account)

See ASHLEY

LYNCH

Elijah and Miranda (Snodgrass) Lynch were married in 1853. Elijah was born July 24, 1829 at 'Hickory Grove,' within a half mile of the Jackson County Courthouse in Independence. He spent all his life in the area, "except during the civil War when he served as a Confederate soldier. He was in the Battle of Lone Jack, and in a later battle was wounded and taken prisoner. After being confined in a prison in St. Louis he was sent to Alabama to be exchanged and had to walk back home at the close of the War.... When Order No. 11 was operative...Mrs. Lynch moved to Lafayette County, Missouri,

[571] Eakin, Joanne Chiles. *Blue & Grey Chronicle*. (December 1998) 2: 2: 5-6. See also Bingham in *Chapter XIV*. Mattie's original writings are also preserved in the Jackson County (Mo.) Historical Society Archives, Independence, Missouri. For more Lykins information see the book, *Missouri Star*, by Rose Ann Findlen.

where she remained 'till the War ended." Elijah and Miranda are buried in the Koger Cemetery.[572]

MADDOX

See ASHLEY

MAJORS

Isham Majors of Cass County went to Shawnee Mission in Kansas. His family later found safe haven in Cooper County, Missouri after Order No. 11 had been issued.[573]

MANN

Christopher Mann came to Jackson County in 1843 and purchased one thousand four hundred and forty acres of land from Nicholas Crenshaw. H. M. Vaile later owned part of the land. Mann's property overlooked the Missouri River to the north. On August 5, 1863, bushwhackers robbed him and told him that they would not burn his home as long as he remained on his property. Mr. Mann then observed rebels setting the torch to other homes in the neighborhood. When the men returned to Mr. Mann's property they asked him to admit that he was a "black abolitionist." Mr. Mann replied "I am a Union man, Sir!" He was then struck over the head and his house was put to the flames. He was compelled to leave his land by Order No. 11, and after the war, he built a second home on the Gilpin farm, one and one-half miles northwest of Independence where he reared a large family and his descendants farmed many years in the Courtney bottoms.[574]

MASTERS

Mary C. (Leader) Masters was born in Independence, Missouri on March 4, 1842. On May 24, 1860, at the age of eighteen, she married twenty-seven year old M. C. Masters who had been born in Jessamine County, Kentucky. M. C. had immigrated into Jackson County in 1853. After their marriage, Mary moved to the farm her husband had purchased in Sni-A-Bar Township in 1859. M. C. and Mary had seven children: Lizzie H., Susie L., Annie M., Charley P., Effie, Nellie, and Gertie B. The Master's remained on their farm until they were forced to move by Order No. 11 and they reported to Independence.

---

[572] "A Native of Jackson County," *Kansas City* (Mo.) *Star*, 22 July 1918.

[573] Rafiner, citing family histories compiled by Cass County Historical Society, Harrisonville, Missouri.

[574] Wilcox, 360, citing *Independence* (Mo.) *Examiner*, 6 June 1906.

During the summers of 1864 and 1865, M. C. became engaged in freighting on the plains, and after the War he returned to his family and farmed until 1871 when he moved the family back to Independence. M. C. operated a mercantile trade in Independence until March 1879 when he relocated his family to Blue Springs and opened a new mercantile business.[575]

MATTIN

S. E. Mattin left Cass County following Order No. 11 and went to Carroll County, Missouri.[576]

MAXWELL

See also ASHLEY

While Thomas F. Maxwell was away fighting for the Confederacy, his wife, Martha A. Maxwell, was banished under Order No. 11. Thomas surrendered at Shreveport, Louisiana, on June 9, 1865, and he and his wife returned to Independence to live; Thomas and Martha are buried at Woodlawn Cemetery in Independence.[577]

MCCREADY

See ASHLEY

McCOOL

James McCool and his son Peter V. McCool lived in Bates County. Peter joined the Missouri State Guard, but never went into the Confederate Army. "His father, James McCool, lived at home until Order No. 11 when he moved to Henry County [Missouri] and died there in 1865."[578]

McFARLAND

T. J. McFarland of Cass County relocated to Henry or Cedar County, Missouri after Order No. 11.[579]

---

[575] Union, 922.
[576] Rafiner, citing Glenn, Allen. 482.
[577] *Independence* (Mo.) *Examiner*, 4 Oct. 1905, as cited in Eakin and Hale, 287. Also, 1900 Census, Independence, Missouri.
[578] Bates 1883.
[579] Rafiner, citing Goodspeed Publishing Company. "History of Hickory, Polk, Cedar, Dade and Barton Counties, Missouri." (Chicago: Goodspeed Publishing Co., 1889), 755.

McFERRIN

Elizabeth McFerrin of Cass County found safe haven in St. Clair County, Missouri after Order No. 11.[580]

McGILL

Mr. McGill, Cass County, went to Boonville, Cooper County, Missouri after Order No. 11 had been issued.[581]

MEADOR

Included within the short vignette of Jackson County pioneers, with emphasis on Lee's Summit families, is the account relating to the affect of Order No. 11 upon the Meador family:

> *With Order No. 11, Sara Angeline Meador, her sister and three children, with other settler's wives and daughters took what they could easily carry and started for safety in Kansas. One day, as they stopped for lunch by a stream, Miss Meador's ox team took a bath, and one of them waded out so far that it was impossible to reach him. The entire company was obliged to wait until evening when the beast decided to conclude his bath and proceed on the journey. Only one elderly man attended this company fleeing from their homes. When they returned after some months, they found the houses but no fences.*[582]

MERS

Andrew Mers a resident of Cass County went to the military post at Pleasant Hill during Order No. 11.[583]

---

[580] Rafiner, citing Glenn, Allen. 464.

[581] Rafiner, obituary.

[582] Short Stories of Pioneer Days by Dollie Bee Breitenbaugh, 1959. 94-leave, typed manuscript. Two copies. Jackson County (Mo.) Historical Society Archives, Document ID 80F6.

[583] Rafiner, National Archives and Records Administration. Washington, D.C., Records of the Quartermaster General. Record Group 92. War Claim: M-167.

MERS
(Additional Account)

W. T. Mers of Cass County stayed at the military post in Pleasant Hill during Order No. 11.[584]

MILAM

Thomas J. Milam of Cass County left Missouri under Order No. 11 and went to Jackson County, Kansas.[585]

MILLER

"Reverend George Miller and family also fled the Order." In letters to his church, Rev. Miller stated, "the Union committed no greater blunder--I feel like saying crime--in its conduct of the war…." Rev. Miller and his family returned to Pleasant Hill, Cass County, in June 1865, after almost two years of exile. He wrote: "My wife and I came up through Johnson, Cass, and Jackson Counties, and for miles and miles saw nothing but lone chimneys to mark the spots where happy homes stood. It seemed like a vast cemetery--not of a living thing to break the silence except the little birds that sang among the leafy branches. Man no longer existed there.'"[586] Miller also mentions an "incident about an old man, John Poyntz, who had been confined in a military prison for his sympathetic expressions on behalf of Southern friends. He was released into the custody of George Miller who kept him in his home until Poyntz was finally banished from Missouri."[587]

---

[584] Rafiner, citing National Archives and Records Administration. Washington, D.C., Records of the Quartermaster General. Record Group 92. War Claim: M-145.

[585] Rafiner, citing Glenn, Allen.

[586] Eakin, "General Order Number Eleven" (First of the Series on Order No. 11). *Kansas City Genealogist;* 23: 1&2: 12.

[587] 1961 article about the Jackson County Historical Society-sponsored exhibit of Bingham paintings at the Nelson Atkins Art Museum. Also, Kansas City (Mo.) Times, 24 Aug. 1963, "Order No. 11, Crux of Civil War Bitterness." And, a third article, "Order No. 11 Drove Families from Border Homes in 1863," Kansas City (Mo.) Star, 9 Apr 1961, lists three families affected by Order No. 11. Jackson County (Mo.) Historical Society Archives, Document ID 144.01F33. For a copy of Poyntz banishment order, see Jackson County (Mo.) Historical Society Archives, Document ID 57F25. Also, Rafiner, citing Miller, George. 100-107.

MIMS

William Leonidas Bone (8th Mayor of Independence) wrote to Professor F. Mims on December 22 the following in which he discussed the treatment of Independence homes and depredations during the Civil War, and after Order No. 11:

> *Your house still stands, not burnt, not many houses have been burnt in the city, but all more or less injured by rough treatment. An Irishman is now in your house, treats it tolerably well, pays no rent. The Stegall brick house, near ours, is now occupied by an old man who cut up some 10 or 15 hogs in the parler* [sic.] *room on last Sunday. That is a fair sample of how dwelling houses are treated in the town. A number of good business houses on the Square are now occupied as house-stables by the Kansas 11th."*[588]

MONTGOMERY

"[The Levi and Elizabeth (Holloway)] Montgomery home was also burned. He was 'called out' by the Union men. The family begged him not to go, for at this time of his life [Levi] was blind. He resisted, and met them anyway.... A young man with the aggressors had been befriended by Levi at some time and prevented his being shot. The family left the area, going to the southeastern corner of Nebraska, staying until it was safe to return."[589]

MOORE

Lawson and Zirlida Moore, and their six children, the youngest 18-months-old, fled Jackson County during Order No. 11. Lawson's brother, Will Moore, "packed the household goods and obtained a team of oxen and a wagon to transport the family and their possessions to the ferry. They crossed the river to Clay County, landing at Baxter's Landing (Liberty Landing). A young couple, perfect strangers, sheltered and fed the refugees until Mr. Moore arrived and they went on to Clinton County. The Moore's did not return to Jackson County after the War."[590]

---

[588] Jackson County (Mo.) Historical Society Archives, Document ID 100F24.

[589] Svobada. 14.

[590] Eakin, "Tears and Turmoil," 84-85, citing Wilcox and *Kansas City* (Mo.) *Star,* 9 Aug. 1942.

MOORE
(Additional Account)

Mr. Moore of Cass County went to Sedalia, Pettis County, Missouri following the issuance of Order No. 11.[591]

MOORE
(Additiona l Account)

See GEORGE

MORRIS

John Calvin Morris of Morristown, Cass County, according to R. L. Morris and W. H. Morris, loaded his milling equipment and personal belongings onto ox-drawn wagons and moved with his family to the military post at Westport as a result of Order No. 11.[592]

MORROW

Jesse Morrow, a farmer by profession, was born on March 4, 1810, in Simpson County, Kentucky, and spent his boyhood days working on a farm. At the age of sixteen, he began learning the blacksmith trade. He married Miss Sylvia Hudspeth, in the spring of 1829. She was also a native of Simpson County, Kentucky, born in 1807. The Morrows were Baptists. He and his wife moved to Jackson County in the fall of 1829 where he farmed and opened a blacksmith shop in the Fort Osage Township of Jackson County. In 1845 he traveled to California, but did not remain there long. Around 1855, Jesse went to Texas.

During the Civil War, Jesse and Sylvia's son Benjamin Morrow, born on December 16, 1840, enlisted in Shelby's Division, and served about five months attaining the rank of lieutenant. Benjamin returned home where he joined William Quantrill and remained for the next three years with the bushwhacker chieftain. Benjamin was involved in the three large guerilla massacres: Lawrence and Baxter Springs, Kansas in 1863 and Centralia, Missouri, October 1864. After the issuance of Order No. 11, Jesse Morrow moved his family to Carroll County, Missouri, where they resided for one winter.

The Morrow family returned to the Fort Osage Township of Jackson County. Jesse and Sylvia had nine children, but six had preceded her in death. At the time of Sylvia' death in 1873, her three surviving children were Benjamin, Nathan and Rufus F.

---

[591] Rafiner; another source said he found safe haven in Kansas City. Belton Star-Herald. "The First Hundred Years: Belton, Missouri 1872-1972." (Belton, Mo.: Belton Star-Herald, 1972).

[592] Jackson County (Mo.) Historical Society Archives, Document ID 152F37.

B.[593] By 1881 Jesse owned three hundred acres of "fine land," part of which was under cultivation. He had also begun deeding most of his farm property to his children. Jesse's daughter, Rufus, kept house for her father. She was described as an "estimable young lady, and one that is deserving of a happy future."[594]

MUIR

Bettie Muir, daughter of Mr. and Mrs. Hezekiah Muir, was born about three miles east of Warrensburg in Johnson County, Missouri. At the start of conflict, Bettie was a young child living with her parents near Raytown. Bettie's father became an officer in the Confederate armies of General Sterling Price and General Jo Shelby. After he enlisted, her mother moved the family to the home of her uncle, John S. Muir, whose home was located on the crossroads between Independence and Westport. Their living conditions were quite difficult for her and her family along her grandmother and three other families resided in the house owned by her uncle. She said that all the children were put to bed crosswise on a long lounge.

Under Order No. 11 they were all banished and joined a wagon train bound for Texas. Bettie said it took about three months to journey to Texas. Her father Hezekiah eventually reached his family in Texas. After several years in Texas and Arkansas, Bettie returned with her parents to Jackson County. She later married Mr. Cabness and they lived in Jackson County. She once wrote that "I do not complain of rationing and food shortages—I think of the hard times of that war of long ago, and know that compared to it, the present has only little inconveniences." Bettie Muir Cabness died in 1948 and was buried in the Blue Springs Cemetery.[595]

Buckner Muir was 84 years old at the time of the enforcement of Order No. 11. Under the Order, Buckner and his 56-year-old son, Sam, were the only two pilots of sixty-one women and children out of Jackson County. Among the women and children under the guidance of Buckner and his son were Laura Harris Flanery, her small son, and her sister, Mrs. Julian Irwin.[596]

Hezekiah Muir's brother was John S. Muir. After Hezekiah went into the service of the Confederacy, his family went to live with John and his family, but they were all banished under Order No. 11 and joined a wagon train bound for Texas. John Muir remained several years in Texas before returning to Jackson County.

---

[593] Union, 922.

[594] Union, 907-908.

[595] Thruston, 269-270; Independence (Mo.) Examiner, 26 Sept. 1930; Kansas City (Mo.) Star, 5 Mar. 1946 (For more information see Flanery above and in *Chapter XII*).

[596] Eakin, Joanne Chiles, *Blue & Grey Chronicle*. (August 2000): 3: 6: 3. See also *Chapter XII* and Flanery).

MUIR
(Additional Account)
See FLANERY

MURRAY

John Murray was born in the County of Galway on the Shannon River on June 15, 1831. He lived with parents in Ireland until eventually moving to the United States in 1848, landing at New Orleans. He worked on the steamboat line on the Mississippi River until he came to Jackson County around March 1852. John married Mary F. (Smith) from the area of Blue Springs on September 6, 1855. John and Mary had one child William F. John worked in the mercantile business in Blue Springs and served as post-master until 1859. After the war broke out John entered the Confederate Army as a private and served until the fall of Vicksburg. He was promoted through various grades up to Provost Marshal. John was fighting for the south when Mary and William were evicted from their property in Blue Springs because of Order No. 11.

When the rebellion ended, the family was reunited and returned to Independence. Mary died in September 1866. As of 1881, John was engaged in the real estate business in Kansas while his son William F. was living in Independence.[597]

MYERS

Harry Myers a citizen of Cass County went to Boone County, Missouri while William H. Myers of Cass County went to the military post at Pleasant Hill during Order No. 11.[598]

NAILOR

Fountain Nailor of Cass County went to Saline County, Missouri because of Order No. 11.[599]

NEAT
See ASHLEY

---

[597] Union, 821

[598] Rafiner, citing National Archives and Records Administration. Washington, D.C., Records of the Quartermaster General. Record Group 92. War Claim: M-47.

[599] Rafiner, citing National Archives and Records Administration. Washington, D.C., Records of the Quartermaster General. Record Group 92. War Claim: 114-1160.

NELSON

Hector Nelson of Cass County traveled to Howard County, Missouri to escape Order No. 11.[600]

NEWMAN

J. N. Newman, Osage Township, Vernon County, was arrested in Vernon by the 3rd Wisconsin Cavalry, August 19, 1863, and sent to Fort Scott to stand trial.[601]

NICHOLSON
See ASHLEY

NOEL

A short vignette of Jackson County pioneers, with emphasis on Lee's Summit families, includes the affect of Order No. 11 on the Noel family: "Under Order No. 11, the [James W.] Noel family took refuge in Carroll County. Mrs. Noel had a large engraving of sorrow and desolation of the country at this time which Mr. Noel bought from the Kansas City Art Institute."[602]

NOLAN
See ASHLEY

NOLAND

Harriett Newton Noland went to Saline County as a result of Order No. 11.

NORMAN

C. P. Norman, Cass County, went to eastern Missouri after Order No. 11 was issued.[603]

---

[600] Rafiner, obituary.

[601] http://freepages.genealogy.rootsweb.ancestry.com/~jrbakerjr/provost3/bushwhackers.htm (viewed 10 June 2013).

[602] Short Stories of Pioneer Days by Dollie Bee Breitenbaugh, 1959. 94-leave, typed manuscript. Two copies. Jackson County (Mo.) Historical Society Archives, Document ID 80F6. A description of the Battle of Lone Jack is provided in this manuscript. Also, *Independence* (Mo.) *Examiner*, 19 Sept. 1919, as cited in Eakin and Hale, 325.

[603] Rafiner, citing National Archives and Records Administration. Washington, D.C. Records of the Quartermaster General. Record Group 92. War Claim: M-1213.

O'BANNON

E. O'Bannon, a Union man from Cass County, moved from his farm to Higginsville in Lafayette County. Another source said he found safe haven in Henry County. Regardless of the location, he definitely was forced off his property because of Order No. 11.[604]

O'DONNELL

See ASHLEY

OLDHAM

The plight of the Oldham family under Order No. 11 provides the following account:

> *Mrs. Oldham and her family left their home as a result of Order No. 11...and they took refuge in Chariton County, Missouri. When they returned two years later it was in the dead of winter. The entire family traveled back to Independence in a wagon drawn by a yoke of oxen. The weather had become very cold by the time they reached Lexington and snow was falling. 'My father,' Mrs. Mary Julia Oldham recalled, 'threw terror into the onlookers when he drove his team and heavily-laden wagon across the frozen river, the ice cracking under every turn of the wheels. We made the other side safely, however.*[605]

According to historian Joanne Chiles Eakin, the members of the Oldham family banished under Order No. 11 consisted of Albert G. Oldham, his wife, Mary Julia Oldham, and their two small children, Althea Oldham and Thomas Oldham along with Mary Julia's mother, Amelia Hickman.[606]

---

[604] Rafiner, citing Glenn, Allen. "History of Cass County, Missouri." (Topeka, Ks. Historical Publishing Co, 1917.), 581; obituary.

[605] Jackson County (Mo.) Historical Society Archives, Document ID Vertical File Newspaper Subject Files: Military/Wars: Civil War: 1863-"Order No. 11": "Death to Woman of 99: Mrs. Julia Oldham had lived in Independence 88 years," *Kansas City* (Mo.) *Star*, 11 Jun 1932. This article mentions the Civil War Battle of Rock Creek in Independence in 1861, and also states that, "rock used in the building of the present ancient county jail at Independence was taken from Rock Creek on the Oldham farm more than 70-years-ago (as of 1932).

[606] Eakin, "Tears and Turmoil," 59.

OLDHAM
(Additional Account)

John Rice Oldham owned property in Blue Township on the Independence-Harrisonville Road about one to two miles south of the provost marshal's post in Independence (today the property would be located on the east side of Noland between 23rd and 35th). At the time of Order No. 11, Oldham was around sixty-two years old, married and had four children. The Oldham family was allowed to return to their farm after John executed on behalf of his family a verification of loyalty on April 5, 1864. In the document, he committed that he and his family would remain loyal, report disloyal activities and disclose guerilla and rebel actions to the provost marshal.[607]

OUSLEY

The story of Matilda (Hunter) Ousely, her daughters and the death of her son Andrew Ousley is fully covered in *Chapter XII* of this book.

OWINGS
See GEORGE

PAGE
See COLCORD

At least one year prior to the issuance of Order No. 11, the Union army was imprisoning (and banishing) women suspected of aiding and abetting the Southern cause. Mrs. Hall reused an envelope to pencil a short note to her sister, Venitia Mary "Nish" Colcord (Mrs. Granville Dyson) Page.

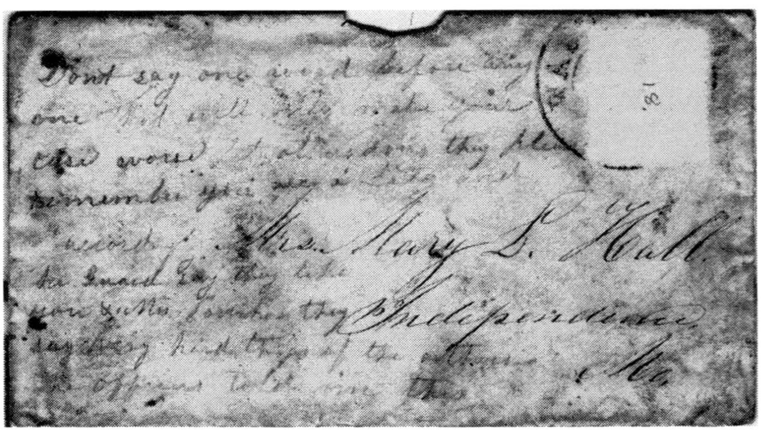

[607] John R. Oldham, Fold 3. NARA Publication M322: Compiled Service Records of Confederate Soldiers Who Served in Organizations from the State of Missouri. Record Group 109, 586957. One of Oldham's wives, Jane Reid Moberly) was the sister of Mary "Polly" Moberly Rhoads (George W. who lived on the site of the Raytown City Hall) and Ann Wood Moberly Hill (Adam).

Nish was imprisoned in a Kansas City jail with other young Southern women. Mrs. Hall warned her sister, Nish, on the face of an envelope, to carefully guard her confidences and conversation. The note is dated in 1862 (the Thomas Property on Grand Avenue had not yet been converted to the jail that would collapse in August 1863), and so Nish was most likely incarcerated in one of those other pest-infected facilities along the Missouri River. She left the county for a short period under Order No. 11.

Nish married Granville Dyson Page in Jackson County on December 31, 1860. Granville purchased the defunct 1840 Jackson County Jail in Independence where he engaged in the slave trade until the eradication of slavery in Missouri. In 1872, Granville became the first, elected Jackson County Marshal.[608]

PALMER

Jacob Teaford Palmer lived in the Hickman Mills area of Jackson County at the time of Order No. 11. A supporter of the southern cause, Palmer had several run-ins with federal troops. He and Pierre Rippetoe were first arrested along the Missouri-Kansas border and ordered to take the loyalty oath. Palmer refused to take the oath and was told he would be executed if he failed to comply. Palmer was allowed to make a final prayer, but he was eventually released after giving his assurance not to take up arms against the Union. He was then hired by Henry Chiles to drive a herd of hogs to the Marcus Gill farm near Santa Fe. While driving the hogs, Kansas Jayhawkers charged into them scattering the hogs and taking the horses. Once again Palmer was released without further harm. Jacob then took several other odd jobs until he landed a position with Mr. Jeffreys, eventually becoming the overseer of Jeffreys' property.

When Mr. Jeffreys returned from Europe on August 25, 1863, he wanted Palmer to remain to supervise his property near Hickman Mills. To do so, Palmer would have had to submit to the loyalty oath; Jacob refused. Jeffreys then offered to assist Palmer and his wife to go to Indiana under Order No. 11. Once again Palmer refused because he would have to submit to the loyalty oath in order to gain a provost marshal pass to make the journey. In the end, Jeffreys took the Palmers to the military post at Hickman Mills.

---

[608] Pictured here is a reproduction of the envelope (Jackson County (Mo.) Historical Society Archives, Document ID 2F16), which was first published with the Jackson County Historical Society's permission in 1960 on page 40 of the publication by Robert S. Townsend, "Decade of Decision," 1855-1865. (Kansas City, Mo., Kansas City Life Insurance Company, 1960). Also, a copy of the second edition printed in 1978. Frederic James provided a variety of drawings throughout the booklet, including Rock Creek School; Battle of Independence (showing the 1859 Jackson County Jail); Price's Last Raid; and, Private Dunlavy Captures a General. Order No. 11 is discussed on pages 39-43. Jackson County (Mo.) Historical Society Archives, Document ID 96F15. More biographical information about Mr. and Mrs. Page may be found in, "LOCK DOWN: Outlaws, Lawmen and Frontier Justice in Jackson County, Missouri" (Independence, Mo.: Jackson County Historical Society, 2009), as meticulously researched and written by David W. Jackson and Paul Kirkman.

The provost post in Hickman Mills was located in a house with a window overlooking a front porch. All citizens had to go to that window and take the loyalty oath. After standing in line, Jacob and his wife approached the window. Once again his employer Jeffreys came to the Palmer's assistance. Jeffreys was sitting on a chair next to the window and when he saw the Palmer's Jeffreys quickly explained to Union officials that Jacob was his foreman and that he and his wife needed a pass to visit her family in Greencastle, Indiana. The pass was issued and neither Jacob nor his wife was compelled to take the oath. The following Monday morning Jacob's father in-law took the family to the Gillis House in Kansas City where they booked passage up the Missouri River to St. Joseph. Jeffreys' benevolence continued as he had one of his carpenters build a box so the Palmers could ship their belongings to Philadelphia.[609]

PALMER
(Additional Account)

Dr. Palmer a resident of Cass County went to Clay County after Order No. 11.[610]

PARISH

Mr. Parish a citizen of Cass County left Missouri after the issuance of Order No. 11 and went to Indiana.[611]

PARK

When Order No. 11 was declared, "The Park family from Independence found refuge at Multnomah with the Dougherty family. Their small daughter of two years, Annie [Park], became friends with John [Dougherty], age six, and years later when they grew up, married."[612]

PARKER

The following account is provided as the impact of Order No. 11 upon the Parker family:

---

609 Eakin, "Tears and Turmoil," 31-32.
610 Rafiner, citing Glenn, Allen, 329.
611 Rafiner, obituary.
612 Jackson County (Mo.) Historical Society Archives, Document ID Vertical File Newspaper Subject Files: Military/Wars: Civil War: 1863-"Order No. 11": "Government Order No. 11 - (1863)," by Vera Haworth Eldridge, in an undated, unattributed newspaper clipping.

With Order No. 11, Mr. [Zachariah] Parker sold some cattle for $400-500 a head, took his family and a few belongings with two ox teams and wagons--their horses had been stolen--and went traveling for safety, not knowing where, and finally arrived in Chariton County, near Keytesville, at the farm of General Sterling Price. Captain Edwin Price, the son, was kind to Mr. Parker and rented him the farm which produced good crops. About a quarter mile away was a store with a room above used for officer's headquarters.

As Mrs. Parker was getting supper, two drunken soldiers arrived, ordered supper and became boisterous with cursing and swearing. Mr. Parker slipped away to the Captain who casually walked up and took a seat on the veranda, waiting to escort the ruffians to camp. They were probably sufficiently reprimanded, and the Parker family was never molested again. Peace restored, Mr. Parker came back with nothing but his family and his farm. "The Negroes were freed, and he was not allowed to bring them even if they wanted to come.

His house was burned and everything gone. Mrs. Parker heard of a stand table in possession of someone in Independence. When she called for it, she virtually had to buy it in "paying for its safekeeping...." Eventually "the family moved into their new home.[613]

PARKER
(Additional Account)

William H. Parker owned a tract of land in Polk Township, Cass County, Missouri and was displaced from his property during Order No. 11.[614]

PARSONS
See AKERS

PARSONS
(Additional Account)

"[Daniel White Parsons] parents lived within the territory that was covered by Order No. 11…and lost all their property."[615]

---

[613] Short Stories of Pioneer Days by Dollie Bee Breitenbaugh, 1959. 94-leave, typed manuscript. Two copies. Jackson County (Mo.) Historical Society Archives, Document ID 80F6. Also, *Lee's Summit* (Mo.) *Democrat,* 7 Nov. 1933, as cited in Eakin and Hale, 79.

[614] Boisvert, Robert. "Order 11 Drove Families Out," *Pleasant Hill* (Mo.) *Times*, 14 Aug. 1991.

PATTERSON

Joel Patterson was born in Surry County, North Carolina on June 19, 1816. He was reared and educated in his native county. He married Miss Sarah Patterson also a native of Surry County on September 8, 1833. In the fall of 1836, Joel and Sarah immigrated to Jackson County. It took them sixty days to reach Jackson County as they traveled by land. During their marriage they had ten children, but only three were living in 1881: Jesse, Joel B. and Mollie.

For several years, Patterson engaged in the saw and grist milling business. His mill was valued at $2,000 but it was burned during the Civil War. By Order No. 11, the family moved to Carroll County, where they remained for about six months before returning to Jackson County. After the war, Joel farmed his one hundred acres farm near the Grain Valley Post Office and worked as a bridge contractor.[616]

PAYNE

E. S. Payne, Cass County, found safe haven in Boone County, Missouri after Order No. 11.[617]

PHILLIPS

James M. Phillips of Cass County went to the provost marshal's post at Harrisonville during Order No. 11 where he signed the loyalty oath and remained living in the city.[618]

PIERCE

See ASHLEY

PLEASANT

Edward Pleasant, Cass County, went to Johnson County, Missouri during Order No. 11.[619]

---

[615] Jackson County (Mo.) Historical Society Archives, Document ID Vertical File Newspaper Subject Files: Military/Wars: Civil War: 1863-"Order No. 11": Unidentified Kansas, newspaper clipping, January 1923, for Daniel White Parsons, son of Samuel and Barbara (Ely) Parsons.

[616] Union, 924.

[617] Rafiner, citing Glenn, Allen, 777.

[618] Rafiner, citing National Archives and Records Administration. Washington, D.C. Records of the Quartermaster General. Record Group 92. War Claim: M-325.

[619] Rafiner's best guess judging from the 1860 and 1870 U. S. Census returns.

POLLARD
See ASHLEY

POTTER

Benjamin Potter, one of the six victims of the Kansas 9th, was living near Lone Jack at the time of the issuance of Order No. 11. Despite being 75 years old, the Kansas troops took him into custody. It is believed that he was seized and killed for having been known as a Confederate sympathizer and for having fed Quantrill and his men two nights before their attack on Lawrence.

Martha Potter was a daughter of Benjamin Potter. She lived with and cared for her aged-father. Potter's four other daughters lived in Indiana and had not moved to Missouri with their father and other siblings. Martha helped her father take care of the children of her late brother, William Potter, who had died years earlier in a trip out west. The small, orphaned children of William lived with their grandfather at the time of his death; they were Amanda, Marion and Thomas, age twelve.[620]

POTTS

Fannie E. (Parsons) Potts was married to Peter J. Potts. She was the daughter of Samuel and Barbara (Ely) Parsons. Fannie "saw the Civil War in all its fury and bloody angles; saw the roof of her home burned over her head by Order No. 11… Through it all, in stress of day and storm of night, she fed her little flock and prayed to Him who is the arbiter of all earthly things."[621]

POWELL

A short vignette of Jackson County pioneers, with emphasis on Lee's Summit families, includes the Powell family story of Order No. 11:

> **Alvis [Powell] was not in the war but stayed to look after the women and do chores while the family went to Saline County under Order No. 11. Cattle roamed at will, one young steer decided to occupy the porch bedroom, walked in and shut the door. How long the creature**

---

[620] *Daily Missouri Republican* (Saint Louis), 11 Sept. 1863; Eakin, "Tears and Turmoil," 6-18; Wilcox, 369; Eakin and Hale, 354. See also *Chapter XII*.

[621] Jackson County (Mo.) Historical Society Archives, Document ID Vertical File Newspaper Subject Files: Military/Wars: Civil War: 1863-"Order No. 11": Unidentified Paola, Kansas, obituary, January 1929, for Mrs. Peter J. (Fannie E. (Parsons)) Potts, daughter of Samuel and Barbara (Ely) Parsons.

> **was imprisoned no one knows, for he had almost perished when found.**[622]

POWELL
(Additional Account)

John Beeler Powell married Parthen Meador before the Civil War and they had six children. They were forced to leave under Order No. 11 and went to Cooper County, Missouri where they lived in a house that had previously been used as a sheepfold—the only shelter they could find. In the late fall of 1863, John returned to Jackson County and harvested a wheat crop which he had left standing the previous June. The "thrifty people" that lived around Independence at the time took everything moveable from their farm.[623]

POWELL
(Additional Account)

Jonathan R. Powell a resident of Cass County traveled to the provost marshal's post at Pleasant Hill during Order No. 11.[624]

POYNTZ
See MILLER

QUICK

Cornelius Quick, Cass County, went to Henry County, Missouri as a result of Order No. 11.[625]

RALSTON

Sam Ralston was born on September 4, 1819. His first wife was Sarah A (Jordan) and from this marriage Sam had two children, John, born in 1838 and Rowena, born in 1839. Sam's wife died on July 20, 1846 and five years later in 1851, his daughter Rowena followed her mother to the grave. Within seven months of the death of his wife, Sam married his second wife Mary Catherine Hill on February 5, 1847; she was the

---

[622] Short Stories of Pioneer Days by Dollie Bee Breitenbaugh, 1959. 94-leave, typed manuscript. Two copies. Jackson County (Mo.) Historical Society Archives, Document ID 80F6. A brief description of the Battle of Lone Jack is provided in this manuscript. Also, Wilcox, 328; Eakin and Hale, 355.

[623] *Lee's Summit* (Mo.) *Journal*, 12 Nov. 1914, as cited in Eakin and Hale, 355-356.

[624] Rafiner.

[625] Rafiner.

daughter of Adam Hill. There were seven children born of this union: Fedora, born Feb 9, 1848, Samuel born in 1850, Anne "Anna" born on January 25, 1853, Adam born on August 29, 1855, William H. born on July 14, 1858, Katherine born on January 4, 1860, and Harry M. born on May 25, 1862 (another daughter, Margaret Woodson was born after the war on January 7, 1867).

The Ralston family was banished before Order No. 11 but still experienced the ravages of the conflict and Order. It appears Samuel spent the rest of the War in Omaha, Nebraska, and went to Texas in the 1870's.[626] In 1875, Samuel learned that his daughter, Anna, wanted to marry the noted guerilla and bandito Alexander "Frank" James. Samuel protested the union of his daughter to the noted guerilla. Despite her father's disapproval, Anna eloped. She and Frank had only one child, Robert James (Frank died in 1915 and Anna passed away in 1944—they were both cremated and their cremains inurned at the Hill Park Cemetery in Independence).

REED

Farlow Reed left Cass County and Missouri and went to Miami County, Kansas to escape Order No. 11.[627]

REID

John W. Reid was born on June 14, 1820 in Lynchburg, Virginia. In 1840 he moved to Jefferson City, Missouri where he taught school and studied law. During the Mexican American War, Reid served as a captain, twice being wounded. After the Mexican War, he moved to Independence and began to practice law. On November 14, 1850, Reid married Marcella Flournoy, a widow with two children. John and Marcella had three children, two of whom died in infancy. Marcella died on April 26, 1855 while John was serving his first of two terms in the Missouri House of Representatives (1854 and 1856). In 1855 Reid and two other legislators was appointed to revise the statutory laws of Missouri.

Reid was actively involved in the days of Bleeding Kansas. On August 30, 1856, Reid led between two hundred and fifty to three hundred pro-slavery men into Osawatomie, Kansas. During this engagement, Reid's men killed one of John Brown's sons. In all ten to twelve men were killed before Reid and his men burned down the Kansas town. From that day forward, Reid was called 'General" and remained a strong pro-slavery advocate during those battling years in Kansas.

---

[626] *Sedalia Weekly Bazoo*--after 1873; additional information was further provided by Nancy M. Ehrlich, a noted historian in Independence and direct descendant of the Hill Family, Sam Ralston's second wife.

[627] Rafiner, obituary.

When Reid was not involved in border war issues, he practiced law in Independence and served in the Missouri House. As a state legislature he developed an interest in the western expansion of railroads, and in 1857 Reid introduced a bill in the Missouri House to incorporate the Hannibal and St. Joseph Railroad Company. His bill passed with considerable opposition. In 1860, Reid was elected to Congress from the 5th district or Missouri (he took his seat on 4 March 1861).

His Congressional stay in the nation's Capitol was short-termed. Although Reid opposed secession, he resigned from Congress and to volunteer on the staff of Sterling Price. He was later taken captive by federal troops and spent about one year in the federal prison in St. Louis. He was released on parole on the condition that he would take no further part in the conflict. Reid did not break the terms of his bond after he was released.

When Order No. 11 was issued, Reid moved to Liberty, Missouri, where he spent a couple of years. At War's end, Reid married Sallie Cochrane Magraw, the daughter of the wealthy pioneer Santa Fe trader, William M. F. Magraw. John and Sallie made their honeymoon home in the mansion at 126 S. Pleasant in Independence. Their residence had been a gift to the newly-weds by Sallie's mother. It was a large six- room, brick home with a large orchard to the south.

Reid's post-bellum life he spent in the rapidly developing community of Kansas City. He became president of the K.C. Savings Association (a forerunner of a number of banks in the state), and he purchased land later incorporated into Kansas City. After he settled in Kansas City, Reid resumed the law practice. He also became actively involved in the banking industry. He died in Lee's Summit on November 22, 1881 and was buried in a large mausoleum Elmwood Cemetery in Kansas City.[628] There is a U. S. veteran's marker from the Mexican War in front of the gravesite.

RICE

Martin Rice was born November 22, 1814 in Campbell County (now Union County, in east Tennessee). He was the eldest of eleven children born to Enoch and Mary Rice. He remained loyal to the Union during the Civil War. Despite his loyalties, he quietly remained at his home, aiding and befriending so far as he could. The impact of General Order No. 11 upon Rice and his family is one of the most haunting and shocking. When Martin, Isaac and six other men were taken into custody by a company of the 9th Kansas on September 6, 1863, Martin and Isaac were released; the other six men Benjamin Potter, age 75, John S. Cave, aged 45-50, William Hunter, aged 47, John David Hunter, age 35, William C. Tate, aged 32 and Andrew Ousley, aged 17 were murdered. Martin, Isaac and seventy-five years old John Hunter had the daunting task of burying the six dead men on the Roupe property—and near the spot where they had been killed. Rice

[628] Eakin, Joanne Chiles. *Blue & Grey Chronicle.* (April 1999): 2: 6: 1-3.

moved his family to Lafayette County, and then removed to Ray County on the north side of the Missouri River before returning home after the War.

Rice's family returned to their farm near Lone Jack. His aged mother lived with him until her death on March 12, 1880 and she was buried next to her husband and father. Rice became known for his poetry and prose writing style. In 1878, his poetry and writings were published. His poem about his experiences during the War, and in particular under Order No. 11, was released under the title of The Exodus of Eighteen Sixty-three; or Order Number Eleven." In 1881, he re-published another edition of his book of poetry, and he provided his account of "What I Saw of Order Number Eleven." Rice's careers included farming, surveying, teaching and operating a nursery. He also commenced the propagation of fruit trees, which he continued throughout his life. He maintained and kept up some of the oldest orchards in Jackson County and in adjoining counties by his grafting procedure. He was often heard say, "when I am dead and gone, the orchards will be my memorial stones." He was described as beginning life poor and never accumulating much of this world's goods. By the time of his death on December 5, 1903, Martin had earned the title of "The Sage of Lone Jack."[629] He is buried in Pleasant Garden Cemetery in Lone Jack, Missouri.

Mary Rice was the daughter of Martin Rice. After the War, Mary married Mr. Butler, a Union veteran. According to several accounts, Mary and her father Martin had been self-appointed spies for the federal government. Martin allegedly gathered information about his southern neighbors and Mary reportedly couriered the data her father obtained to the Union provost marshal in Pleasant Hill. These allegations were spread by one of Rice's nieces who told it to pro-southern neighbors. In his poem, Rice extolled the patriotism of a girl who had given reports to Union soldiers who "gave three cheers for that young lady when she went riding home." Those who later held Rice culpable for the death of the six men on September 6, 1863 pointed to this line in the poem as evidence to their contentions against Rice (see *Chapter XII* for further details of the Rice story and the death of the six men).

William Calvin Tate, one of the six victims murdered on September 6, 1863 was a son-in-law of Martin Rice. William was born in North Carolina in 1832 and married Martha J. Rice, the daughter of Martin Rice, in Lafayette County. In 1928 David Tate who was only five at the time of his father's death was told that his grandfather Martin Rice had been responsible for the death of his father during Order No. 11.[630] It is

---

[629] Martin Rice, "Rural Rhymes, and Poems from the Farm." (Kansas City, Mo.: Millett & Hudson, 1877); Martin Rice, "Rural Rhymes, and Tales of Olden Times." (Kansas City, Mo.: Ramsey, Millett & Hudson, 1882); "Martin Rice Remembers: What I saw of Order Number Eleven." *Missouri Historical Review* (April 2012): 106: 3; Union, 940-941; and *Daily Missouri Republican*, Saint Louis, Missouri, 11 Sept. 1863. In 1879, Martin was remembered as one of the Old Settlers of Cass County, and was named an honorary member of the Association.

[630] Eakin, "Tears and Turmoil," 7.

unknown whether David believed the story or not. It is undisputed that his grandpa Rice if he had been alive at the time would have vigorously denied the allegation.

RICE
(Additional Account)
See ASHLEY

RICE (Additional Account)

A recollection of a grandchild of Martin Rice included the following account of the family's experience under Order No. 11:

> *On August 15, 1863, a meeting was held in Van Buren Township. The purpose of the meeting was to take into consideration the guerrilla warfare existing in the Sni Hills, and the rumors of Order No. 11. Resolutions requesting General Ewing to refrain from issuing Order No. 11 were adopted. The resolutions were to be taken by Grandfather Rice to General Ewing, who was stationed in Kansas City. By the time Grandfather could get to Kansas City, General Ewing had gone to Leavenworth. His secretary, Mr. Plumb, would not 'move an inch.' Order No. 11 was to be put into force. Grandpa arrived home from his fruitless mission on the evening of August 20. Quantrill and his men had been there and had ordered large amounts of food. On Tuesday, August 25, General Ewing issued his famous Order. On Monday, August 31, 1863, Grandfather went to Pleasant Hill and secured a Certificate of Loyalty along with permission to move his family to an area of safety. Lexington was the chosen site. September 6, 1863, found the Rice family ready to leave their home. A squad of soldiers from the Kansas 9th Regiment rode up. Captain Coleman was in charge. Martin Rice, his son, Isaac, a son-in-law, William Tate, David Hunter, Andrew Ousley, a youth of 17, John S Cave, William Hunter and Benjamin Potter--all neighbors of Mr. Rice's--were marched to the Roupe farm, about one mile south of the Rice home. During a conference which followed, Grandpa showed his Certificate of Loyalty to Mr. Barney Dempsey, an officer.*
>
> *In a few moments Captain Coleman called Grandfather to one side, 'You will take your son and travel,' was the order given. Grandpa and Uncle Isaac left at once. A few moments later shots were heard. As soon as was possible, Grandpa, his daughter, Martha (Mrs. Tate), whose husband had been shot, Miss Jane Cave, and some of the smaller children returned to the scene of the tragedy.*

Six bullet-ridden bodies were found. After a hasty burial, proceeded by a committal service performed by my Great Grandfather, the remaining few families left the community in compliance with Order No. 11.[631]

RICE
(Additional Account)

See also, *What I saw of Order No. 11 and Poems by Martin Rice,* reprinted by Fabius Martin Butler in 1965.[632]

RICHARDSON

Mary I. Richardson was born in Cass County and married Joseph Richardson. They were forced off their property and "experienced all of the hardships of the Civil War" and of Order No. 11. Joseph died on November 15, 1893. After his death, Mary married James S. Berry in January, 1909, and they lived in Sapulpa, Creek County, Oklahoma. Mary died on May 25, 1911, at the home of her daughter, Mrs. P. W. Oldham, 3520 Flora Avenue, Kansas City and she was buried in Freeman, Cass County.[633]

RIDLESBERGER
See ASHLEY

RITTER

Henry T. Ritter was born around 1823. He came with his father's family from Virginia to Missouri in 1835. They settled near what is today Lee's Summit. In 1849, Ritter married Miss Taylor in Jackson County. He worked making boards from logs and operated a blacksmith shop in Kansas City. Kansas Jayhawkers continuously harassed the Ritter family because of their Southern sentiments. Ritter fought in the battle of White Oak Creek near Raytown on August 18, 1862 and two days after the Battle of Lone Jack. White Oak Creek is a small tributary of the Little Blue River. Ritter described the battle as follows:

---

[631] Manuscript written for the Jackson County Historical Society Contest, entitled "Some Interesting Ancestors Who Lived in Jackson County." Jackson County (Mo.) Historical Society Archives, Document ID 77.01F29. The story given by one of Rice's grandchildren was inaccurate about Order No. 11. The reference to the adopted resolutions delivered by Rice actually referred to the orders issued by Ewing under Order No. 9 and Order No. 10 and not Order No. 11—see *Chapter VIII* above relating to Rice delivering the committee's resolutions about Order No. 9 and Order No. 10 to Ewing's headquarters in Kansas City.

[632] Jackson County (Mo.) Historical Society Archives, Document ID 163.01F26.

[633] "Death of Mrs. J. S. Berry," *Kansas City* (Mo.) *Star*, 25 May 1911.

> *During the fight I saw a Jayhawker fall within twenty-five yards of where I stood. Another ran to pick him up and with my shotgun I piled him on top of his comrade. Another came and I piled them three deep. I was somewhat shielded by a rock and discovered one of the enemy shooting at me from behind a tree. I bided my time and finally got a reasonably fair shot at him. He immediately broke cover and ran, falling before he went far. Those were the days indeed.*

Under Order No. 11, he and his family left Missouri and made their way to Texas were they settled on a farm. They returned to Jackson County after the War and settled on their property. By 1900 H. T. Ritter was seventy-eight living in Lee's Summit.[634]

ROBINSON

See also DEHONEY

J. J. Robinson was known as Parson Robinson. He was a Baptist preacher who had settled in the area of Raytown after arriving from Kentucky in 1846. Neither the house of Parson Robinson nor that of neighbor Leander DeHoney was destroyed under Order No. 11. It was believed that this uncommon fortune was because the Red-leg's captain knew that the preacher and DeHoney were Masons. Parson Robinson actually delivered a copy of Order No. 11 to Leander DeHoney. Like the DeHoney's and others the Robinson's left Raytown under Order No. 11.[635]

ROBINSON

John Robinson a resident of Cass County left Missouri under Order No. 11 and went to Jacksonville, Illinois.[636]

ROSS

See ASHLEY

ROUPE

The Order No. 11 story of blacksmith John Roupe and his son John Henry is covered in *Chapter XII*.

---

[634] *Lee's Summit* (Mo.) *Journal*, 2 Feb. 1900; Eakin, Joanne Chiles. *Blue & Grey Chronicle.* (February 1999): 2: 3: 4.; Hickman, 738-739; Eakin and Hale, 371.

[635] Thruston, 21-23. An elementary school in the Raytown School District is named for the Robinson family—Robinson Elementary (the authors two daughters attended the school).

[636] Rafiner, citing Hon Family Papers, Pleasant Hill Historical Society, Pleasant Hill, Missouri.

ROUT
See CLEMENTS

ROWLAND

The Rowland's of Cass County went different directions after Order No. 11 was issued. Jesse Rowland went to Carroll County, Missouri.[637] William Rowland went to the military post at Harrisonville where he remained after he took the loyalty oath.[638]

RUNYAN
See ASHLEY

RUSSEL

Mrs. Elizabeth Southworth (Williams) Russel was born July 3, 1820 to Ezra and Mary (Fisher) Williams, and died in Breckenridge, Macon County, Missouri, in April 1911. She lived in Lone Jack during the Battle of Lone Jack. The next year she and her family were "driven out by Order No. 11" as were Mr. and Mrs. A. C. Ellez (later of West Line, Cass County, Missouri).[639]

SAINT CLAIR
See CLARKSON

SAINT CLAIR
(Additional Account)

Joseph V. St. Clair "was born June 3, 1844, on a farm near Blue Springs." During the early years of the Civil War he served under General Sterling Price and later Confederate General Kirby Smith. "He returned to Jackson County where he spent the rest of his life, except in the years when Order No. 11 was in force and he lived in Lafayette County." After the War, he became a farmer and married Alice Broaddus. They are buried in Perdue Cemetery near Blue Springs.[640]

---

[637] Rafiner, citing Glenn, Allen, 502.
[638] Rafiner, citing National Archives and Records Administration. Washington, D.C., *Records of the Quartermaster General. Record Group 92. War Claim: L-3306.*
[639] "As to the Battle of Lone Jack," *Kansas City* (Mo.) *Star,* 26 April 1911; and Russel's Missouri death certificate.
[640] "Deaths in Kansas City—St. Clair," *Kansas City* (Mo.) *Star*, 18 Dec. 1918.

SANDERS

Mr. Sanders of Cass County traveled south to Taney County, Missouri when Order No. 11 was issued.[641]

SAUNDERS

During the Civil War, John B. and Druscella (Stayton) Saunders lived in a brick dwelling, later known as 17536 R. D. Mize Road. On one occasion, a Union soldier who had been shot from his horse and left for dead was given aid and assistance by the Saunders.

> *Although of Southern sympathies and in danger of reprisals, the Saunders carried the wounded trooper into their home and nursed him back to life and strength, sufficient to enable him to rejoin his command. Months later, when, in the grim struggle, it was decided that this home should be included among those to be burned, this former trooper, now risen in the ranks to a command, used his authority to order that whoever touched the Saunders' house would have to answer to him in person. So, the old house still stands, after more than a century.*

The Saunders family "left for Richmond, Missouri, their only possessions were those which could be carried in a farm wagon drawn by a 'span' of oxen, horses being out of the question." According to historian Joanne Chiles Eakin, "eleven-year-old Susan Saunders rode on an overturned oak drop-leaf table stuffed with a featherbed."[642] "On their return after the conflict, the Saunders found their house still standing but stripped of its doors and windows and of all furnishings, except the top of a sewing machine and a broken rocking chair."[643]

SCRIVNER[644]

See also CLEMENTS

---

[641] Rafiner, citing Creighton Centennial Commission. *History of Creighton, Missouri*. Clinton, Mo.: The Printery, 196). 269.

[642] Eakin, "Tears and Turmoil," 73-76. Two pieces of the Saunders' furniture are pictured, courtesy Louis K. and Jeanie White, who restored the Saunders' home at 17601 R. D. Mize Road.

[643] Jackson County (Mo.) Historical Society Archives, Document ID Vertical File Newspaper Subject Files: Military/Wars: Civil War: 1863-"Order No. 11": "Mercy Shown Union Soldier Saved Old Saunders' Home," undated, non-attributed newspaper clipping by Harry R. Nelson. The Saunders' home passed to the Latimer and then to the Haas family.

[644] Burton, Angela M. and Hannah Clements Montgomery. "Order No. 11 in Hickman Mills." *Kansas City Genealogist;* 24: 1: 16.

SCOTT

Mr. Scott a resident of Cass County in order to escape Order No. 11 bordered a Missouri riverboat that was completely filled to capacity.[645]

SCRUGGS

See ASHLEY

SECHREST

Charles Sechrest and his wife Katie had fourteen children, twelve of whom grew to adulthood; three served in the Civil War. The Sechrest's suffered many depredations by Kansas Red Legs who stole their livestock, eight head of mules and other family belongings. Mrs. Katie Sechrest stood at the gate, bravely trying to save some of the livestock, but the leader of the Red Legs threatened to burn the house if she refused to let them remove the mules. They took refuge at the government post at Hickman's Mill during Order No. 11. [646]

SETTLE

J. F. Settle from Cass County went to Arrow Rock, Missouri after Order No. 11.[647]

SEVIER

"Henry Clay Sevier…was one who was financially broken by Order No. 11, and removed his family to Clay County."[648]

---

645 Rafiner, citing "The Elvira Scott Diary," State Historical Society of Missouri, Collection 1053, 22.

646 Hickman, 664-665. Eakin And Hale, 388.

647 Rafiner. Arrow Rock had been the child hood residence of George Caleb Bingham.

648 Jackson County (Mo.) Historical Society Archives, Document ID Vertical File Newspaper Subject Files: Military/Wars: Civil War: 1863-"Order No. 11": "Government Order No. 11 - (1863)," by Vera Haworth Eldridge, in an undated, unattributed newspaper clipping.

SHALER

N. T. Shaler of Cass County went to a military post in Jackson County under Order No. 11.[649]

SHARP

The Sharp family legend claims that under Order No. 11, Matt Sharp of Cass County went to Harrisonville at Harris' spring. Another source said he went to Adams County, Illinois.[650]

SHELEY

The Nelson and Lucinda (Robertson) Warren "family witnessed the Battle of Independence from a hilltop. When Order No. 11 was issued, the Warren family went to Wayne City and took steamboat to Nebraska City."[651]

SHELTON

William H. Shelton and his father's family from Bates County went to Madison County, Illinois, for a year after Order No. 11 was issued. In the fall of 1865 they returned to Johnstown in Bates County.[652]

SHELTON
(Additional Account)

Minerva Shelton who lived in rural Cass County reported to the provost marshal at Harrisonville during Order No. 11.[653]

---

649 Rafiner, Family histories compiled by Cass County Historical Society, Harrisonville, Missouri.
650 Rafiner, citing National Historical Company.
651 Jackson County (Mo.) Historical Society Archives, Document ID Vertical File Newspaper Subject Files: Military/Wars: Civil War: 1863-"Order No. 11": "Parents of Mrs. Jane Sheley, wife of Dr. Oliver Sheley." Independence (Mo.) Examiner, 8 Nov. 1937. The First Battle of Independence occurred on August 11, 1862. See, Monaco, "Blood on the Streets," 25.
652 Eakin and Hale, 391.
653 Rafiner, obituary.

SHIPLEY

William Shipley sent his wife Amanda to live in Harrisonville, Cass County, during Order No. 11. Unfortunately, Amanda was arrested for singing Confederate songs and was imprisoned in the jail and where she died in 1863. Before her death, their son Preston Shipley had gotten into trouble with the children of northern parents. Amanda sent her son to Preston to be with Cole Younger, but he was too young, so they found him a home with widow Mrs. Brackon near Peculiar in Cass County. Two years later when Preston was fourteen he was sent to work and live on the Nathan Harrelson ranch near Peculiar.[654]

SLAUGHTER

O. V. Slaughter and his wife Elizabeth lived on property that was in the southern part of Jackson County on the Blue Ridge. Eight children were born of their marriage. Their land was contiguous and directly to the north of the farm owned by Solomon Young and his family. When Order No. 11 was issued, they all abandoned their homes. Like their neighbors the Slaughter's returned to their properties when the restrictions of Order No. 11 were lifted. In the succeeding years, their children would attend college and had the opportunity to watch Young's grandson Harry S. Truman grow-up working on the Young farm.[655]

SLOAN

F. X. Sloan of Cass County left Missouri and went to Miami County, Kansas in compliance with Order No. 11.[656]

SMART

See ASHLEY

SMART
(Additional Account)

Historian Joanne Chiles Eakin provides the insight of the Civil War experiences of Mrs. Susan Smart and her sufferings under Order No. 11. Susan was the wife of Judge Robert Glover Smart. Judge Smart settled in Independence at an early age. He went on to become an attorney practicing in the city for several years with Russell Hicks. He served

[654] Osborn, Donald L. *Tale of Amarugia Highland*, as cited in Eakin and Hale, 393.
[655] McCullough, 32 and 76.
[656] Rafiner, citing National Historical Company, 471.

in the Missouri Senate during the 12th General Assembly. By the time of the Civil War he served as the elected Circuit Court Judge for Jackson County. However, he resigned from his judicial post in 1861 as he refused to submit to the Loyalty Oath. He and Susan owned ten slaves and had four children: Virginia L., Robert A., Mary C. and Susie. Before Judge Smart and Susan moved the family to Saline County, Missouri during the winter of 1862, they buried their silverware and other valuable article.

In July 1863 while residing in Saline County, Judge Smart was accused of harboring bushwhackers. Although he had been sympathetic to the southern cause, he never took part in the War. On July 20, 1863, the Federals of the 7th Mo. Volunteers went to his house. At first Judge Smart intended to flee capture, but soon chose to surrender. He was immediately executed by Federal troops. Mrs. Smart was officially banished from Jackson County. Her youngest child Susie was born after the death of her husband.[657]

SMITH

See ASHLEY

SMITH
(Additional Account)

Eliza (Wallace) Smith's home was located about three miles south of Independence. When she was a child, her father loaded a small table onto a raft where he floated down the Tennessee River to Ohio River; then down the Ohio to the Mississippi River; then up the Mississippi and then west on the Missouri River to their home near Independence. "The pioneer housewife's feather beds were not only her most highly prized possessions but in an emergency were always to be saved first and never abandoned until all else was lost." When Order No. 11 was enforced, she carried her family's small table and her feather bed that she had saved from her burning home to a rowboat. She and her precious heirlooms were then ferried to safety to Clay County."[658]

[657] Wilcox, 359, Eakin, "Tears and Turmoil," 83; and Branded as Rebels.

[658] Jackson County (Mo.) Historical Society Archives, Document ID Vertical File Newspaper Subject Files: Military/Wars: Civil War: 1863-"Order No. 11": "Government Order No. 11 - (1863)," by Vera Haworth Eldridge, in an undated, unattributed newspaper clipping. Also printed in: Withers, Robert Steele. Historical Notes and Comments (Memoirs and a Gospel Testimony of A. H. Wallace). Missouri Historical Review (1946-1947), Vol. 41, 226.

SMITH
(Additional Account)

At the time of Order No. 11, Franklin Smith Jr. lived in Cass County. The following account of the affect of the Order on Smith was retold years later.

> *When Franklin Smith, Jr., was born, Franklin, Sr., was in the Legislature. As a birthday gift he had a full-length portrait of himself painted. There is some doubt as to whether it is an authentic Bingham portrait, but it figures in the tale of Order No. 11. The Smith home was burned in the wake of the order and someone remembered the portrait and entered the burning home, cut it out of its frame, and rolled it up to save it, only to find that because of the heat, it had been damaged. Only the bust portion of the portrait was salvaged. It hung in the home of Franklin Smith, Jr., for as long as he lived.*[659]

SMITH
(Additional Account)

Dicey Smith described leaving Cass County under Order No. 11 and surviving the hardship of winter.[660]

SMITH
(Additional Account)

Isaac Smith from Cass County went to Cooper County, Missouri because of Order No. 11[661]

SMITH
(Additional Account)

John Smith left his home in Cass County under Order No. 11 and went east to Centralia, Missouri.[662]

---

[659] Svobada, 14.
[660] Rafiner, citing "Reminiscences of Aunt Dite: Diary of Dicey Smith." (Harrisonville, Mo.: Cass County Historical Society), 4.
[661] Rafiner, citing Glenn, Allen, 769; and, obituary.
[662] Rafiner, citing Glenn, Allen, 817.

STAYTON

Moses Stayton, born on December 6, 1828, was the son of John Stayton. He came to Jackson County as a young boy with his parents. In 1837, his father built a fine brick home some two miles east of Independence (the home was razed in 1920). Moses later married Lucy Ann (Hite) Stayton who had been born in Kentucky on September 18, 1838. Her father James Coleman Hite had settled in Miller County, Missouri in 1853. Moses and Lucy were the parents of Rebecca Stayton. During Order No. 11, Moses and his family went to stay with Lucy's parents in Miller County. They did not return to Jackson County until after the War.[663]

Rebecca Stayton was a young girl at the time of Order No. 11. She and her parents went to stay with her maternal grandfather James Coleman Hite in Miller County to escape the Order. Rebecca later married John Alfred Bills Adcock a native of Independence (see Adcock).

STAPLES

Abner Ford Staples was born in 1793. Abner and his wife Elsie traveled from Tennessee to Missouri by wagon with their infant child Mary in 1826. The following year the family settled in Independence. Abner built a home for his growing family in 1835. The family was financially secure at the outbreak of the Civil War. They owned a farm that consisted of three hundred and thirteen acres on the Lexington Road and about sixty slaves. During the War, Abner took the slaves to Cooper County, Missouri where he freed them. His wife and children followed him to Cooper County during Order No. 11. They returned to their property after the War and found everything destroyed, except their house.[664]

STEELE

Albert Gallatin Steele was born on January 8, 1810 in Kentucky. He was a descendant of the famous pioneer Daniel Boone. In 1854 Albert moved to Jackson County and eventually married Emily Hampton where they resided in Sibley, and near the former factory, Ft. Osage (the fort had stood from 1808 until the 1820s).

Albert and Emily had three sons. Not only did the Steele's lose their home under Order No. 11, they lost their boys within a one-year time period. Two of their sons Joseph and George served in the Missouri State Guard in 1861 and enlisted in the Confederate Army in 1862. Their eldest son Joseph E. died in Alabama on July 7, 1864

---

663 Eakin, "Tears and Turmoil," 48.

664 Eakin, "Tears and Turmoil," 83, citing Bill Curtis and *Kansas City (*Mo.) *Star*, 19 Jan 1930.

as a result of battle wounds. He was buried in Hamburg, Alabama. Their son George served under General John Pemberton in Hardee's Corp., Bledsoe's Battery of Smith's Regiment of the Light Artillery during the Battle of Champion Hill, Mississippi in May 1863. George and some two hundred other men were taken prisoners during the battle, and incarcerated in the Ft. Delaware prison where George died on October 9, 1863. Their third son, James Thadeus Steele, the twin brother of George, was the only brother who did not serve in the War. Instead, James attended school in Mopville, Illinois where he tragically died sometime around the issuance of Order No. 11.

Albert and Emily were banished during Order No. 11. Emily wrote a letter to one of her sons on July 24, 1863 in which she described the horrific conditions they were experiencing in Sibley.

> *...We have had several horses jayhawked and we had 8 tons of hemp burned, it was all bailed ready for shipment. Your Pa could not get a boat to land at Sibley because other Bushwhackers had been firing on them all along this part of the river. Nearly all of Sibley is burned, a great many of the citizens have been banished. Mrs. Hudspeth has been burnt out twice, first her dwelling was burnt out last spring and a few days ago her cabin, corncrib and meat house was burnt with their contents, all because Frank Sheppard would stay around there. He is a bushwhacker. Old Aunt Rach and little Rachael are dodging about over the River to keep from being put in prison . . . Priss and Rach are parted, old Aunt Rachel's house is burned and everything has gone to rack.*[665]

Following the war, Albert Steele was prohibited from voting under the terms of the Drake Constitution because he had raised the "secesh flag at Sibley."[666]

STEGALL

See MIMS

---

[665] Jackson County (Mo.) Historical Society Archives, Document ID 51F1. For more information on the Steele letters, see *Tear & Turmoil*; Eakin, Joanne Chiles. *Our Far Away Boys: The Civil War Letters of the Emily Steele Family*. (Independence, Mo.: Two Trails Publishing, 2004); *Kansas City Genealogist*; 23: 1&2: 11-12; and, Eakin and Hale, 409-410. Albert Gallatin Steele Family Papers in the Jackson County Historical Society Archives. Old Aunt Rachael and little Rach referenced in Emily's letter were slaves the family owned. Included in the Steele Papers are photocopies of letters pertinent to this era in Jackson County's history (See also, *Chapter XII* for other letters of Mrs. Steele). Letters of Emily Steele.

[666] Eakin, "Tears and Turmoil," 19-29.

STEEN

Elizabeth Steen left Cass County and Missouri under Order No. 11, and she relocated to Indiana.[667]

STEPHENS

John L. L. Stephens and family moved from their farm on South Fork, Cass County and went to Kentucky after the issuance of Order No. 11. The family did not return to their property in Cass County until 1866.[668]

STONESTREET

James Stonestreet was born in Kentucky in 1827 and died in Pleasant Hill, Cass County on September 1919 at the age of ninety-two. He came to Jackson County in 1852, "bringing with him the second herd of Shorthorn cattle ever brought to Jackson County. He had driven them 700 miles overland. During the Civil War he and his wife were compelled to leave the country under Order No. 11 and returned to Kentucky, where his wife died. The farm had been swept clean of improvements, except the house, when he returned."[669]

STORMS

Mr. Storms, Cass County, went to the military post at Pleasant Hill during Order No. 11.[670]

SUMMERS

See ASHLEY

SWEENEY

See ASHLEY

---

[667] Rafiner.
[668] Eakin and Hale, 413.
[669] "A Missouri Pioneer Dead," *Kansas City* (Mo.) *Star*, 3 Sept. 1919.
[670] Rafiner, citing Glenn, Allen, 473.

SYMINGTON

Robert Smith Symington was a well-known, respected Presbyterian minister. He was of the "Old School Presbyterian Faith." During 1848 he preached inside the courtroom of the Independence Square Courthouse until 1850 when a Presbyterian church was built in Independence on the southwest corner of Lexington and Osage. During his pastorate in Independence his sister Myra married Samuel A. Lucas, a long-time resident of Independence and former general in the Missouri State Militia. Rev. Symington was involved in establishing several churches including the Six-Mile Presbyterian Church near Sibley, a congregation in Pleasant Hill in Cass County and the Old School Church in Westport. As many other Presbyterian pastors, he earned his living by teaching school and operating his churches.

Rev. Symington gave as many as three hundred and sixty sermons and developed a "minister's sore throat" in 1850. In the same year, he became the financial agent for the newly organized Synodical College in Fulton, Calloway County, Missouri. He traveled the state until 1854 when he was stationed at Pleasant Hill and Six Mile. He moved from Pleasant Hill to Westport in 1855 where he lived in a log cabin on John Calvin McCoy's farm until his own house was built in Westport. His time was divided between the Westport Church and the new one in Kansas City. The congregation in Kansas City, organized in 1857, initially met inside the Courthouse at Third and Main until it purchased a lot at on the south side of Third near Walnut in 1860. With the new church, the preacher became solely connected with the Kansas City congregation. In 1862, Rev. Symington's throat was again afflicted and he was advised to rest for a year. Being unable to preach, he moved to Pleasant Hill.

During Order No. 11, Symington decided to relocate to St. Louis. A Presbyterian congregation in Mexico, Missouri learning of Symington's travel plans asked for his assistance in getting federal troops removed from the church they occupied. Symington telegraphed President Lincoln and the soldiers were removed. Before his departure from Pleasant Hill, Symington was arrested on September 3. Provost Marshal Comingo a long associate and friend intervened by writing a petition that secured Symington's release. After his discharge, Symington temporarily lived with Rev. Kerr near Mexico in Audrain County.[671]

---

[671] Rev. Kerr had been the administrator of the "Deaf & Dumb Asylum" in Fulton. It is believed that during a period of the War Rev. Kerr took patients from the asylum to live in his home in Mexico. Joseph D. Liles and Robert Smith Symington, NARA Publication M322: Compiled Service Records of Confederate Soldiers Who Served in Organizations from the State of Missouri. Record Group 109, 586957; Fold3 papers, Hill's History of the Presbytery of Kansas City; and Rafiner, citing National Historical Company, 521. Rev. Symington's brother-in-law Sam Lucas had warned General James T. Buel of the rebel attack on Independence the night before the battle, but Buel ignored the warning. See, Monaco, "Blood on the Streets," 24. The author wishes to personally thank historian, Nancy M. Ehrlich, for contributing much of the information on Rev. Symington.

TALBOT

Joshua Talbot of Cass County reported to the provost marshal's post at Pleasant Hill during Order No. 11.[672]

TALLEY

A short vignette of Jackson County pioneers, with emphasis on Lee's Summit families, includes the impact of Order No. 11 upon the Talley family:

> *In war times, Union soldiers arrested David C. Talley and banished him to Kansas City. The mother [Mary (Fristoe) Talley] was sometimes ordered to prepare meals for 100-200 who tied their horses in the timber.*
>
> *Order No. 11 came in August, when Mr. Talley had 150 hogs, cattle and 30 sheep roaming at large, and 50 acres of corn that would yield 50 bushels per acre. All this he was required to leave within five days. Stock ranged on Blue and Cedar Creek, each animal was branded by owners' marks in each ear which was recorded in Independence. Hogs were fed enough in summer to entice them home every night, and came at call. Cattle used to lie on beds of leaves on slopes beside streams, and were corralled in the fall.*
>
> *With two old horses left by soldiers in place of two good ones, and two young steers, the Talley family, leaving practically everything except bedding and clothing, set out for Saline County where they rented a farm. There were the father, mother, and eight children, the oldest son wounded in the war, two grandchildren, the colored woman, her four children, and the daughters' negro maid, all transported in two wagons. John Fristoe, the son-in-law, was banished to Council Grove, Kansas. There was not a house on the prairie south of the Dr. [Pleasant Graves] Lea home, now McKeighan (1959), and the Missouri Pacific was started from St. Louis when Lincoln was assassinated, and the Talley family returned home.*[673]

---

[672] Rafiner, citing National Historical Company, 473.

[673] Short Stories of Pioneer Days by Dollie Bee Breitenbaugh, 1959. 94-leave, typed manuscript. Two copies. Jackson County (Mo.) Historical Society Archives, Document ID 80F6.

TATE

William Calvin Tate was one of the six victims murdered by the 9th Kansas on September 6, 1863. He was a son-in-law of Martin Rice and was 32 years old at the time of his death. William was born in North Carolina in 1832 and he married Martha J. Rice, the daughter of Martin Rice, in Lafayette County, Missouri in 1856. At the time of his death, William and Martha had three children, John, age 5, Mary, age 3 and the baby Nancy Rosalie. The Tate's lived near Martin Rice and Benjamin Potter at the time of Order No. 11 (see *Chapter XII* for further account of the tragic death of William Tate).

John Tate was only five years old at the time of his father's murder on September 6, 1863. In adulthood John became a minister. In 1928 he was told that his grandfather, Martin Rice, had been responsible for the death of his father during Order No. 11.[674] It is unknown whether John believed the story. It is undisputed that his grandpa Rice if he had been would have vigorously denied the allegation.

On December 14, 1864, Martha Jane "Mattie" (Rice) Tate wrote a letter in which she expressed the anguish and turmoil she and her children had experienced during Order No. 11. The letter also provides additional insight to the background of her late husband and family (original style is presented as written by Mattie).

> *I am the wife of Calvin Tate who was in Indian Terr[itory] ten years ago when you was quite small I have often hear him speak of you and your Pa when he was there. But alas I never shall hear his lovely voice again on earth for he is gone gone from me forever and I am left with three small children to take care of and not able a great part of my time to take care of myself and what is to become of them I cannot tell But the Lord has said I will be a husband to the widow and a father to the fatherless and I hope he is such to me and mine for if ever a poor family needed his help it surely is mine alone for there are many these war times that need a great deal of help from above such is and has been the [case] with me and six other familys all related to my husband save one and all living within one mile of each other and although I say it myself a better set of men never live or die then they were in September 1863.*
>
> *"Ewing made the Order No 11 to devastate the Countys of Jackson Cass Bates and part of Vernon and people had but nine days to get out of those Countys a Military Post on [--?--] were all trying to get out [--?--] County and the men would have started in one hour on Sunday Morning But before they got off some soldiers from Kansas came and taken My Father Brother and my Husband and five others two Hunters Brothers and*

[674] Eakin, "Tears and Turmoil," 7.

> *one by the name of Cave Brother in law to the Hunters and one boy 17 years old named Owsley Nephew to the Hunters and one old man 60 years old by the name of Potter Father in law to one of the Hunters and taken them something over half mile from home and shot them all But father and brother they were released and came home and had just got home when the guns were heard that made four widows and 26 fatherless children and Oh what a sight to behold they were shot all to pieces [most?] and left on the ground by them to stay there Forever for what they [--?--] there was but three men left to bury them my Father Brother and Uncle John Hunter 70 years old the Father of the two that was killed everybody was gone but those familys and one other woman and we had to just dig a hole and put them all six in without any coffins or boxes nothing but blankets and right where they was Massacred by a band of Murders and a set of Blood thirsty__________ None of those men had ever been in arms on either side except David Hunter who was in the Molisha at Kansas City about two weeks they was all at home and had been and was minding their own Business they were killed in the morning about 10 o'clock and we left home about 4 in the evening that home that was to be home no more for me we went to Ray Co[unty] on the North side of the River and stayed all winter and came back in the spring when we left home we all had to leave all our stock and left with one wagon a piece and what could be put in them Mr. Hunters family and Caves have not come home yet they live near Dover Lafayette Co[unty] there is four families of them and but one man among them and that is Uncle John Hunter the women has all the work to do in and out of the house...." Mattie Jane Tate P.S. address me at Pleasant Hill Cass Co, Mo, (my childrens' names John Martin / Mary Matilda / Nancy Rosalie).*[675]

TATUM

Thomas Jefferson and Elizabeth (Clark) Tatum and family (including John W. Tatum) of Blue Springs, Missouri "removed to Carroll County where they remained until 1865. When the war closed his slaves were gone, his horses stolen and everything portable on the place had been carried off, and his buildings were in ruins."[676]

---

[675] The original Tate letter was donated to the Lone Jack Historical Society by Mr. and Mrs. Jane Foltz (Independence, Wisconsin). A copy of the letter, plus the transcript by Sharon Snyder and edited by David W. Jackson are filed in the Jackson County (Mo.) Historical Society Archives Document ID 140F16.

[676] Hickman, 535.

TAYLOR

Elizabeth Taylor left Cass County and Missouri during Order No. 11 and relocated to Illinois.[677]

THATCHER

Vera Haworth Eldridge, in an undated newspaper clipping, gave the following account of Mrs. J. M. Thatcher's [Fanny?] war experience with federal troops under Order No. 11:

> *Mrs. J. M. Thatcher tells of living in Westport with her father and four children while her husband served with the state guard. Colonel Pennick, commander of a Union force of 100 bloodthirsty men were stationed nearby. Her home there was raided and the family was subjected to such indignities that the husband planned to take them away. He was detained for two weeks, and with others was taken out and shot. A baby born later never saw his father. She fled to Liberty with her children. The splendid library of her father, which they had to leave, later was found with all books ruined.*[678]

THOMAS

Griffin Thomas from rural Cass County went to live within the military district of Harrisonville during Order No. 11.[679]

THOMPSON

George M. Thompson was born on January 12, 1816 in Lincoln County, Kentucky. His father Nelson A. Thompson was a native of Virginia, and a direct descendant of Wattie Thompson, a noted politician from South Carolina. On November 24, 1840, George married Susan A. Pemberton of Barren County, Kentucky who had been born on July 29, 1822. Susan's father was Charles Pemberton of Barren County,

---

[677] Rafiner, obituary.

[678] Jackson County (Mo.) Historical Society Archives, Document ID Vertical File Newspaper Subject Files: Military/Wars: Civil War: 1863-"Order No. 11." There is a 35-year-old, Kentucky native, Fanny Thatcher in the 1870 U. S. Census for Westport, Missouri, with four children; James, 15; Mary, 14; William, 13; Tiny, 9; and Milton, 7, all born in Missouri. None of the children appear to have been married in Jackson County. See, United Daughters of the Confederacy.

[679] Rafiner, obituary.

Kentucky. George and Susan were members of the Presbyterian Church. In March 1853, they immigrated to Jackson County and settled in Van Buren Township. Seven children were born of the marriage, but only three survived: James C., born January 29, 1847; David, born June 14, 1849; and Lizzie H., born April 4, 1853.

When the Civil War broke out, George and Susan were owners of three hundred acres of finely cultivated land. They also owned a large herd of stock. This all changed under Order No. 11 as George and his family were forced to leave behind a great deal of stock, grain and household goods, all of which was confiscated or destroyed. Before leaving their property, the Thompson's were forced to endure many insults and indignities by federal troops. Among their losses was one very fine racehorse, valued at one thousand dollars. In 1864, the Thompson's large residence was burned and all of their fences and improvements were destroyed. When the family returned to their property after the close of the rebellion, they found nothing but the naked ground. They rebuilt their property in Section 22 of Van Buren Township and the family returned to the business of farming and raising stock.

THOMPSON
(Additional Account)

William and Lucinda (Webb) Thompson's home was burned to the ground and the livestock stolen and driven away. When the family left for Illinois under Order No. 11, they took with them all the livestock they possessed: one old mare and a yoke of oxen. Remarkably, "during the war he had many narrow escapes from death. Three times he was stood up to be shot, but each time escaped."[680]

THORNTON

C. T. Thornton who lived in Cass County sought refuge in Chariton County, Missouri after the issuance of Order No. 11.[681]

THRUSTON
(see also DEHONEY)

Captain W. Creighton Thruston and his family lived in the area of Raytown. Following the issuance of Order No. 11 he and his family along with the Leander DeHoney family and other neighbors went to Clay County. The Thruston's and

[680] Hickman, 513. Eakin and Hale, 429.

[681] Rafiner, citing Cass County Historical Society. "Cass County Missouri Families." (Harrisonville, Mo: Cass County Historical Society, Inc., 1976).

DeHoney's did not return to Raytown until after the rebellion had ended. Thurston's wife Lina Lander Thruston remained in the Raytown community until her death in 1922.[682]

TRIPLETT

Eli Triplett, his wife and son Dan were all impacted by Order No. 11 and left their property along with their neighbors.[683] The account of how Bird Hamilton assisted the Triplett's during the Order follows:

> *Bird Hamilton...pulled a clumsy, squeaky old wagon on which rode his family and their few most needed possessions. Five days of the ten given persons to leave the county had already expired, and he was on his way out.*
>
> *Mr. Hamilton knew his neighbor, Eli Triplett, was without a team to move his family away so he invited them to go along and share alike with his family in his wagon. Mr. Triplett's son said he followed his father down in the wood pasture where he hid his gun in a hollow tree. He pushed his old cap and ball pistol up in a hollow log and crammed dead grass over the top of it....*

Dan Triplett, the Triplett's son remembered watching his parents prepare to leave their property under Order No. 11. His mother "hastily" emptied "the milk out of the crocks." His father dug a large hole in the ground to bury dishes and other small household articles. They gathered together a few pieces of bedding and their clothes and climbed into the wagon with the Hamilton family to start on the long journey.

The first night the Triplett's exited Jackson County under Order No. 11 they camped on the banks of Fire Prairie Creek, just east of Levasy. Dan Triplett recalled the campfire and the women preparing supper. He played with the other children on the approach to a bridge that had been burned. The next day about noon, they reached the town of Wellington in Lafayette County.

> *While they were resting the tired team, for the load was heavy and the day was hot, several soldiers rode up, with big brass buttons and gold braid on their coats. They looked at their team and after a whisper or two with one another, ordered Mr. Hamilton to unhitch the good horse and*

---

[682] Thruston, 24.

[683] Wilcox, 364-365, citing *Independence* (Mo.) *Examiner*, May, 1906. Triplett's homestead is pictured in the "Illustrated Historical Atlas of Jackson County, Missouri," originally published in 1877, and indexed and reprinted for sale in 2007 by the Jackson County (Mo.) Historical Society.

they led it away. That left them with the old gray mare with the knocked down hip and with many miles yet to go.

Dan did not know how they located another horse. Anyhow they joined the many travelers all going east. The next day they drew near the town of Lexington where the wagons and teams were stretched out one after another for over a mile waiting their turn to be ferried across" the Missouri River. "Most of the people were going to Richmond where they planned to make their new home for a while at least. The wagon in front of theirs was drawn by a team of oxen" with the family cow tied behind the wagon. A little boy sat in the rear holding on to a big bulldog.

In December 1863 the Provost Marshal issued the Triplett's a permit to return to their farm. They discovered found some hogs and butchered them for the winter. Under the term of their provost pass, they were permitted to remove from their land six hogs. The wild hogs had been fattened on acorns from the oak trees in the nearby timbers.

After 18 months more of hardship, the families were allowed to return to their homes in Jackson County. Blackened chimneys standing alone in the high weeds along the roadside gave evidence of the destruction all about them. Every so often new graves were seen, persons were buried here who died of the cholera on their way home.[684]

TRUMAN
See HOLMES

TWYMAN
See also FRISTOE

L. W. Twyman

Dr. Lyddall Wilkerson "L. W." Twyman and his wife Frances C. (Fristoe) lived at Blue Mills in northeast Jackson County. Frances, born in Jackson County on April 20, 1829, was the daughter of Richard M. Fristoe, one of the first Jackson County Court justices. Dr. Twyman was a well-known physician and surgeon with years of experience in Jackson County. He and Frances had two children, sixteen-year old Julia and George Twyman, age ten, at the time of Order No. 11.

Before the Order was issued, Frances had gone to Independence, but through intelligence she was told that federal

[684] Jackson County (Mo.) Historical Society Archives, Document ID Vertical File Newspaper Subject Files: Military/Wars: Civil War: 1863-"Order No. 11": "Order No. 11 Era," *Independence* (Mo.) *Examiner*, 2 June 1939. Also, Wilcox, 364-365, citing *Independence* (Mo.) *Examiner*, May, 1906.

troops were going to search her. Knowing this, Frances rearranged the interior of her baby's diaper satchel. When a sentry stopped her, he demanded to search the satchel; she consented. Grabbing the baby satchel the soldier rammed his hand in without looking. He quickly withdrew his hand and holding it at arm's length and turning his head in the other direction, he curtly said to Ms. Twyman, "Madam you may pass on." One can only imagine the laugh or smile she held within herself following this episode. Unfortunately, it proved to be a short-term moment of happiness.

As a consequence of the Order, their house was stripped and the family was forced off their property. At the time of their departure, they only had a wagon with three wheels. They borrowed an extra wheel from Jesse Morrow and escaped to Howard County. Frances provided numerous accounts of her family's plight under Order No. 11. They joined a column heading out of the county. Forty-seven years after the tragedy, Frances still remembered how many children cried for bread or to "be taken back to their homes." Frances recounted two memories: A mother had stowed her children in an ox-drawn wagon and as the wagon neared a bridge over the stream, the oxen, sensing water, started running and the mother lost control of the team. Frances feared that the children would be killed, but "Providence watched over them." No one was injured but the mother began weeping and cursing the Federals. The other tale she recalled was seeing another woman who "had two cows hitched to a wagon; a little boy was leading them."

Frances C. "Fannie" (Fristoe) Twyman

> *There were some boughs on the wagon, and old-time coverlid stretched over them. Inside the wagon was a very sick child. The wagon halted, the mother got out with her sick babe in her arms and seated herself under the friendly shade of a tree. It was apparent to all that the child was dying. There sat the mother with her child dying in her lap; her husband had been killed, she was forced to leave her home, broken hearted mother as she sat there, with tears streaming down her pale cheeks, knowing she was powerless to save her child. Some kind-hearted people of the neighborhood came to her assistance. The crowd surged on, women and children dragging their weary limbs through the dust and heat.*

The Twyman family continued east and settled in Howard County. There were three adults in the Twyman family, including Frances' seventy-two year-old mother, and six children. They were piled into a two-horse buggy and traveled west across three counties to Missouri City. There was near-constant snow and sleet and no one would give them shelter from the bad weather. They were unable to find anyone who would even rent rooms or outbuildings to them. By November 1863, they had settled in Missouri City in Clay County north of the Missouri River, and only three miles north of their farm in Jackson County. Tragedy soon followed when Julia, their sixteen year old daughter, became very ill and died. Julia was the only daughter of the family. Her grieving mother described Julia as having been "too pure for this earth."[685]

Near the close of the War, the Twyman's returned to their home and found a scene of desolation. Everything had been destroyed and their hogs having subsisted on acorns had gone wild. The herd of hogs was rounded-up and butchered. Dr. Twyman exchanged the meat for much needed staples so he could reestablish his family at their new temporary property.[686] They continued to live at Blue Mills, which later was known as Twyman, until later moving to Independence.[687]

Frances Tywman later told a descendant of the family experience under Order No. 11. The following is that account:

> *The road from Independence to Lexington was crowded with women and children. Babies were carried in their mothers' arms. Packs were on the mothers' backs and four or five children following, some crying for food and wanting to go back to their homes. The torch had been applied to all homes, with nothing standing but the chimneys.' All the horses and wagons had been taken by the soldiers. They went to Howard County for a while, but returned in November to Missouri City as it was directly across the Missouri River from their farm. 'My mother [Mary L. (Sullivan) Fristoe], she said, 'was 72 years old and ill, and my husband, the doctor, with our six children, had a one-horse buggy then. We traveled through the snow and sleet to get as near home as we could. No one was willing to give us shelter for the night. We tried to rent a room or some outhouse for the night, even the school. I finally became desperate and told one man we had to have shelter, even if we were rebels and would just as soon be shot as to be tortured. He then offered them a room in his home*

---

685 Leslie, 262-263.

686 Wilcox, 364, citing *Independence* (Mo.) *Examiner*, October 25, 1901. An image of their homestead in 1877 appears in, "Illustrated Historical Atlas of Jackson County, Missouri," reprinted for sale in 2007 by the Jackson County (Mo.) Historical Society.

687 Jackson County (Mo.) Historical Society Archives, Document ID Vertical File Newspaper Subject Files: Military/Wars: Civil War: 1863-"Order No. 11": "Mrs. Fannie Twyman," Jackson (Independence, Mo.) Examiner, 2 Apr. 1909. A lengthier account was also published in United Daughters of the Confederacy, 263-266. Hickman, 436; Eakin and Hale, 439; and, Bartels, Carolyn M. "Bitter Tears: Missouri Women and Civil War, Their Stories." (Independence, Mo. Two Trails Press, 2002), 146-152.

> *for he had mistaken them for Yankees.' After many trials they reached Missouri City. They were arrested several times and their horses taken, but the worst was the death of their 16-year-old daughter who did not survive the hardships.*[688]

Frances Tywman died in 1909 and her husband L. W. had predeceased her six years earlier. The obituary provided a description of the France's character:

> *[She was] unusually bright mind and of high literary attainments; and in her younger days she contributed to many newspapers and magazines. In her later years she has always loved to tell of her early days in Independence. She was living here during the stirring times of border life and her mind was enriched with many graphic memory pictures of the early days. A few years ago she was awarded a prize at the Independence fair as being the oldest woman present on the grounds. In spite of her extreme age at the time, she was also one of the brightest minded and happiest.*[689]

Their son George T. Twyman was born in Jackson County on December 16, 1853. Both his father and paternal grandfather Dr. Leo Twyman was physicians and his maternal grandmother was Fannie (Fristoe) Twyman. George received a common school education. He went with his family after Order No. 11. Following the Civil War, George studied medicine at William Jewell College in Liberty, Clay County and studied medicine with his father. George attended the Louisville Medical College, of Kentucky, in the year 1877, for five months, and then attended the Kentucky Medical College for another five months. He completed his course at the Louisville Medical College, where he graduated on February 28, 1879, and returned to Jackson County to practice medicine with his father.[690] He practiced medicine until he unexpectedly died in April 1910. Prior to his death, George had been the private physician of Col. Thomas Hutton Swope, the Great Benefactor of Kansas City. At the time of George's death, he was on the prosecutor's witness list to testify against Dr. Clyde Bennett Hyde for the alleged murder of his wife's uncle, Col. Swope.[691]

---

688 Jackson County (Mo.) Historical Society Archives, Document ID Vertical File Newspaper Subject Files: Military/Wars: Civil War: 1863-"Order No. 11": "Government Order No. 11 - (1863)," by Vera Haworth Eldridge, in an undated, unattributed newspaper clipping. A lengthier account was also published in United Daughters of the Confederacy, 263-266. Also, Eakin, "Tears and Turmoil," 96-101.

689 Obituary of Mrs. Frances Tywman; *Kansas City Star* "Mrs. Frances C. Twyman Dead," 2 April 1909.

690 Union, 909-910.

691 Monaco, Ralph A., II. "The Strange Story of Col. Swope & Dr. Hyde." (Independence, Mo.: Two Trails Publishing, 2010).

UNDERWOOD

John Underwood of Cass County went to Johnson County, Missouri under Order No. 11.[692]

VAN KIRK

Mr. Van Kirk left Cass County and Missouri under Order No. 11 and went to Douglas County, Kansas.[693]

VAN NESS

See ASHLEY

VAUGHN

Historian Donald R. Hale relayed the family history of Reuben and Rebecca (Harper) Vaughn, who in 1832 settled in New Santa Fe, on the western border of Jackson County. Their son James Newton Vaughn was in the Battle of Lone Jack on August 16, 1862, and their home was burned in 1863 as a result of Order No. 11.[694]

VERGIN

After Order No. 11, William G. Vergin received a provost pass on November 25, 1863 to go five miles south of Harrisonville to round-up hogs for the winter.[695]

WADE

W. Wade of Cass County went to Sedalia, Pettis County, Missouri under Order No. 11.[696]

---

692 Rafiner's best guess judging from the 1860 and 1870 U. S. Census returns.

693 Rafiner, citing Glenn, Allen, 467.

694 Manuscript written by Donald R. Hale, Lee's Summit, Missouri, 1961, for the Jackson County Historical Society Contest, entitled "Some Interesting Ancestors Who Lived in Jackson County." Jackson County (Mo.) Historical Society Archives, Document ID 77.03F8.

695 Jackson County (Mo.) Historical Society Archives, Document ID 36F6.

696 Rafiner, citing Glenn, Allen, 827.

WALKER

Andy Jackson Walker was born in Jackson County on September 1, 1836. He married Miss Elizabeth Braden of Missouri on April 12, 1860. He was the son of James Morgan Walker and Pollcy (Cox) Walker. A. J. was eight of twelve children born of the marriage of his parents. A. J. farmed on land near the Blue Springs, but after Order No. 11, he moved his family in October 1863 to Texas. They lived in Texas until the spring of 1864. A. J. traveled to Mexico where he stayed about nineteen months before going to California and Oregon before returning to Jackson County in March 1867. He and his wife had four children before she died in 1874. After Elizabeth died A. J. remarried Eliza Hannon on November 25, 1875. He and Eliza had four children, three who survived, including twins, Morgan and Monroe. After his second wife died in 1879, he remarried Fanny Warren of Missouri on February 18, 1880.[697] By 1910, A. J. was living in Weatherford, Parker County, Texas. He sat for a series of interviews with a reporter from the *Weekly Herald* and recounted his experiences during the rebellion. The newspaper then in a series of weekly reports published their A. J. stories, which were later published in book form.[698]

> *When Order No. 11 went into effect many refugees from Jackson County came into Clay. I remember that a niece of Uncle Bob Walker, daughter of Morgan Walker, came to live with him. She was a fine looking woman, and visited us with my cousins. There were also many Confederates and bushwhackers, who had not reported and been paroled, lurking around in our county. This cousin of my cousins proved to be a notorious spy for Quantrell [sic.]*[699]

WALLACE

Rev. Joseph William Wallace was a Presbyterian minister who had been born in Fayette County, Kentucky in 1821. His father Colonel John C. Wallace, a native of Virginia, had been a farmer, and an officer in the Kentucky militia long before the Civil War. His paternal grandfather Captain John Wallace born December 18, 1748, in Virginia served under General Washington throughout the Revolutionary War.

---

[697] Union, 928; and Wilcox, 364, citing *Independence* (Mo.) *Examiner*, September 13, 1957. Also see, *Chapter III* for historic account of Quantrill and the Morgan Walker Raid in December 1860.

[698] Eakin, "Recollections of Quantrill's Guerrillas," Introduction, 93-94.

[699] Frick, John Henry. "Recollections of the Civil War." *Missouri Historical Review* (1924-1925), 19: 630-654 (634).

Rev. Wallace married Ann Elizabeth Hockaday Wallace in Kentucky. She was the eldest daughter of Edmund Warwick Hockaday. Their marriage produced only one child, William Wallace. Ann died when William Wallace was about four-years old. Five years after his first wife died, Rev. Wallace remarried Jessamine Ryley.

Jessamine (Ryley) Wallace was born in Jessamine County, Kentucky—named for the county of her birth. Her first husband was Dr. Alfred Ryley of Fulton, Missouri. Their marriage produced two children, Emma and Archie. Although both children were alive at the time of Order No. 11, they did not survive to adulthood.

After Rev. Wallace and Jessamine were married, they moved to Jackson County in early 1857. The couple and their blended family disembarked from the steamboat at Wayne City Landing near Independence, and settled on an open prairie between what is today Blue Springs and Lee's Summit. Rev. Wallace and Jessamine had four sons: Charles H., John C. and Addison A. Wallace, an infant (their son Theodric was not born until after the war). Rev. Wallace was financially secure as a minister. The family farm was well stocked with cattle, hogs and sheep and twenty horses, most of which Rev. Wallace had brought to Missouri from Kentucky. Combined Rev. Wallace and Jessamine owned twelve slaves.

During the early years of the Civil War, the Wallace's experienced many privations at the hand of Kansas' troops. A band of Jennison's Kansas Jayhawkers came to the Wallace farm and sacked their home, taking everything they could carry of value. One of the Wallace slaves, Alfred, went with the Kansas men, but their other slaves refused to leave with the troops. On another occasion, the Wallace family saw Kansas' bushwhackers burn over twenty farms in the area. Rev. Wallace himself was threatened at gunpoint to surrender his wife gold watch, but he refused.

The Wallace family along with Aunt Mary and her two daughters, Mandy and Patsy and two family slaves, Marshall and Louisa, left their property under Order No. 11 and reported to the provost marshal's post located in the county jail at Independence. They signed the loyalty oath and received a pass to move to Callaway County, Missouri. The Wallace family remained two years in Callaway, where Rev. Wallace taught college. When hostilities ended, Rev. Wallace brought his family back to their farm in Jackson County. Despite the destruction of the property, he was able to rebuild the farm.[700]

Infant son Addison A. Wallace eventually became a Presbyterian minister in Mexico, Missouri.[701]

Charles H. Wallace who was about fourteen at the time of the issuance of Order Number 11 provided a detailed account of how their farm had been found completely

[700] Wallace, 247-260.

[701] Withers, Robert Steele. "Historical Notes and Comments (Memoirs and a Gospel Testimony of A. H. Wallace)." *Missouri Historical Review* (1946-1947); 41: 226; and Wallace, 247-260.

waylaid following their return. Charles grew-up to become a physician in St. Joseph, Missouri.[702]

John C. Wallace was a young boy at the time of the issuance of Order No. 11. John became an attorney, but died before 1909.[703]

William H. Wallace was born on a farm in Clark County, Kentucky on October 11, 1848. He was almost fifteen when Order No. 11 was issued. He described it as a "tragic hour." He traveled to Calloway County with his father, stepmother and half-siblings. Wallace did not return to Jackson County with his family but remained in the house of Judge T. B. Nesbit to obtain a college education.

William went onto have a very productive legal career. He was elected Jackson County Prosecutor in 1878 and became the first prosecutor to garner a conviction against the James Gang in his successful prosecution of Bill Ryan, the Whiskey Head, for the Blue Cut Train Robbery. He also unsuccessfully prosecuted Frank James in the Winston Train Robbery in 1883. When William left the prosecutor's office, he continued to practice law in Kansas City.[704] In October 1898, William Wallace was visiting Jesse James, Jr., the son of the bandit Jesse James, at the young man's tobacco shop inside the Jackson County Courthouse at Kansas City. The bandit's son that afternoon was taken into custody for allegedly robbing the Missouri Pacific Train in the Leeds portion of Jackson County. William Wallace vehemently protested the treatment of young James and joined many other leading citizens in Jackson County who came to the aid and defense of Jesse James, Jr.[705]

WALLACE
(Additional Account)

The house at 8607 Wilson Road, Independence, was formerly the site of "a big white house with columns...the first Wallace home. It was burned during Order No. 11,' Mrs. [Robert (Nell Wallace)] Brown said. 'The Union soldiers ripped up the rugs and cut them into pieces for saddle blankets." Mrs. Brown related further accounts of her grandparents experience under Order No. 11.

> *My mother fled to Clay County in an ox wagon, with the belt from the saw mill hid under the bundles of blankets.*

---

[702] Wallace, 247-260.

[703] Wallace, 247-260.

[704] Wallace, 247-260.

[705] Monaco, Ralph A. II. "Son of a Bandit, Jesse James and the Leeds Gang." (Independence, Mo.: Monaco Publishing, L.L.C., 2012); the train robbery occurred south of present day Truman Sports Complex—home of the Kansas City Royals and Chiefs.

Earlier, John Wallace had been imprisoned by Union troops. Sentenced to be hanged, he was rescued by Quantrill's band on the eve of his execution.... They returned to their ruined home, lived in a one-room cabin left standing on the place...and did a land office business at the sawmill.[706]

WALLACE (Additional Account)

A story relating to the John Wallace's home was related years later:

Another memory dear to me is that of the 'Ford' house, one of two left standing after the ravaging of homes as the result of Order No. 11. The other house was the 'Wallace' house. They were left for use as headquarters. At one time one of my Uncles lived there.... I have been told that the house was built with wooden pegs instead of nails. It had slaves' quarters in the upstairs back of the house which were completely closed off, so there was no approach except from the outside.[707]

WALTON

At the time of Order No. 11, Mrs. Walton's husband served as a lieutenant colonel in the Trans-Mississippi department of the Confederate Army. She and her husband had one son Thomas, who was three years old at the time of Order No. 11. Mrs. Walton and her son were forced off their property by the Order. Mrs. Walton and little Tom were among the sixty-one women and children who were piloted out of Jackson County to Clinton, Henry County, Missouri by eighty four year old, Buckner Muir, and his 56-year-old son, Sam. In route to Texas, baby Tom grew ill with membranous croup. His condition soon worsened. An Indian woman doctor attempted to administer treatment to him, but it was of no avail and the little boy soon died. Mrs. Walton became "prostrated with grief" over her son's death. When the party reached Boggy Depot in Texas, they met Colonel Harris with a detachment of about fifteen hundred Confederate

---

[706] Jackson County (Mo.) Historical Society Archives, Document ID Vertical File Newspaper Subject Files: Military/Wars: Civil War: 1863-"Order No. 11": "House with a History to Go on Block," *Kansas City* (Mo.) *Times*, 20 Sept. 1967. John Davis willed his house to Miss Lizzie Wallace. Mrs. Brown's father, Jefferson Davis Wallace, the youngest of nine children, lived in the house with Lizzie, along with his wife, Mrs. Gertrude Boothe Wallace. Mrs. Robert (Nell Wallace) Brown was the last survivor of the four children of Jefferson and Gertrude Wallace. Her grandmother referenced in the account above was Mrs. John D. (Nellie Duncan) Wallace.

[707] Svobada, 14.

troops. When Harris and his men learned of the death of Little Thomas they provided him a coffin and buried him with full military honors in respect to the boy's father.[708]

WALTON
(Additional Account)
See also FLANERY

WARD

Cyrus Ward from Cass County went to Johnson County, Missouri after Order No. 11 was issued.[709]

WARREN

Nelson Warren was one of the earliest settlers in Jackson County, coming to the county in 1831. Nelson worked for the federal government as a blacksmith for the Wyandotte Indians. He also worked for the Chouteau Fur Company and later was associated with Robert Weston during the gold rush days.

Nelson married Lucinda (Robinson) Warren. They had one daughter Lucinda Jane Warren who was born in Jackson County on October 16, 1852. The family was banished from their property by Order No. 11 and boarded a steamboat at Wayne City Landing and traveled up the Missouri River to Nebraska City where they stayed until after the War when they returned to Independence.[710]

Lucinda Jane Warren was born October 16, 1852, in Jackson County and she was about eleven years old when Order No. 11 forced her and her parents off their property. On January 17, 1878, Lucinda Jane married Dr. Oliver C. Sheley. They lived at Pink Hill in Jackson County. Later they moved to Independence and in 1889, she was living at 142 East Short Street.[711]

WARREN
(Additional Account)

Sarah Warren left Cass County and went to Johnson County, Missouri after Order No. 11 had been issued.

---

[708] Eakin, Joanne Chiles. *Blue & Grey Chronicle*. (August 2000): 3: 6: 3. (See also, plight of the Walton's and death of Thomas under Flanery story in *Chapter XII*).
[709] Rafiner's best guess judging from the 1860 and 1870 U. S. Census returns.
[710] Wilcox, 359-360, citing *Jackson* (Independence, Mo.) *Examiner*, 1906.
[711] Wilcox, 359-360, citing *Jackson* (Independence, Mo.) *Examiner*, 1906.

WEST

Mrs. Edward (Canine) West was married to Edward West. She was the daughter of Cornelius Canine. She and Edward were the parents of George Cornelius West. Little George had been born on July 15, 1856, about one mile east of Raytown. The West's were residing in the Raytown community when the War began.

Edward West joined General Sterling Price and served as a military guide through Jackson, Henry and Cass Counties. For the first two years of the conflict, Mrs. West remained in contact with her husband, but after Order No. 11, she heard nothing from him until hostilities had been over for about six months. During enforcement of Order No. 11, Red Legs robbed their property and took all family provisions, including her husband's guns he had left for her protection. When the Red Legs vandalized her property, she and her son moved to her father's house. Her father Cornelius Canine moved the family to a timbered farm where they remained until Edward came riding home one October morning in 1865.

In the spring of 1866, Edward moved his family south of Westport (today 35th and Prospect) into a double log house where they farmed one hundred acres, harvesting a large crop of hay. The West's had three other children born after the rebellion, including a daughter named Shelby. Edward West went on to serve as a member of the first school board in the geographic area from state line on the west, Westport on the north, Swope Park on the east and about 80th Street on the South.[712]

George Cornelius West, the oldest of the four children born to Mr. and Mrs. Edward West, married the daughter of Judge I. M. Hale. George became a teacher in the school district in which his father Edward served on the school board. When the Kansas City Convention Hall burned down in 1900, George volunteered his services to rebuild it. George died in Kansas City at the age of ninety-four.[713]

WEST
(Additional Account)

Lysander West, Cass County, went to Harrisonville during Order No. 11 and then traveled to Kansas City.[714]

---

[712] Thruston, 44-46. The property in which George was born and the West's were evicted under Order No. 11 was located at what today is 62nd & Woodson, Raytown.

[713] Thruston, 44-46. See also, Canine above.

[714] Rafiner, citing obituary and National Archives and Records Administration. Washington, D.C. Records of the Quartermaster General. Record Group 92. War Claim: 124-164.

WHITE

Mr. White of Cass County first went to Iowa then resettled in Illinois after Order No. 11 was issued.[715]

WHITE

William White of Cass County returned to Virginia following the issuance of Order No. 11.[716]

WHITSITT

J. B. Whitsitt was the father of Mrs. Jacob Palmer. His son-in-law, Jacob Teaford Palmer, was a southern sympathizer and after they reported to the nearest post headquarters in Jackson County at Hickman Mills, J. B. assisted his daughter and son-in-law and took them to the Gillis House in Kansas City where they booked passage up the Missouri River to St. Joseph.[717]

WILEY

See ASHLEY

WILHITE

Jefferson Wilhite had joined Quantrill and his guerrillas sometime before his relatives was killed by members of the 9th Kansas near Lone Jack, Missouri on September 6, 1863 during Order No. 11.[718]

WILLETT

J. N. Willett of Cass County left the state and went west into Colorado after Order No. 11.[719]

---

[715] Rafiner, citing "Drexel: The First 100 Years." 1989, 379.
[716] Rafiner, citing Belton Star-Herald. "The First Hundred Years: Belton, Missouri 1872-1972." (Belton, Mo.: Belton Star-Herald, 1972).
[717] Eakin, "Tears and Turmoil," 32-23. See also, story of Jacob Teaford Palmer above.
[718] Eakin, *Wellington, Missouri: Diary of a Town,* and *Western Journal of Commerce*, 23 July 1863, as cited in Eakin and Hale, 467.
[719] Rafiner, obituary.

WILLIAM(S)

David William(s) a resident from Cass County went to Harrisonville during Order No. 11 to report to the provost marshal.[720]

WILLIAMS
(Additional Account)

J. R. Williams of Cass County first went to Clay County and then to Jackson County during Order No. 11.[721]

WILLIAMS
(Additional Account)

Mary Mildred Zick DeWitt provided an extract of an August 29, 1863, "order to pass for one J. W. Williams, Independence, Missouri."[722]

WILLIS

Alpheus Willis a resident of Cass County went to Saline County, Missouri when Order No. 11 was issued.[723]

WILSON

A short vignette of Jackson County pioneers, with emphasis on Lee's Summit families, includes the story of the affect of Order No. 11 upon the Wilson family:

> *Mr.* [R. G.] *Wilson said that Order No. 11 was for the purpose of starving out the Bushwhackers, but that they paid no attention to pickets on the road. They were excellent horsemen, and ate supper in Independence every night and had the protection of all southern sympathizers. [Mr. and Mrs. and Louisa (McMurray) Wilson] often fed the Kansas Red Legs, as*

---

720 Rafiner, citing National Archives and Records Administration. Washington, D.C. Records of the Quartermaster General. Record Group 92. War Claim: 87-464.

721 Rafiner, citing National Archives and Records Administration. Washington, D.C. Records of the Quartermaster General. Record Group 92. War Claim: 97-535.

722 "Story Hour Manuscripts," by the Independence Young Matrons from 1964. "Uprooting of Families of Order No. 11." Jackson County (Mo.) Historical Society Archives, Document ID 66F40.

723 Rafiner, citing National Archives and Records Administration. Washington, D.C., Records of the Quartermaster General. Record Group 92. War Claim: L-3628.

many as 150 at a time. The first crowd always took the knives and forks, and the rest ate with their fingers. All provisions were consumed. At one time, the family had nothing but parsnips, they made a substitute for coffee....

With horses and everything gone, someone in the neighborhood would get an old horse, another a harness, another an old wagon and all the money available and set out for Independence to buy groceries for the families....

When they vacated their home, the Wilson family left 150 Chester White hogs and 10,000 bushels of corn in cribs that reached for a quarter a mile....

When the Wilson family returned, after the war, from Liberty, Missouri, where they finally went for safety, with two loads of goods, a cow and one horse and wagon, they stretched their necks in sought of their home for fear their house was an ash heap as others. They saw the roof only. Fearing a cyclone had laid the building flat, they arrived at twilight to find the home in a forest of weeds.

Hogs had eaten the grass in the yard and rutted the ground around the house and meat house, so that fire consuming everything else on the place, had passed by and left the house in high weeds.

However, the hogs were gone, so was the corn. With the horse they broke a way to the door, tied the horse to the wagon, took the beds out and lay down on the floor in the darkness, not knowing whether snakes and creeping things might disturb them. In the morning the family discovered what a good housekeeper nature had been. With the doors and windows all gone, probably for three years, the wind had swept the place free of dust.[724]

WILSON
(Additional Account)

Mrs. Henry C. (Sarah Louisa Wilson) Mayes, "vividly recalls when Order No. 11 was issued…She was but a small girl but the scene was too vivid to forget. Her people being southern sympathizers went to Lexington for safety instead. She says that when they returned two years later the country was desolate." They left in such haste,

They had time "to take only part of their pigs, chickens, and crops in the field. They took their negro slaves, however. Most of their pigs were

[724] Short Stories of Pioneer Days by Dollie Bee Breitenbaugh, 1959. 94-leave, typed manuscript. Two copies. Jackson County (Mo.) Historical Society Archives, Document ID 80F6. Other accounts of Quantrill, Bushwhackers and Captain Pennick are mentioned in the Wilson memoirs

> *left behind and when they returned after the war the pigs were wild and were roving the countryside at will. When the war was over their farm was up in weeds, fences could hardly be seen. It meant starting over again as her parents had done years before when the country was new.*[725]

WOODRUFF

Miss Julia Kern (later Woodruff) experienced the ravages of the rebellion. She and her family suffered at the hands of Jennison and his Jayhawkers. They were banished from Jackson County after Order No. 11. Their home was raided and robbed. Her family hid their old box of silver, but it was unearthed and taken by the Kansas raiders. Other items they lost included a "locket containing a dead child's hair." By Order No. 11 her mother took Julia and her siblings to St. Louis and then proceeded down the Mississippi River until finding "a haven of rest in Southern Texas." The Wodfruff family lost everything. Their home was burned; their plantation was devastated; in the end their land was confiscated or sold for back taxes. After "Reconstruction" they found their once beautiful "homestead" desolate and in waste.[726]

WOODSON

See ASHLEY

WOODSON
(Additional Account)

Mrs. M. A. Woodson, formerly Mrs. Ervin and originally Miss Bowers, was compelled to leave Jackson County in 1863 by Order No. 11.[727]

WOOLERY

Caroline Woolery of Cass County sought safe haven in Johnson County, Missouri because of Order No. 11.[728]

---

[725] Jackson County (Mo.) Historical Society Archives, Document ID Vertical File Newspaper Subject Files: Military/Wars: Civil War: 1863-"Order No. 11": "Recalls Order No. 11," undated, unattributed newspaper.
[726] United Daughters of the Confederacy, 124-125.
[727] *Lee's Summit* (Mo.) *Journal*, 29 Apr 1915, as cited in Eakin and Hale, 479.
[728] Rafiner's best guess judging from the 1860 and 1870 U. S. Census returns.

WOOLERY
(Additional Account)

J. J. Woolery of Cass County traveled to Cooper County, Missouri during Order No. 11.[729]

WORRELL

Mr. Worrell a Cass County resident went to Cooper County, Missouri under Order No. 11.[730]

WRIGHT

Mrs. Emory C. Wright of Blue Springs, Missouri provided the following account of her grandfather's experience during Order No. 11:

> *When my grandfather's home in Independence was burned he was given only a few hours at night to get out. He and his family went to the River and managed to get aboard a steamboat. They went as far upriver as Cincinnati, Ohio, where they had relatives. They were unable to return for many years because all they had was burned and looted.*[731]

WYATT

A short vignette of Jackson County pioneers with an emphasis on Lee's Summit families there is included the following story about the Wyatt family:

> *Order No. 11 allowed Mrs. A. M. (Loraine) Wyatt to load all she could in a wagon to start to Clay County. Under the table was a feather bed which was undisturbed by officers making an investigation of the wagon before the start. They found nothing, but when the exiles reached Clay County, Mr. Wyatt cautiously released himself from the feather bed. Still pursued, Mr. Wyatt and a friend were hidden in a thicket when two*

---

[729] Rafiner, obituary.
[730] Rafiner, obituary.
[731] "Order No. 11," a speech Alberta Wilson Constant delivered to the Independence Young Matrons, ca. 1960, Jackson County (Mo.) Historical Society Archives, Document ID 154.01F30. [NOTE: The author has grouped this recollection under WRIGHT because Mrs. Emory C. Wright did not name her paternal or maternal grandfather, and this was the only reference there was to classify the account.]

officers rode down the road. Mr. Wyatt aimed at one, and his friend aimed at the other, and said Mr. Wyatt, 'That was the only man I ever killed.'[732]

WYATT
(Additional Account)

During Order No. 11 the wife of Wiley Wyatt hid her husband in a wagon box and drove to Clay County. From there Wiley Wyatt went to Caldwell County, Missouri, before returning to Johnson County, Kansas. He moved back to Washington Township, Jackson County in 1879.[733]

YANCEY

George Yancey left Cass County and traveled to Indiana to escape Order No. 11.[734]

YANKEE

Mrs. David (Orlena Beeler) Yankee of Lone Jack was born in 1832 and lived in Jackson County all her life "with the exception of the time she was evicted by Order No. 11."[735]

YANKEE
(Additional Account)

Wesley Yankee of Cass County went to Lafayette County under Order No. 11.[736]

YANKEE
(Additional Account)

Mr. and Mrs. Jacob Yankee lived in the area where Martin Rice and all six victims of the September 6, 1863, massacre resided. According to Martin Rice, when he returned from taking a load after the tragedy, he found Mrs. Yankee and her niece in total

---

[732] Short Stories of Pioneer Days by Dollie Bee Breitenbaugh, 1959. 94-leave, typed manuscript. Two copies. Jackson County (Mo.) Historical Society Archives, Document ID 80F6.

[733] *Memorial and Biographical Record*, as cited in Eakin and Hale, 481.

[734] Rafiner, obituary.

[735] Jackson County (Mo.) Historical Society Archives, Document ID Vertical File Newspaper Subject Files: Military/Wars: Civil War: 1863-"Order No. 11": "Untitled," undated, unattributed typewritten transcript.

[736] Rafiner, citing Glenn, Allen. 489.

"loneliness, entirely ignorant of all that occurred in the vicinity during the last forty-eighty hours not having seen a person since Saturday." Mrs. Yankee and her niece told Martin Rice that they had spent the previous two days removing farming tools, implements and other items that they could take with them or take out of their house and barn in the event they were burned.[737] The Yankees like their neighbors left their property in the Lone Jack area under Order No. 11.

YODER

Solomon Yoder left Cass County and Missouri under Order No. 11 and went to Ohio in exile.[738]

YOUNG

Solomon Young was born on April 24, 1815, in Simpsonville, Shelby County, Kentucky, one of nine children born to Jacob Young and Rachel Goodnight. Solomon's eight siblings were Landis, John, Jackson, Elizabeth, Michael G., Rachel Anna, Sarah, Juliet and a young unnamed child who died in infancy.

Harriett Louisa Gregg was born on October 15, 1815, in Shelby County, Kentucky. On January 9, 1838, in Shelby County, Solomon married Harriett and they had seven children: William, Sarah A., Elizabeth, Harrison, Laura, Martha E (the mother of future President Harry S. Truman) and Adah. Four of their children were living at home at the time of Order No. 11: Sarah Ann Young, born in 1843; Laura Jane, born 1850; Harrison and Martha "Mattie" born on November 25, 1852.

Solomon Young was a successful farmer who also owned and operated wagon trains that departed from Westport and Independence. One of the son's of Solomon and Harriett, William, served under Upton Hays until Hayes was killed in September 1862 and then joined guerrillas Virgil Miller, Cole Younger, Dick Yeager and Boon Muir.

On one occasion Kansas Red Legs came to the Young farm. Solomon's wife Harriett fed the soldiers, but afterwards the men stole the family silverware, killed over one hundred hogs, and burned their barns and haystacks.

The Young's were evicted from their property under Order No. 11 even though Solomon had signed the loyalty oath in Kansas City. After the War, Solomon bought property near Grandview, Jackson County where he farmed the land. Solomon died on January 26, 1892 and was buried in Forrest Hill Cemetery in Kansas City. The family home he had built was destroyed by fire the year after he died.[739]

---

737 United Daughters of the Confederacy; and, "Martin Rice Remembers."

738 Rafiner, citing National Historical Company, 678.

739 McCullough, 32; Joanne Eakin & Donald Hale, "Branded as Rebels," 484, and Eakin, "Tears and Turmoil," 52-55. Today the Federal Parks' Truman historic site in Grandview, Missouri includes the home

Harriet Louisa (Gregg) Young always hated what the federal troops did to her family during the Civil War. She most vehemently grew to disdain anything relating to the Union Blue. This even applied to that eventful occasion over forty years after Order No. 11 as told in *Chapter XII* above). Harriett Gregg Young died in Grandview on December 9, 1909 and was buried in Forrest Hill Cemetery in Kansas City.[740]

Laura Jane Young was the daughter of Solomon and Harriett Young. She was born in 1850 in Jackson County. She later married Mr. Everhart and died in Jackson County in 1933.

Martha "Mattie" Young was the daughter of Solomon and Harriett Young. She was born on November 25, 1852 in Jackson County and was only eleven at the time of Order No. 11. She later married John Anderson Truman who had been born in 1851. The first child born of their marriage was simply Baby Truman as he died in infancy in 1882. Their son and future president, Harry was born in 1884 and died in 1972. Their third child was John Vivian Truman, born in 1886 and died in 1965; and the fourth child was Mary Jane Truman born in 1889 and died in 1978. Mattie died in July 26, 1947 in Grandview and was buried in Forrest Hill Cemetery in Kansas City.

Sarah Ann Young was the daughter of Solomon and Harriett Young. She was born in 1843 and later married the guerilla Jim Crow Chiles. Her parents did not approve of their daughter's marriage to Chiles. After the war, her husband and son Elijah were killed in a shoot-out with Sheriff Peacock in Independence. Sarah died in 1907.

YOUNG
(Additional Account)

Fred L. Young, grandson of Daniel W. Young, provided some recollections about his mother, Rebecca Jane (Smith) Young, during Order No. 11:

> [Rebecca] *lived together with other women in the family during the Civil War. Their farm was close enough to the site of the Battle of Lone Jack that they could hear the shooting.* [Daniel] *said he was sitting on a rail fence and seeing a soldier deserting the battle and was clearing the fences with ease as he ran from the fighting.*
>
> *When Order No. 11 was issued,* [Rebecca] *proceeded to move to Independence. Their slave, Jesse, helped with the move. On one trip Rebecca urged Jesse to stay in Independence, but he wouldn't. That*

built after Solomon Young died, and it is the same dwelling in which grandma Young had berated little Harry S. Truman in 1904 for wearing the federal blue.

740 Eakin, "Tears and Turmoil," 52-55 (also see Chapter XII for more details on the Solomon Young family).

evening some men came to the farm and asked to borrow their mules. Rebecca knew what they wanted and sent Jesse to the barn to get the animals.

Shortly after he left the house, the family heard a shot. No one dared to go investigate.

The next morning Jesse was found dead standing between the gate and the fence post. Daniel reached into Jesse's pocket and took his pocket knife, which he kept for many years. Jesse's wife and children disappeared and nothing was ever heard of them again.[741]

YOUNGS

Mr. Youngs left his home in Cass County and went south into Texas after the issuance of Order No. 11.[742]

ZION
(see also FRANSE)

Nancy Zion of Cass County traveled with Celia Franse to Boone County, Missouri as a consequence of Order No. 11.[743]

[741] Jackson County (Mo.) Historical Society Archives, Document ID Vertical File Newspaper Subject Files: Military/Wars: Civil War: 1863-"Order No. 11."

[742] Rafiner, citing Belton Star-Herald. "The First Hundred Years: Belton, Missouri 1872-1972." (Belton, Mo.: Belton Star-Herald, 1972), 26F.

[743] Rafiner, citing Cass County Historical Society. "Cass County Missouri Families." (Harrisonville, Mo: Cass County Historical Society, Inc., 1976).

ILLUSTRATIONS

Cover "Order No. 11," or "Martial Law," by George Caleb Bingham, from an original John Sartain engraving in the private collection of Ralph A. Monaco. Two original copies are conserved in the Jackson County (Mo.) Historical Society Art Collection (one, a signed mezzotint proof by the artist, has been reproduced for resale) (JCHS021291AM).

13 The Kansas-Missouri Border, 1863. This map, originally drafted under the supervision of James C. Malin, was modified in 2013 by David W. Jackson at the instruction of author Ralph A. Monaco, II. It shows the "Border District," aka. the vicinity of the 6th Congressional District in Missouri.

15 General Robert E. Lee (C.S.A.) by Vic Arnold, courtesy Library of Congress Prints & Photographs Division, Washington, D.C., Call Number: PGA-Seer-Robert E. Lee (D size).

16 U.S. President Abraham Lincoln by Anthony Berger, courtesy Library of Congress Prints & Photographs Division, Washington, D.C., Call Number: PRES FILE-Lincoln, Abraham-Portraits-Meserve no. 85 [item]. This is the portrait used on the U.S. five dollar bill.

18 John C. Fremont, ca. 1855-1865, courtesy Library of Congress Prints & Photographs Division, Washington, D.C., Call Number: LC-BH82-523 C Lot 4192.

21 James Henry "Jim" Lane, ca. 1865-1880, courtesy Library of Congress Prints & Photographs Division, Washington, D.C., Call Number: LC-BH832-595.

23 David Rice Atchison, by Mathew B. Brady, courtesy Library of Congress Prints & Photographs Division, Washington, D.C., Call Number: DAG no. 134.

26 Stephen A. Douglas, ca. 1860, courtesy Library of Congress Prints & Photographs Division, Washington, D.C., Call Number: LOT 13301, no. 232.

30 John Brown, by Small, Maynard & Co., courtesy Library of Congress Prints & Photographs Division, Washington, D.C., Call Number: BIOG FILE-Brown, John, 1800-1859.

33 Brigadier General Thomas Ewing, Jr., ba. 1860-1865, courtesy Library of Congress Prints & Photographs Division, Washington, D.C., Call Number: LC-B813-2054 A Lot 4192 (corresponding print).

35 Charles Ransford "Doc" Jennison, courtesy Kansas State Historical Society, Kansas Memory, Item Number: 211289.

40 General Sterling Price (C.S.A.), ca. 1860-1870, courtesy *New, David F. Tales of the Trail: The Personal Correspondence of Soldiers and Men Who Made the Long Journey to Santa Fe in 1846 to 1849.* (Anacortes, Wa.: David F. New, 2013).

45 Daniel B. Anthony, ca. 1854-1874, courtesy Kansas State Historical Society, Kansas Memory, Item Number 221895.

46 General Henry Halleck, ca. 1861, courtesy Library of Congress Prints & Photographs Division, Washington, D.C., Call Number: LOT 9934, p. 26.

47 James "Jim" Lane, standing with gun, by Brady's National Photographic Portrait Galleries, courtesy Library of Congress Prints & Photographs Division, Washington, D.C., Call Number: LOT 14022, no. 329.

49 George Caleb Bingham, photograph, courtesy The State Historical Society of Missouri, Columbia, Missouri (also reproduced as JCHS004007L).

52 James S. Rollins, courtesy Ruth Rollins Westfall Photographic Collection (P20; Accession 006110), The State Historical Society of Missouri, Columbia, Missouri.

59 William Clarke Quantrill, portrait in charcoal on paper, which was prominently displayed in a frame for years at Quantrill reunions (JCHS009241XX).

60 William Clarke Quantrill birthplace on Third Street, Dover, Ohio, and later the site of Holmes Grocery Store (pictured in the Kansas Capitol newspaper, Topeka, Kansas, August 26, 1888 (JCHS022671AL). This image is of the house after it was moved from its original location and remodeled, per Mrs. John Cooley, Dover, Ohio. Images courtesy, Donald R. Hale Papers, Jackson County (Mo.) Historical Society Archives (JCHS022671BL).

60 Caroline Cornelia Clarke Quantrill, image courtesy, Donald R. Hale Papers, Jackson County (Mo.) Historical Society Archives (JCHS022664AM).

60 Thomas Henry Quantrill, image courtesy, Donald R. Hale Papers, Jackson County (Mo.) Historical Society Archives (JCHS022664BM).

61 Morgan Walker (JCHS021989XX)

62 Sketch of the 1859 Marshal's Home (the 1859 Jackson County Jail adjoins to the rear of this residence facing Main Street), 217 N Main Street, Independence, Missouri, by Virginia Jennings (attributed) (JCHS003144L).

64 Colonel Upton B. Hays (JCHS007765M).

67 Major General John M. Schofield, by Brady National Photographic Art Gallery, courtesy Library of Congress Prints & Photographs Division, Washington, D.C., Call Number: LC-B813-1944.

71 Abram Comingo, mayor of Independence, Missouri, 1852-1857, and Provost Marshall of 6th Congressional District, June 1863. This image courtesy Historic Elmwood Cemetery, Kansas City, Missouri (http://www.elmwoodcem-kc.org/family_photos-c.htm)

74 William Chrisman, later in life (JCHS000937L).

74 Samuel H. "Sam" Woodson, later in life; Mayor of Independence, Missouri. This image from 1901 is cropped from a group of men comprising the City of Independence Mayor and City Council at Wayne City Landing at the Missouri River (JCHS011762L).

75 1853 Jackson County Courthouse on Independence Square, Independence, Missouri, as taken from the 1868 *Bird's Eye View of Independence*, by A. Ruger (JCHS004816AS).

79 Abram Comingo, in later life (JCHS001367S).

81 Brigadier General Thomas Ewing, Jr., ca. 1860-1870, courtesy Library of Congress Prints & Photographs Division, Washington, D.C., Call Number: LC-B813-6535 C.

83 William Tecumseh Sherman, courtesy Library of Congress Prints & Photographs Division, Washington, D.C., Call Number: BIOG FILE – Sherman, William Tecumseh <item> [P&P].

87 Pacific House, 6th and Delaware, Kansas City River Market, where "General Orders No. 11" was issued in August 25, 1863 (JCHS000702L).

94 The makeshift Union Women's Prison (a building inherited by George Caleb Bingham's wife referred to as the 'Thomas property,' and commandeered by the Federal Army) was about the middle of the Metropolitan Block on Grand Avenue between 13th and 14th Streets, Kansas City, Missouri (JCHS000725L).

95 Charity Kerr, on right. Nancy Elizabeth "Nannie" Harris is on the left; she survived the prison collapse tragedy and soon thereafter walked with her family to Howard County, Misosuri. See also HARRIS biography in Appendix H. Image courtesy Donald R. Hale Papers, Jackson County (Mo.) Historical Society Archives (JCHS006174M).

101 Martha Ann "Mattie" (Livingston) Lykins (later Bingham) by S.M. Eby & Son, Ambrotype and Photographic Artists, Kansas City, Missouri, 1865 (the only year S. M. Eby & Son were listed in annual Kansas City city directories, at the corner of 3rd and Main Streets, 3rd floor). One undated newspaper clipping claimed that Mattie was the first woman to have a photo made in Kansas City. Mr. Robert Dewit Owen, a great-great grandnephew of Mattie, donated this original image to the Jackson County (Mo.) Historical Society Archives, where it accompanies his previous gift of Mattie's original, "Recollections of Old Times in Kansas City," and unpublished manuscripts of his grandmother, Mrs. Dewit Livingston (Ada Armintha Campbell) Owen (JCHS01315M).

110 William Clarke Quantrill, with mustache (JCHS022665M), sometimes reprinted with a "Quantrell" label (JCHS005501M).

111 "On to Lawrence," from Jay Donald's book, *Life and Adventures of the James Brothers*, 1882, , image courtesy, Donald R. Hale Papers, Jackson County (Mo.) Historical Society Archives (JCHS022692BL).

112 "Destruction of Lawrence," from Jay Donald's book, *Life and Adventures of the James Brothers*, 1882, image courtesy, Donald R. Hale Papers, Jackson County (Mo.) Historical Society Archives (JCHS022692CL).

112 "Quantrell's Raid on Lawrence, Kansas," from Augustus C. Appler's book, *The Life, Character and Daring Exploits of the Younger Brothers*, 1875, and Croy's book, *Last of the Great Outlaws*, image courtesy, Donald R. Hale Papers, Jackson County (Mo.) Historical Society Archives (JCHS022692DL).

112 Quantrill's Raid on Lawrence, as sketched in pencil by Sherman Enderton, Co. E, 11th Kansas Volunteers, who reached the ruined city hours later. Courtesy Kansas State Historical Society.

113 Quantrill's Retreat, by Ernst Ulmer, painted on commission for a mural in the private dining room of Mr. and Mrs. Bud and Dee Morris, Prairie Village, Kansas, who generously allowed David W. Jackson to photograph the artwork for inclusion in *Scattered to the Four Winds*. The imagery is based on Mrs. Morris' family history tradition. Mrs. Morris' mother's grandfather, George N. Simpson, is depticted in the mural, protecting his wife from Quantrill's fleeing raiders, by hiding in a corn field.

114 *Harper's Weekly*, 5 Sept. 1863, Jackson County (Mo.) Historical Society Archives, Document ID OV9.

117 David Lee Campbell, courtesy David W. Jackson.

126 Reproduction of an original, handwritten "General Orders No. 11," from the
127 National Archives and Records Administration.

138 BEFORE. Homes of productive citizens were burned to the ground, resulting in a landscape domintéd by an array of solitary chimneys across the horizon. Original artwork courtesy Bill Wehner, Planning Committee, *Smoke and Ashes* living history event co-sponsored by Jackson County Parks + Rec and Jackson County Historical Society at Missouri Town 1855, Saturday, September 14, 2013.

139 AFTER. Homes of productive citizens were burned to the ground, resulting in a landscape domintéd by an array of solitary chimneys across the horizon. Original artwork courtesy Bill Wehner, Planning Committee, *Smoke and Ashes* living history event co-sponsored by Jackson County Parks + Rec and Jackson County Historical Society at Missouri Town 1855, Saturday, September 14, 2013.

144 Luttrell cabin, restored at Missouri Town, 1855, Fleming Park at Lake Jacomo, Blue Springs, Missouri, courtesy Jackson County Parks + Rec.

145 Map of Jackson County, Missouri, 1860, by Nancy M. Ehrlich.

148 Harry S. Truman, 1906, in his National Guard uniform, courtesy Harry S. Truman Presidential Library and Museum, Independence, Missouri, photograph ID 79-22.

149 Laura Elizabeth (Harris) Flanery (later, Bridges), as she was in her youth as Mrs. Martin Van Buren Flanery, *Kansas City* (Mo.) *Journal-Post*, 16 May 1926.

152 Laura Elizabeth (Harris) Flanery (later, Bridges), grandmother of former Missouri Congressman, William J. Randall (JCHS005807AL).

155 Rice Cabin, undated newspaper clipping, Jackson County (Mo.) Historical Society Archives, Vertical file subjects in newsprint, "Biography: Rice."

161 Six Man Cemetery burial marker, Lone Jack, Missouri, courtesy the Lone Jack Historical Society, historiclonejack.org/whatisaw.html.

167 George Caleb Bingham home, Pacific and Osage, Independence, Missouri, as it appeared when he lived there in 1865. It has since been enlarged and today is the Lewis-Bingham-Waggoner Estate house museum. From the Bingham-Waggoner Estate Historical Society Collection in the Jackson County (Mo.) Historical Society Archives (JCHS004805X).

168 1853 Jackson County Courthouse on Independence Square, Independence, Missouri, as taken from the 1868 *Bird's Eye View of Independence*, by A. Ruger (JCHS004816BS).

170 1859 Marshal's Home (and Jackson County Jail to the rear), as represented in the *Illustrated Atlas of Jackson County, Missouri,* 1877. This is the earliest view of the 1859 Marshal's Home. (JCHS011133L).

171 Rev. Abner Holton Deane, as painted by George Caleb Bingham when Deane was incarcerated in the 1859 Jackson County Jail after the Civil War. Bingham's original was copied by muralist Charles Goslin and donated to the Jackson County (Mo.) Historical Society Art Collection; it is on permanent exhibit at the 1859 Jail Museum, 217 North Main Street, Independence, Missouri (JCHS020420L).

172 Jackson County Jailor, Henry Bugler, post mortem, 1866 (JCHS003807L).

175 "Order No. 11," or "Martial Law," by George Caleb Bingham, from an original John Sartain engraving in the private collection of Ralph A. Monaco. Two original copies have been donated and are conserved in the Jackson County (Mo.) Historical Society Art Collection (one, a mezzotint proof signed by the artist and donated in 1961 by a descendant of the Browning family of Jackson County; and, another engraving signed by the artist, and donated in 2013 by a descendant of the Captain and Mrs. Leslie Smith family of Jackson County). This artwork has been reproduced and is available for resale at the Historical Society's Bookshop (JCHS021291AM).

176 George Caleb Bingham's Independence studio where "Order No. 11," or "Martial Law," was painted. The original studio building was later razed. A new building, used as a carriage house and museum orientation center, has been erected in close proximity to the foundations of Bingham's studio (JCHS4016L).

178 "Order No. 11," or "Martial Law," by George Caleb Bingham, when it was displayed in the reception hall of the Colonel Joseph W. Mercer home, 116 South Pleasant, Independence, Missouri (JCHS000240L).

182 George Caleb Bingham, a self portrait in oil, ca. 1877, courtesy The State Historical Society of Missouri, Columbia, Missouri (JCHS000663L).

186 James Lane tombstone, courtesy contributor, Tom Denardo, findagrave.com (Lane), viewed 8 May 2013.

190 George Caleb Bingham's Kansas City studio, after the Civil War, possibly taken at the time of his death. Image may be from the Native Sons and Daughters of Greater Kansas City collection at the State Historical Society of Missouri, Kansas City Research Center (also, JCHS003943M).

192 George Caleb Bingham tombstone, courtesy contributor, Bill Walker, findagrave.com (Bingham), viewed 8 May 2013.

197 William Clarke Quantrill lock of hair, image courtesy, Donald R. Hale Papers, Jackson County (Mo.) Historical Society Archives (JCHS022686BL).

197 William Clarke Quantrill leg and arm bones, courtesy Kansas State Historical Society (JCHS022686AL).

198 William Clarke Quantrill skull when on display at the Dover Historical Society, image courtesy, Donald R. Hale Papers, Jackson County (Mo.) Historical Society Archives (JCHS022682AM).

199 William Clarke Quantrill tombstone in Higginsville, Missouri, courtesy contributor "SMG," findagrave.com, viewed 8 May 2013.

1995 Quantrill family burial plot, Dover, Ohio, image courtesy, Donald R. Hale Papers, Jackson County (Mo.) Historical Society Archives (JCHS022683AM). Quantrill's subsequent tombstone courtesy contributor, Paula Lorenz, findagrave.com (Quantrill), viewed 8 May 2013.

204 John Schofield tombstone, courtesy contributor, Ryan Gleason, findagrave.com (Schofield), viewed 8 May 2013.

207 Abram Comingo mausoleum, Elmwood Cemetery, Kansas City, Missouri, courtesy the author.

211 Thomas Ewing tombstone, courtesy contributor "EddieM ," findagrave.com (Ewing) viewed 8 May 2013.

224 *Kansas City Daily Journal*, 25 Aug. 1863, Jackson County (Mo.) Historical Society Archives, Marian Franklin and family, who donated the original Colonel Upton Hays and Margaret Jane (Watts) Hays Overstreet Family Collection, 1849-1865 [early 1900s] to the Jackson County (Mo.) Historical Society. This item is located in the Society's Archives at Document ID OV9.

298 Colonel Upton B. Hays (JCHS007763M).

299 Margaret Jane (Watts) Hays, wife of Colonel Upton Hays. They married in 1852 and had four children: John Nathan; Mary Elizabeth; Elfred; and Jane Upton. Margaret later married William Overstreet and they had one son, Claude C. Overstreet. Colonel Upton Hays and Margaret Jane (Watts) Hays Overstreet Family Collection, 1849-1865 [early 1900s], Jackson County (Mo.) Historical Society Archives. This collection is also available online at wattshaysletters.com, courtesy Marian Franklin and family, who donated the originals to the Society for permanent conservation (JCHS022918S).

306 Nancy (Tyler) Holmes, great grandmother of Harry S. Truman (JCHS000010L).

311 1834Hudspeth Bible, gift of Dora Slaughter, Jackson County (Mo.) Historical
312 Society Archives, Document ID FOLIO4F1; image courtesy David W. Jackson, Archives and Education Director, 2007.

312 Civil War Passes; Jackson County (Mo.) Historical Society Archives Document ID 237F4 and 237F5.

336 Note written to Venitia (Colcord) Page, while incarcerated in the Women's Prison at Kansas City, Missouri. Jackson County (Mo.) Historical Society Archives Document ID 2F16.

366 Dr. L. W. Twyman (JCHS004204M)

367 Mrs. Frances C. "Fannie" (Fristoe) Twyman (JCHS020602S)

BIBLIOGRAPHY

Archives

Cass County Historical Society, Harrisonville, Missouri.

Tom A. Rafiner Papers, containing research for his book, "Caught Between Three Fires: Cass County, Mo., Chaos, & Order No. 11, 1860-1865." (Xlibris, Corp., 2010).

Fold3

Hill's *History of the Presbytery of Kansas City*.

Joseph D. Liles, John R. Oldham, and Robert Smith Symington. NARA Publication M322: Compiled Service Records of Confederate Soldiers Who Served in Organizations from the State of Missouri. Record Group 109, 586957.

Jackson County Records Center (and on microfilm at the **Midwest Genealogy Center**, Independence, Missouri)

Jackson County (Mo.), County Court Records.
July Term, 1863
July Term, 1866 (Record Book No. 13)

Jackson County (Mo.) Historical Society Archives, Independence, Missouri.

Document ID 2F16. Jacob Hall Family Papers, 1831-1904.

Document ID 3F12. Alberta Wilson Constant Papers, Constant, Alberta Wilson, compiler. Reverend Abner Holton Deane (1828-1912) material.

Document ID 10F15. Akers, Joseph Harold, tape-recorded interview and transcript of Joseph Harold Akers on August 27, 1985.

Document ID 11F10. J. Calvin and Fannie Iserman Family Letters, 1858-1863.

Document ID 36F6. Pass issued to William G. Vergin to go 5 miles south of Harrisonville for hogs, 25 Nov. 1863 (photocopy).

Document ID 51F1. Albert Gallatin Steele Family papers (photocopies). For more information and transcriptions of additional letters to and from Emily (Hampton) Steel (dated 24 Jul 1863; 31 July 1863; 20 Sept. 1863; and 5 Dec. 1863) which are NOT yet in this collection, see the Fall 1980 Jackson County Historical Society JOURNAL and the 2004 book by Joanne Chiles Eakin and Annette and Bill Curtis, "Our Far Away Boys."

Document ID 57F25. Special order to John Poyntz, banishing him from Missouri, 19 Sept. 1863 (copy). For more see Jackson County (Mo.) Historical Society Archives, Document ID 144.01F33.

Document ID 62F3. Mrs. William J. (Mary Agnes Johnson) Bland Collection.

Document ID 66F40. Independence Young Matrons Records, 1941-, "Story Hour Manuscripts," 1964, "Uprooting of Families of Order No. 11."

Document ID 68F24. undated newspaper article titled, "An Interesting Momento: List of Those Banished by Order No. 11," reprints from The Columbia Herald, two documents from 1863. One is an extract of an evacuation order, "Special Orders No. 6," from the Headquarters District of the Border, Kansas City, Missouri, August 29th 1863, "A. O. Runyan and family, residents of Independence, Missouri, are ordered to remove from this district within ten days from the date hereof. They will not go to the counties of Platte, Clay, Ray, or Carroll, Missouri, to reside nor return to this district, during the rebellion, without previous express permission from competent military authority." The second clipping reprinted gives the names of those banished (as compiled and re-published in this book).

Document ID 68F25. An unpublished article titled, "A Panoramic View of the Lost Townships," by Mrs. Greenberry (Jesse Crosby) Ragan, ca. 1965.

Document ID 76F28. Lambert Moon Printing Company. "Centennial Greetings, Independence, 1827-1927: Wishing Our City Prosperity in its Second Century."

Document ID 77.01F16. Jackson County Historical Society Contest Entries, "My Most Interesting Ancestor Who Lived in Jackson County," 1959-1960, manuscript relaying family history of the Hudspeth and Hambright families that had been passed down by family members, written by Thomas Lorraine (Mrs. Paul M.) Campbell, Minden, Louisiana.

Document ID 77.01F21. Jackson County Historical Society Contest Entries, "My Most Interesting Ancestor Who Lived in Jackson County," 1959-1960, manuscript about Narcissa Hudspeth that includes a recollection of Order No. 11.

Document ID 77.01F29. . Jackson County Historical Society Contest Entries, "My Most Interesting Ancestor Who Lived in Jackson County," 1959-1960, manuscript about Martin Rice, relaying family history, including his involvement in Order No. 11.

Document ID 77.03F1. Jackson County Historical Society Contest Entries, "An Interesting Ancestor Who Lived in Jackson County," 1960-1961, manuscript about Harmon Gregg, written by Maude Lee (Burgess) Immele, Kansas City, Missouri.

Document ID 77.03F8. Jackson County Historical Society Contest Entries, "An Interesting Ancestor Who Lived in Jackson County," 1960-1961, manuscript written by Donald R. Hale, Lee's Summit, Missouri, relaying family history of Reuben and Rebecca (Harper) Vaughn, who settled in 1832, in New Santa Fe, on the western border of Jackson County, Missouri. Their son, James Newton Vaughn, was in the Battle of Lone Jack in August 1862, and the Vaughn home was burned in 1863 as a result of Order No. 11.

Document ID 77.03F13. Jackson County Historical Society Contest Entries, "An Interesting Ancestor Who Lived in Jackson County," 1960-1961, manuscript written by Maude Lee (Burgess) Immele, Kansas City, Missouri, relaying family history of her grandfather, William Sanders Jones, including an incident of Order No. 11.

Document ID 77.03F14. Jackson County Historical Society Contest Entries, "An Interesting Ancestor Who Lived in Jackson County," 1960-1961, manuscript written by Maude Lee (Burgess) Immele, Kansas City, Missouri, relaying family history of her grandmother, Jemima (Best) Jones, including an incident of Order No. 11.

Document ID 77.04F6. Jackson County Historical Society Contest Entries, "A Story of an Early Jackson County Home and the People Who Occupied It," 1962-1963 , manuscript written by Ethel Milton Hughes realying a history of the Samuel Luttrell log house (which stood just west of the intersection of 39th Street and Lee's Summit Road as late as February 1962, and was re-located to Missouri Town, 1855), including an incident of Order No. 11.

Document ID 80F6. Dolly Bee Breitenbaugh Papers, 1848-1958, "Short Stories of Pioneer Days," by Dollie Bee Breitenbaugh, 1959. 94-leave, typed manuscript.

Document ID 82.06F12. Jackson County Historical Society Contest Entries, "History of the City, Town or Village in Jackson County," 1966-1967, manuscript, "History of Old Westport: Phillip [sic.] Becker and Order No. 11," by Adrienne Christopher after an interview with Mrs. August (Rosa "Rose" (Becker)) Hahn, who provided a claim about the origin of Order No. 11.

Document ID 87.01F2. Haukenberry Noland Collection. A letter from Mrs. G. T. Twyman to a Mrs. Green, 19 Apr. 1926, relating history of Richard Marshall Fristoe.

Document ID 96F15. Townsend, Robert S. Decade of Decision, 1855-1865. (Kansas City, Mo.: Kansas City Life Insurance Company, 1960).

Document ID 100F24. Photocopy of a letter from William Leonidas Bone (8th Mayor of Independence) to Professor F. Mims, 22 Dec. 1863, discussing the treatment of Independence homes and depredations during the Civil War, and after Order No. 11.

Document ID 110F7, Bingham, Martha A. "Mattie" (Livingston) Lykins. "Recollection of Old Times in Kansas City," 67-page handwritten manuscript, ca 1883-1900. Gift of Robert Dewitt Owen, in memory of his grandmother, Mrs. Dwit Livingston (Ada Armintha Campbell) Owen.

Document ID 113F21. The Truth About Order No. 11 by Flora E. Stevens, from an undated newspaper clipping from, "The Sunday Republic Feature Section" (see also *Confederate Veteran* (April 1917): 25: 154).

Document ID 121F38. Balfour Collection of Duvall Family Papers, 1846-1935. Duvall genealogy, and a 22 Aug 1927 affidavit by W. A. Duvall recollecting Order No. 11.

Document ID 140F16. Photocopy of a letter to Mary from cousin Mattie Jane (Rice) Tate, 14 December 1864, from Jackson County, Missouri, describing Order No. 11 and its enforcement. The original is in the collection of the Lone Jack Historical Society, Lone Jack, Missouri, and should be credited as such. Transcript included.

Document ID 144.01F4. Donald R. Hale Papers. Extract from a letter written by Frances Fitzhugh (George) Kabrick about the George and Owings family's ordeal through Order No. 11.

Document ID 144.01F33. *Kansas City* (Mo.) *Times*, 24 Aug. 1963, "Order No. 11, Crux of Civil War Bitterness." "Order No. 11 Drove Families from Border Homes in 1863,"

Kansas City (Mo.) *Star,* 9 Apr 1961, lists affected families: Henry Cordell and Rev. George Miller. For more on Miller, see his book, Missouri's Memorable Decade (Columbia, Mo., 1898), 100-107, who also mentions an "incident about an old man, John Poyntz, who had been confied in a military prison for his sympathetic expressions on behalf of Southern friends. He was released into the custody of George Miller who kept him in his home until Poyntz was finally banished from Missouri." See Jackson County (Mo.) Historical Society Archives, Document ID 57F25.

Document ID 152F37. John Calvin Morris and Order No. 11, by R. L. Morris and W. H. Morris.

Document ID 154.01F30. Alberta Wilson Constant Papers. "Order No. 11," a speech Alberta Wilson Constant delivered to the Independence Young Matrons, ca. 1960.

Document ID 163.01F26. "What I saw of Order No. 11 and Poems by Martin Rice, reprinted by Fabius Martin Butler, 1965.

Document ID 167.5F32. Dorothy (Pierson) Butler Oak Grove Collection. Butler, Dorothy (Pierson) Butler, "In Early Days," *Oak Grove* (Mo.) *Banner*, undated: A newspaper clipping describing James Russell Hudson and his 1852 trip across the plains on the California Trail for the California Gold Rush.

Document ID 223.03F16. Mrs. Con (Ethel Massie) Withers Papers, 1900-1935, enclosing a copy of, "A Half Century Since Order No. 11," *Kansas City* (Mo.) *Star*, 25 Aug 1913.

Document ID 237F4. Thomas Jefferson Hudspeth Papers, 1824-1848. Paralee Hudspeth, Pass permitting travel in border counties of Missouri, 18 November 1863; Order No. 11.

Document ID 237F5. Thomas Jefferson Hudspeth Papers, 1824-1848. Cynthia Hudspeth, Grant of protection from U.S. Army, 4 February 1862.

Document ID 229. The Colonel Upton Hays and Margaret Jane (Watts) Hays Overstreet Family Collection, 1849-1865 [early 1900s], gift of Marian Franklin and family. This collection includes letters by Margaret and Upton Hays; John S. Watts; John S. Watts, Jr.; Elizabeth Ewing Berry Yocum Watts; Thomas A. Yocum; a short 1872 diary by Berry Yocum; and materials about allied families including, but not limited to: Boone, Scholl, and Kimberlin. These materials have been digitized and are available at wattshaysletters.com. Transcriptions of all of these letters, as mirrored from the Internet postings (wattshaysletters.com), are assembled at the end of this collection. Earlier transcriptions of three of these originals (plus three of which are NOT included in this collection) appear at Jackson County (Mo.) Historical Society Archives Document ID 11F18. The John S. Watts family moved from Kentucky to Van Buren (Cass) County, Missouri, ca. 1839. His daughter, Margaret Jane Watts (1836-1923) married Upton Hays in 1852, and they lived on a farm near Westport, Jackson County, Missouri. Upton Hays was a son of Boone Hays who bought the farm from his uncle, Morgan Boone, in the 1830s. The letters describe life in Jackson County in the mid-1800s and also contain stirring descriptions of the Civil War and Jayhawker raids in the area. Colonel Upton Hays, 2nd Missouri Cavalry, Confederate States of America, was killed in a skirmish in 1862, and Margaret Hays was exiled from Jackson County by "Order No. 11" in 1863 [an original page from the August 1863 Kansas City Daily Journal is included in this collection]. Margaret Hays lived in Callaway County, Missouri, until the end of the Civil War, then returned to Jackson County and lived there until 1872 when she migrated with her children to California, where she homesteaded in Tulare County (later Kings County) in 1873, and married William B. Overstreet in 1877.

Document ID 269F9. Reproductions of secondary sources relating to Reverend Abner Holton Deane, who was incarcerated in the 1859 Jackson County Jail.

Document ID L62F3. Miss Cammie Johnston Collection. Genealogical material relating to Miss Cammie Johnston, focusing on the Doneghy family. Included are copies from the Holy Bible of Mrs. James (Kate S.) Doneghy.

Vertical File Newspaper Subject Files
Biography: "Bridges" (formerly Flanery)
Military/Wars: Civil War: 1863-"Order No. 11"

Library of Congress
John McAllister Schofield Papers, Diary of Events in Missouri, 1863, Box 1

Missouri State Archives
Box F 14495
Missouri Death Certificate Collection: Chouteau; Doneghy; Russel ;

National Archives and Records Administration

Records of the Quartermaster General. Record Group 92. War Claims.

United States Census 1810: Harrisonburg, Rockingham County, Virginia.

United States Census 1830: Arrow Rock Township, Saline County, Missouri.

United States Census 1840: Arrow Rock Township, Saline County, Missouri.

United States Census 1850: Arrow Rock Township, Saline County, Missouri.

United States Census 1850: Jackson County, Missouri.

United States Census 1860: Jackson County, Missouri.

United States Census 1870, Jackson County, Missouri.

United States War Department by an Act of March 3, 1863 (12 Stat. 731) abolished on August 28, 1866, by General Order No. 66, War Department, August 20, 1866, implementing an act of July 28, 1866 (14 Stat. 337).

Books

Anderson, Galusha. "The Story of a Border City During the Civil War: The Missouri Oath of Loyalty of 1865." Excerpted and Introduced by G. E. Rule. (Little Brown and Co., 1908).

Bartels, Carolyn M. "Bitter Tears: Missouri Women and Civil War, Their Stories." (Independence, Mo. Two Trails Press, 2002).

Barton, O. S. "Three Years with Quantrill: A True Story by His Scout John McCorkle." Originally published 1914. (Norman, Ok.: University of Oklahoma Press, 1992).

Belton Star-Herald. "The First Hundred Years: Belton, Missouri, 1872-1972." (Belton, Mo.: Belton Star-Herald, 1972).

Brownlee, Richard S. "Gray Ghosts of the Confederacy: Guerrilla Warfare in the West 1861-1865." (Baton Rouge, La.: Louisiana State University Press, 2011).

Burch, John P. "A True Story of Charles W. Quantrell and his Guerrilla Band." (Vega, Tx.: J. P. Burch, 1923).

Casey, Albert Eugene. "Amite County, Mississippi 1699-1890." Volume 3: The Environs. (Birmingham, Al.: Amite County Historical Fund, 1957).

Cass County Historical Society. "Captain Alexander Robinson's Co.: Enrolled Mo. Militia Muster Roll." (Harrisonville, Mo.: Cass County Historical Society, 2002).

------. "Cass County, Missouri, Families." (Harrisonville, Mo.: Cass County Historical Socity, 1976).

Castel, Albert. "A Frontier State at War: Kansas 1861-1865." Second Edition. (Topeka, Ks.: Kansas Heritage Press, 1992).

-----. "Winning and Losing in the Civil War, Essays and Stories." (Columbia, Sc.: University of South Carolina, 1996).

Castel, Albert, and Tom Goodrich. "Bloody Bill Anderson: The Short, Savage Life of a Civil War Guerilla." (Lawrence, Ks.: University Press of Kansas, 1998).

Collins, Robert. "Jim Lane Scoundrel: Statesman, Kansan." (Gretna, La.: Pelican Publishing Company, 2007).

Connelley, William E. "Quantrill and the Border Wars." Originally published 1910. (New York: Pageant Book Co., 1956).

Cummins, Jim. "Jim Cummins' Book, Written By Himself." Originally published 1903. (Independence, Mo.: Two Trails Publishing, 1999).

Christensen, Lawrence O., William E. Foley, Gary R. Kramer and Kenneth H. Winn, eds. "Dictionary of Missouri Biography." (Columbia, Mo.: University of Missouri Press, 1999).

Dean, Virgil W., ed. "Kansas Territorial Reader." Feature by Brian R. Dirk, "By the Hand of God." (Topeka, Ks.: Kansas State Historical Society, 2005).

Donald, David Herbert. "Charles Sumner and the Coming of the Civil War." Originally published 1960. (Naperville, Il.: Sourcebooks, Inc., 2009).

Eakin, Joanne Webb Chiles, and Donald R. Hale. "Branded as Rebels: A List of Bushwhackers, Guerrillas, Partisan Rangers, Confederates and Southern Sympathizers from Missouri During the War Years." Volume 1. Second Printing. (Lee's Summit, Mo.: Print America, 1995).

Eakin, Joanne Webb Chiles. "Five Years of the Blue and Grey Chronicle, 1997-2002." (Independence, Mo.: Two Trails Publishing, 2002).

-----. "Our Far Away Boys: The Civil War Letters of the Emily Steele Family." (Independence, Mo.: Two Trails Publishing, 2004).

-----. "Recollections of Quantrill's Guerrillas as Told by A. J. Walker of Weatherford, Texas to Victor E. Martin." Originally published in 1910. (Independence, Mo.: Two Trails Publishing, 1996).

-----. "Tears and Turmoil: Order No. 11." (Shawnee, Ks.: Two Trails Genealogy Shop, 1996).

-----. "The Little Gods: Union Provost Marshals in the Civil War." (Shawnee, Ks.: Two Trails Genealogy Shop, 1996).

Edwards, John N. "Noted Guerrillas: Being a History of the Lives and Adventures of Quantrell, Bill Anderson, George Todd, Dave Poole, Fletcher Taylor, Peyton Long, Oil Shepherd, Arch Clements, John Maupin, Tuck and Woot Hill, Wm. Gregg, Thomas Maupin, the James brothers, the Younger brothers, Arthur McCoy, and Numerous Other Well Known Guerrillas of the West." Originally published in 1877. Republished as an indexed edition. (Shawnee, Ks.: Two Trails Publishing, 1996).

Etcheson, Nicole. "Bleeding Kansas: Contested Liberty in the Civil War Era." (Lawrence, Ks.: University Press of Kansas, 2004).

Fellman, Michael. "Inside War: The Guerrilla Conflict of Missouri During the American Civil War." (New York: Oxford University Press, 1989).

Findlen, Rose Ann. "Missouri Star: The Life and Times of Martha Ann "Mattie" (Livingston) Lykins Bingham." (Independence, Mo.: Jackson County Historical Society, 2011).

Garwood, Darrell. "Crossroads of America." (New York: W. W. Norton & Co., 1948).

Gentzler, Lynn Wolf, and Roger E. Robinson, eds. "The Letters of George Caleb Bingham: "But I Forget That I am a Painter and Not a Politician." (Arrow Rock, Mo: The State Historical Society of Missouri and Friends of Arrow Rock, Inc., 2011).

Gerteis, Louis S. "The Civil War in Missouri: A Military History." (Columbia, Mo.: University of Missouri Press, 2012).

Glenn, Allen. "History of Cass County, Missouri." (Topeka, Ks.: Historical Publishing Co, 1917).

Goodrich, Thomas. "Black Flag: Guerrilla Warfare on the Western Border, 1861-1865." (Bloomington, In.: Indiana University Press, 1999).

Goodrich, Thomas. "Bloody Dawn: The Story of the Lawrence Massacre." (Kent, Oh.: Kent State University Press, 1991).

Goodrich, Thomas. "War to the Knife: Bleeding Kansas 1854-1861." (Mechanicsburg, Pa.: Stackpole Books, 1998).

Guelzo, Allen C. "Lincoln: Emancipation Proclamation: The End of Slavery in America." (New York: Simon & Schuster, 2004).

Hale, Donald R. "We Rode With Quantrill: Quantrill and the Guerrilla War as Told by the Men and Women Who Were with Him: A True Sketch of Quantrill's Life." Originally published in 1975. (Independence, Mo.: Blue and Grey Book Shoppe, 1998).

Hansen, Duncan E. "A Reunion in Death: Gravesites of Quantrill's Men and the James Gang." (Independence, Mo.: Two Trails Publishing, 2002).

Harrold, Stanley. "Border War: Fighting Over Slavery Before the Civil War." (Chapel Hill, Nc.: University of North Carolina Press, 2010).

Houts, Joseph K., Jr. "Quantrill's Thieves." (Kansas City, Mo.: Truman Publishing Company, 2002).

Jackson, David, W. "A History of Pink Hill Park and Veterans Way Memorial." (Independence, Mo.: Jackson County Historical Society, 2013).

Jackson, David W., ed. "Illustrated Historical Atlas of Jackson County, Missouri." Originally published in 1877. Indexed edition. (Independence, Mo.: Jackson County Historical Society, 2007).

Jackson, David W., and Paul Kirkman. "LOCK DOWN: Outlaws, Lawmen and Frontier Justice in Jackson County, Missouri." Sixth Edition. (Independence, Mo.: Jackson County Historical Society, 2012).

Johnson, Robert Underwood, and Clarence Clough Buel. "Battles and Leaders of the Civil War: Being for the most part contributions by Union and Confederate officers." From Sumter to Shiloh, Vol. I (New York: Castle Books, 1956).

Johnson, Robert Underwood, and Clarence Clough Buel. "Battles and Leaders of the Civil War: Being for the most part contributions by Union and Confederate officers Retreat From Gettysburg, Vol. III. (New York: Castle Books, 1956).

"Journal and Proceedings of the Missouri State Convention Held at Jefferson City and St. Louis." (St. Louis: George Knapp & Co., March 1861).

"Journal of the Missouri State Convention Held at Jefferson City." (St. Louis: George Knapp & Co., June 1862).

"Journal and Proceedings of the Missouri State Convention Held at Jefferson City." (St. Louis: George Knapp & Co., June 1863).

Kelly, Alfred H., and Winfred A. Harbison. " The American Constitution: It's Origin and Development." Fourth Edition. (New York: W. W. Norton & Company, Inc., 1975).

Larkin, Lew. "Missouri Heritage." (Columbia, Mo.: American Press, Inc., 1968).

Leslie, Edward E. "The Devil Knows How to Ride." (New York: Random House, 1996).

Long, E. B., and Barbara Long. "The Civil War Day By Day: An Almanac 1861-1865." (New York: Da Capo Press, Inc., 1971).

McCollough, David. "Truman." (New York: Simon & Schuster, 1992).

Miller, George. "Missouri's Memorable Decade: 1860-1870." (Columbia, Mo: Press of E. W. Stephens. 1898).

"Missouri Constitution, 1865."

Monaco, Ralph A., II. "Blood on the Streets: The Civil War Comes to Jackson County, Missouri, August 1862." (Independence, Mo.: Monaco Publishing, L.LC. and Jackson County Historical Society, 2012).

Monaco, Ralph A. II. "Son of a Bandit: Jesse James & The Leeds Gang." (Raytown, Mo.: Monaco Publishing, L.L.C., 2012).

Monaco, Ralph A., II. "The Strange Story of Col. Swope & Dr. Hyde." (Independence, Mo.: Two Trails Publishing, 2010).

Monaghan, Jay. "Civil War on the Western Border, 1854-1865." (Lincoln, Ne.: University of Lincoln, Press, 1955).

Moore, LaVonne Belew. "Missouri Town-1855: A Program in Architectural Preservation." (Kansas City, Mo.: Jackson County Parks and Recreation in cooperation with Lowell Press, Inc., 1987).

Nagel, Paul C. "George Caleb Bingham: Missouri's Famed Painter and Forgotten Politician." (Columbia, Mo.: University of Missouri Press, 2005).

National Historical Company. "History of Cass and Bates Counties, Missouri." (St. Joseph, Mo.: National Historical Company, 1883).

Neely, Mark E., Jr. "Lincoln and the Triumph of the Nation Constitutional Conflict in the American Civil War." (Chapel Hill, Nc.: University of North Carolina Press, 2011).

Norton, Richard L. "Behind Enemy Lines: The Memoirs and Writings of Brigadier General Sidney Drake Jackman." (Springfield, Mo.: Oak Hills Publishing, 1997).

Parrish, William E. "A History of Missouri." Vol. III, 1860-1875. (Columbia, Mo.: University of Missouri Press, 1973).

-----. "Frank Blair: Lincoln's Conservative." (Columbia, Mo.: University of Missouri Press, 1998).

Petersen, Paul R., and David W. Jackson. "Lost Souls of the Lost Township: Untold Life Stories of the People Buried in the Davis-Smith Cemetery, Kansas City, Missouri." (Kansas City, Mo.: The Orderly Pack Rat, 2011).

Petersen, Paul R. “Quantrill of Missouri: The Making of a Guerrilla Warrior.” (Nashville, Tn.: Cumberland House, 2003).

Poppino, Hattie E., abstracter. “Census of 1860: Population Schedules of Jackson County, Missouri.” Index. (Kansas City, Mo.: Hattie E. Poppino, 1964).

Rash, Nancy. “The Painting and Politics of George Caleb Bingham.” (New Haven, Ct.: Yale University Press, 1991).

Rice, Martin. “Rural Rhymes, and Poems from the Farm.” (Kansas City, Mo.: Millett & Hudson, 1877);

-----. “Rural Rhymes, and Tales of Olden Times.” (Kansas City, Mo.: Ramsey, Millett & Hudson, 1882).

Scott, Robert N., H. M. Lazelle, George B. Davis, Leslie J. Perry, Joseph W. Kirkley, Fred C. Ainsworth, John S. Moodey, and Calvin D. Cowles. “Official Records of the War of the Rebellion: A Compilation of the Official Records.” (Washington, D.C.: Government Printing Office, 1880-1901).

Sears, Stephen W. “Chancellorsville.” (New York: Houghton Mifflin Company, 1996).

Siddali, Dilvana R., ed. “Missouri’s War: The Civil War Documents.” (Athens, Oh.: Ohio University Press, 2009).

Smith, Ronald D. “Thomas Ewing, Jr.: Frontier Lawyer and Civil War General.” (Columbia, Mo.: University of Missouri Press, 2008.)

Stiles, T. J. “Jesse James: Last Rebel of the Civil War.” (New York: Alfred A. Knopf, 2002).

Truman, Harry S. “Autobiography of Harry S. Truman.” (Boulder, Co.: Colorado Associated University Press, 1980).

Truman, Margaret. “Harry S. Truman.” (New York: William Morrow & Company, Inc., 1972).

Thruston, Ethylene Ballard. “Echoes of the Past.” (Kansas City, Mo.: The Lowell Press, 1973).

Union Historical Company. "History of Jackson County, Missouri." Origianlly published by Birdsall, Williams & Co., Kansas City, Missouri, 1881. Indexed edition by Felix Eugene Snider. (Cape Girardeau, Mo.: Ramfre Press, 1966).

United Daughters of the Confederacy, Missouri Division. "Reminiscences of the Women of Missouri During the Sixties." (Jefferson City, Mo.: The Hugh Stephens Printing Co., 1920). This book has been reprinted by and is available through the United Daughters of the Confederacy, Missouri Division (http://www.rootsweb.ancestry.com/~moudc/index.html).

United States House of Representatives. "Inaugural Addresses of the Presidents of the United States." Document, 93-208. (Washington, D.C.: United States Government Printing Office, 1974).

Ventimiglia, Jack "Miles." "Jesse James in the County of Clay," *The Friends of the James Farm*, Kearney, Missouri. 2001

Wallace, William H. "Speeches and Writings of William H. Wallace, With Autobiography." (Kansas City, Mo.: Western Baptist Publishing Co., 1908).

Ware, E. F. "The Lyon Campaign, And the History of the 1st Iowa Infantry 1861." Originally published in 1907. (Iowa City, Ia.: Press of the Camp Bookstore, 1991).

Webb, W. L. "Battles and Biographies of Missourians," or, "The Civil War Period of Our State." Originally published in 1900. Indexed edition. (Springfield, Mo.: Oak Hills Publishing, 1999).

Wilcox, Pearl. "Jackson County Pioneers." Reprinted. (Independence, Mo.: Jackson County Historical Society, 1990).

Yeatman, Ted P. "Frank and Jesse James: The Story Behind the Legend." (Nashville, Tn: Cumberland House Publishing, Inc., 2000).

Younger, Cole. "The Story of Cole Younger, By Himself. Originally published in 1903. (Springfield, Mo.: Oak Hills Publishing, 1996).

Journals

American Heritage
Leslie, Edward E. "Quantrill's Bones." (July-August 1995).

Blue and Grey Chronicle
Eakin, Joanne Chiles. "Five Years of the Blue and Grey Chronicle: 1997-2002." (Independence, Mo.: Two Trails Publishing, 2002).

Civil War History: A Journal of the Middle Period
Hardeman, Nicholas P. "The Bloody Battle That Almost Happended: William Clarke Quantrill and Peter Hardeman on the Western Border." (September 1977): 23: 3.
Hattaway, Herman, and Archer Jones. "Lincoln as Military Strategist." (December 1980): 26: 4.

Confederate Veteran
(April 1917), Vol. 25.

Jackson County Historical Society JOURNAL (Jackson County Historical Society)
Cogswell, Burdette. "Aunt Nannie [Nancy Brawner (Rogers)] Cogswell's Life Colorful" (Fall 1966).
George, Todd. "The Mystery of Ezekiel Carlton" (March 1963)
The Buckner Family: Miss Johnston Proud of Illustrious Ancestors" (March 1973)
Noland, Mary Ethel. "Nancy Tyler Holmes' Motto Fitting: 'Difficulties No Dismay" (Fall 1967)
Jackson, David W. Order No. 11: The Lawlessness of Martial Law (Autumn 2003)
Shaw, Mary Gentry, "My Most Unforgettable Character." *Jackson County* (July 1965)

Journal of Supreme Court History
White, Jonathan W. "The Lincoln Administration and the Supreme Court during the Civil War: A Letter by Attorney General Edward Bates." (2012): 37: 3.

Kansas City Genealogist (Heart of America Genealogical Society)
Brigforth, Charles B., contributor. "How Order No 11 Affected the Jim Knight Family of Cass County, Missouri." 23: 1&2.
Burton, Angela M. and Hannah Clements Montgomery. "Order No. 11 in Hickman Mills." 24: 1: 16.
Eakin. "General Order Number Eleven (First of the Series on Order No. 11)." 23: 1&2.
Svobada, Mary Frances. "Stories My Mother Told Me of Order No. 11" 23: 1&2.

Kansas History: A Journal of the Central Plains
de la Cova, Antonio Rafael. "Samuel J. Kookogey in Bleeding Kansas: A Fearless Vindicator of the Rights of the South." (Autumn, 2012): 35: 3: 146.

North and South
Matthews, Matt, and Kip Lindberg. "Shot All to Pieces, The Battle of Lone Jack, Missouri, August 16, 1862." (January 2004): 7: 1.

Missouri Historical Review (State Historical Society of Missouri)
Barclay, Thomas S. "*The Test Oath for the Clergy in Missouri*." (April 1924): 18: 3: 345-381.
Castel, Albert. *"Order No, 11 and the Civil War on the Border,"* (July 1963): 57: 4: 357-368.
Curran, Thomas F. *"On the Road to Dixie": A Missouri Confederate's Review of the Civil War at its Midpoint."* (January 2002): 96: 2: 69-92.
Gentzler, Lynn Wolf. "*Martin Rice Remembers: What I Saw of Order Number Eleven*." (April 2012): 106: 3: 152.
Harris, Charles F. "Catalyst for Terror: The Collapse of the Women's Prison in Kansas City." (April 1995): 89: 3: 290-306.
McKee, George A. "*Boyhood Impressions of the Lexington, Missouri, Area, 1858-1863."* (July 1979): 73: 4: 16-24.
Niepman, Ann Davis. "General Orders No. 11 and Border Warfare during the Civil War." (January 1972): 66: 2: 185-210.
Unattributed. "Commemorating George Caleb Bingham: An Exhibit." 73: 4: 407-425.
Unattributed. "Famous Personages of the Civil War in Missouri, Thomas Ewing." (July 1962): 56: 4.
Withers, Robert Steele. "Historical Notes and Comments (Memoirs and a Gospel Testimony of A. H. Wallace)" (1946-1947).

Pioneer Wagon (Jackson County Genealogical Society)
Vol. 12, No. 3.

Schiller Institute
Carr, Steve. "200th Birthday of George Caleb Bingham: American Artist of the American System," (June 2011); http://schillerinstitute.org/educ/reviews/2011/george_bingham.html.

The Trail Guide (Kansas City Posse of the Westerners)
Ismert, Martin E. "Quantrill: Man and Myth: An Examination of W. E. Connelley's Historianship." (June 1961): 6: 2.

Newspapers

The Advocate (Topeka, Kansas)
January 29, 1896
Daily Missouri Republican (St. Louis, Mo.)
October 3, 1861
September 11, 1863
Daily Times (Leavenworth, Kansas)
August 18, 1863
August 19, 1863
August 22, 1863
*Evening Star (*Washington, D. C.)
January 21, 1896
The Goodland Republic (Goodland, Kansas)
January 24, 1896
Jefferson City People's Tribune
July 4, 1866
Kansas City Daily Journal of Commerce
"Guardhouse" August 14, 1863
August 15, 1863
August 18, 1863
August 26, 1863
August 29, 1863
June 14, 1866
Kansas City Daily Journal
January 22, 1896
Kansas City Post
May 2, 1912
Kansas City Star
February 22, 2009
"General Order No. 11" February 26, 1911
"As to the Battle of Lone Jack," April 26, 1911
"Collapse of a Guardhouse," November 19, 1911
"Mrs. Cheautu Dead," February 7, 1912
August 21, 1912
August 25, 1913
"Deaths in Kansas City," October 18, 1919
"Made Old Home His New," September 4, 1921
December 27, 1925
"Order No. 11 in 1863" April 9, 1961

November 6, 1963
Kansas City Times
"Civil War Ordeal for Hickman Mills," *Kansas City* (Mo.) *Times*, 24 Aug. 1969
Kansas City Times
"Order No. 11," August 24, 1963
"A Piece of Quantrill," December 21, 1977
Liberty Tribune (Liberty, Missouri)
August 14, 1863
August 21, 1863
August 28, 1863
September 4, 1863
September 11, 1863
December 24, 1863
June 22, 1866
February 5, 1875
February 12, 1875
New-York Daily Tribune
May 29, 1856
New York Times
"The War in the Far West," July 8, 1863
"Falling of a Female Prison," August 16, 1863
September 24, 1863
May 16, 1868
New-York Tribune
January 22, 1896
Rome (New York) *Daily Sentinel*
February 27, 1911
The State Journal (Jefferson City, Missouri)
February 5, 1875
February 12, 1875
Warrensburg Standard (Warrensburg, Missouri)
June 15, 1866
Western Journal of Commerce
August 14, 1863

Supreme Court Decisions

Cummings v. The State of Missouri, 71 U.S. 277 (December Term 1866)
Ex Parte Milligan, 71 U.S. (4 Wall.) 2 (1866)
Scott v. Sanford, 60 U.S. 393 (1857)

Internet Ressources

archives.gov/historical-docs/todays-doc/NG?dod-date=516
archives.gov/.../emancipation_proclamation/transcript.html
artcyclopedia.com/artists/bingham_george_caleb.html.
cdm.sos.mo.gov/cdm4/browse
chroniclingamerica.loc.gov/
civilwar.org/education/history/biographies/johnschofield
civilwarsaga.com/william-quantrills-three-graves
civilwarstlouis.com/History/Oathofloyalty.htm
civilwar.org/education/history/biographies/john-schofield
ebooks.library.cornell.edu/m/moawar/waro.html
findagrave.com
genforum.genealogy.com/ralston/messages/1873.html
historiclonejack.org/order11.html
historiclonejack.org/whatisaw.html
history.com/this-day-in-history/pinkertons-maim-frank-and-jesse-james-mother
history.lawrence.com/project/community/quantrill/victims
jurist.law.pitt.edu/trials13.htm
law.cornell.edu/wex/due_process
law2.umkc.edu/faculty/projects/ftrials/impeach/proceedings
lecomptonkansas.com/page/samuel-j-jones
mcadamreport.org/Acton.html
mocivilwar150.com/history/figures/291
newadvent.org/cathen/01114a.html
nyti.ms/Y6G3b4
opinionator.blogs.nytimes.com/2013/03/23/another-name-
schillerinstitute.org/educ/reviews/2011/george_bingham
shs.unsystem.edu/newsaper/newspapercatalog/pdf/jackson
stlouis.com/history
sos.mo.gov/archives/provost/#history
sos.mo.gov/archives/provost/#history
wattshaysletters.com
wesclark.com/jw/quantril.html

INDEX

Adams Family 243
Adams, James ....Apx G
Adams, John Quincy.................................52
Adcock Family 243
Adcock, John Alfred Bills ....Apx G
African-American soldiers..................10, 46
Aiken, Celeste (Houk) 125
Akers Family 243
Akers, Henry Joseph ....Apx G
Akers, Henry, Jr. ....Apx G
Akers, James "Jim" (child) ....Apx G
Akers, Joseph ....Apx G
Akers, Mary "Belle" ....Apx G
Akers, Sylvester "Vest" 195
Aldridge Family......................................244
Aldridge, Ben ....Apx G
Alley, (Major) ....Apx G
Amend, Matthias 51
Ament Family 244
Ament, John ....Apx G
Anderson, Bill...............93, 97, 98, 110, 194
Anderson, Jim..................................93, 110
Anderson, Josephine.....................96, 98, 99
Anderson, Margaret Jane "Janie"...97, 98, 99
Anderson, Mary..97
Anderson, Mollie 98, 99
Anderson, Robert......................................37
Anthony, Daniel J. "Dan" 38, 45, 63, 89
Arkansas State Guard 40
Armstrong Family..................................244
Armstrong, Katherine H. A. (age 9) (later, married J. D. Kreeger)........... Apx G
Army of Northeast Virginia.......................72
Army of the Frontier.....................68, 85, 86
Army of the Potomac.......................9, 10, 15
Arnold, Samuel.......................................209
Arrow Rock, Missouri ..49, 51, 53, 153, 351, 403, 405
Ashley Family 244
Aspling, Thomas and Family .....Apx G, 246
Atchison, David Rice....................23, 29, 55
Atchison, Kansas 44
Atzrodt, George A. 210
Austin Family 246, 247
Austin, Eliza (child) ....Apx G
Austin, Eliza Jane ....Apx G
Austin, George ....Apx G
Austin, Jessie G. ....Apx G
Austin, John H. ....Apx G
Austin, Robert ....Apx G
Austin, Sarah Ann ....Apx G
Austin, Thomas T. ....Apx G
Austin, William F. ....Apx G
Baker Family 247
Baker, Maggie ....152, Apx G
Ballinger, John.. 158
Barkley, W.G. and Family ....Apx G, 247
Barnett, Stephen ....Apx G, 248
Barrett Family.. 248
Barrett, Jack (child) ....Apx G
Barrett, John ....Apx G
Barrett, Mollie (child) ....Apx G
Barrett, Nellie ....Apx G
Barrett, Robert "Bob" ....Apx G
Bassham, Bill.. 195
Battle of Cane Hill 86
Battle of Chattanooga 11
Battle of Chickamauga............................. 11
Battle of Ft. Franklin (Kansas) 31
Battle of Ft. Titus (Kansas)...................... 31
Battle of Hemp Bales.................................. *See* Battle of Lexington, Missouri
Battle of Lone Jack . 64, 251, 255, 271, 275, 279, 298, 322, 325, 334, 342, 347, 349, 370, 385, 399, 411, 413
Battle of Oak Hills *See* Battle of Wilson's Creek
Battle of Prairie Grove........................ 85, 86
Battle of Westport.................................. 208
Battle of White Oak Creek............. 150, 281
Battle of Wilson's Creek40, 41, 43, 45, 46, 68
Beagles, Edward ....Apx G, 249
Becker, Phillip 124
Beets Family 249
Beets, Elizabeth (Tuggle) ....Apx G
Beets, James ....Apx G
Behan, Hugh (Father) 199

Bell, John 56, 76
Bennett Family 250
Bennett, Elizabeth (Cave) ....Apx G
Bennett, George (age 5) ....Apx G
Bennett, Jacob ....Apx G
Bennett, Jacob (Mrs.)..........161
Bennett, Price (age 2) ....Apx G
Best Family..........250
Best, John ....Apx G
Big Blue River..........109, 223, 317
Big Cedar, Missouri *See* Lee's Summit, Missouri
Bingham studio..........56, 93, 100, 179
Bingham, Clara..........53, 55, 167
Bingham, Elizabeth (Thomas)..........54, 189
Bingham, Elizabeth 167
Bingham, George Caleb 3, 4, 8, 10, 47, 49, 50, 51, 52, 53, 54, 55, 56, 57, 58, 63, 76, 91, 93, 99, 100, 101, 102, 106, 122, 133, 135, 136, 137, 141, 142, 143, 153, 166, 167, 168, 169, 170, 171, 173, 174, 175, 176, 177, 178, 179, 180, 181, 182, 187, 188, 189, 190, 191, 192, 193, 203, 210, 212, 255, 268, 325, 329, 351, 355, 400, 405, 408, 409, 412, 432
Bingham, Henry Vest 51, 53
Bingham, Horace..........55, 167
Bingham, Isaac Newton..........51
Bingham, James Rollins 167, 187
Bingham, Joseph Hutchison 54
Bingham, Martha Ann "Mattie" (Livingston) Lykins 100, 101, 106, 122, 190, 192, 233, 246, 324, 325, 400, 405
Bingham, Mary Christina (Amend)..........51
Bingham, Nathaniel..........53
Bingham, Newton..........53
Bingham, Sara Elizabeth (Hutchison) 53
Black Friday 3, 109, 110, 155, 183
Black Jack, Kansas 31
Blackwater Creek 110
Blair, Frank..........39, 132
Blankenbaker, Abraham ....Apx G, 241
Bleeding Kansas . 17, 25, 31, 32, 35, 38, 41, 42, 44, 63, 82, 84, 114, 202, 249, 343, 405, 406, 411
Bloody Bill Anderson......*See* Anderson, Bill
Blue Mills, Missouri..........156, 295, 366, 368
Blue Township..........146, 249, 250, 336
Blunt, James G.... 11, 44, 46, 69, 82, 85, 86, 89, 184, 194, 201
Boggs Family.......... 251
Boggs, John O. ....Apx G
Boggs, Mrs. (formerly Mrs. Peace McGuire) ....Apx G
Boone, Daniel 52, 298, 356
Booth, John Wilkes.......... 34, 208
Boren, M. F. ....Apx G, 251
Bouse, Susan ....Apx G, 251
Boyd Family 252
Boyd, William ....Apx G
Bradbury, Leonard ....Apx G, 252
Bradin, Martha A. ....Apx G
Bradley Family 252, 253
Bradley, (Mrs.) ....Apx G
Bradley, Levin W. ....Apx G
Bradley, Margaret M. (Parish) ....Apx G
Bradley, Thomas..........159, Apx G, 253
Brannock Family 253
Brannock, A. B. ....Apx G
Brannock, George W. ....Apx G
Branson, Jacob.......... 27
Breckenridge, John 34
Bridges, Laura (Harris) Flanery *See* Flanery, Laura (Harris)
Briscoe, Ruth ....Apx G, 254
Britt Family.......... 254
Britt, Jefferson Waller ....Apx G
Broadhurst, James O. 141
Bronaugh, William ....Apx G, 254
Brookhart Family..........Apx G, 254
Brooking Township 273
Brooks, Preston.......... 29
Brown Family 254
Brown, Egbert E. 87, 128
Brown, John 30, 31, 32, 34, 35, 44, 61, 84, 343
Brown, Mary ....Apx G
Brown, Robert A. ....76, Apx G
Browning Family.......... 255
Browning, O. N. ....Apx G
Browning, Orville 209
Buchanan, James.......... 31
Bugler, Henry 171, 172, 173, 433
Bugler, Mary.......... 172
Bullock Family ....Apx G, 255
Burdette Family ....Apx G, 255
Buried belongings ...75, 146, 210, 267, 280, 286, 310, 314, 324, 365 *See also* Feather beds
Burns Family 255, 256

Burns, Dick 195
Burns, Kate (Mrs.) ....Apx G
Burns, Rice ....Apx G
Burnside, Ambrose 9, 15
Burris, F. M. ....Apx G
Burrus Eva (child) ....Apx G
Burrus Family 256, 257
Burrus, A. (Kimberlin) ....Apx G
Burrus, James M., Sr. ....Apx G
Burrus, Kate ....Apx G
Burrus, T. K. (child) ....Apx G
Bushwhackers 12, 19, 47, 64, 69, 81, 82, 86, 88, 89, 90, 93, 94, 105, 107, 108, 109, 110, 115, 118, 121, 129, 131, 132, 154, 158, 159, 163, 164, 165, 172, 173, 194, 195, 196, 205, 251, 252, 266, 277, 279, 326, 334, 354, 371, 372
Butler, Andrew Preston 28
Butler, Missouri 45
Bynum Family 257
Bynum, Charlotte (Cave) (age 44) .....Apx G
Bynum, Galen C. (age 14) ....Apx G
Bynum, James P. (age 13) ....Apx G
Bynum, John (age 41) ....Apx G
Bynum, Robert (age 18) ....Apx G
Byrum's Ford.............*See* Battle of Westport
Cabness Family 258
Cabness, J. B. ....Apx G
Cabness, James ....Apx G
Cabness, Rachel ....Apx G
Cagin Family 258
Cagin, John and Family ....Apx G
Calhoun, John 83
Calloway, Jimmy 153
Cameron, Simon 38
Campbell children ....Apx G
Campbell Family 258
Campbell, Courtney ....Apx G
Campbell, David Lee 116
Campbell, Harriet (Buckner) ....Apx G
Campbell, James 116
Campbell, John B. ....Apx G
Campbell, Rachel A. (Grubb) ....Apx G
Campbell, Washington and Family ...Apx G
Canine Family 259
Canine, Cornelius ....Apx G
Canine, Cornelius (Mrs.) ....Apx G
Capell Family 259
Cappell, Bennetter A. (child) ....Apx G
Cappell, Britton M. ....Apx G
Cappell, John ....Apx G
Cappell, Martha (child) ....Apx G
Cappell, Mary (child) ....Apx G
Cappell, Sarah (child) ....Apx G
Cappell, Susan (child)........................Apx G
Carleton Family 259
Carlton, Ezekiel ....Apx G
Carlton, Mary Jane (Smith) ....Apx G
Carney, Thomas 47, 116, 117, 118, 123, 183
Carter Family ....Apx G, 260
Carter, Al ....Apx G
Carter, James ....Apx G
Carter, Michael (Father) 199
Cary, Edmon ....Apx G, 260
Case, Theodore S. 211
Casey, Philip ....Apx G, 260
Cave Family.. 261
Cave, Allguis (age 20) ....156, Apx G
Cave, Amanda (Potter) (age 20s) 155, Apx G
Cave, Berilda T. (age 7) ....156, Apx G
Cave, Ella E. 156
Cave, Henry C. (age 6) ....Apx G
Cave, Jane ....160, Apx G, 346
Cave, John (age 20s) ....Apx G
Cave, John B. (age 4) ....Apx G
Cave, John S. . (age 45) 155, 156, 159, 161, Apx G, 261, 344
Cave, John 155, 156, 159, 253, 261, 314
Cave, Lillian L. (child) ....155, Apx G
Cave, Marion ....Apx G
Cave, Martha "Matilda" A. (age 12) 156, Apx G
Cave, Mary J. 156
Cave, Nancy (age 28) ....Apx G
Cave, Nancy (Hunter) (age 44) 155, 161, Apx G
Cave, Pleasant F. (age 4) ....Apx G
Cave, Susan (age 56) ....Apx G
Cave, Susan F. (age 17) ....156, Apx G
Cave, William (age 33) ....Apx G
Charleston, South Carolina 10
Chicago, Illinois 23, 24, 28, 173, 187, 244, 327
Chick, Washington H. 124
Chilcote, John ....Apx G, 262
Chiles Family.. 262
Chiles, Alexander 152
Chiles, Anna Azubah (child) ....Apx G
Chiles, Azubah (Skinner) ....Apx G
Chiles, Dick 128

Chiles, Elijah Young (age 3) ....Apx G
Chiles, H. C. (child) ....Apx G
Chiles, Jim "Crow"........................147, 152
Chiles, Joel Franklin (age 15) ....Apx G
Chiles, Mary Jane (child) ....Apx G
Chiles, Sarah Anne (Young) ......147, Apx G
Chiles, Sarah Margaret (age 17) ....Apx G
Chipman Family 263
Chouteau, Cyperian ....Apx G, 263
Chouteau. Nancy (Francis) ....Apx G
Chrisman, Lucy (Lee)...............................74
Chrisman, William..............55, 74, 173, 324
Clark Family 263
Clark, Charles S.....................................160
Clark, David M. and Family ....Apx G
Clarkson Family 263
Clarkson, Virginia (Saint Clair) ....Apx G
Clay, Henry (portrait) 55
Clay, Henry 52, 53, 54, 82, 83, 235, 262, 351
Clements Family................Apx G, 264, 265
Clements, Arch 110
Clements, Hannah H. (Scrivner)
(Mrs. John Ambrose) ....Apx G
Clements, Milton (Mrs.) ....Apx G
Cockerel, Elizabeth...................................95
Cogswell Family.....................................265
Cogswell, Nancy Brawner
"Nannie" (Rogers) ....Apx G
Cogswell, Oscar H. ....Apx G
Coleman, Charles F. 159
Collins Family 266
Collins, "Buck" ....Apx G
Columbia, Missouri .. 33, 43, 49, 50, 51, 52, 80, 83, 122, 125, 182, 191, 244, 268, 329, 398, 401, 404, 406, 407, 408, 409
Comingo, Abram 3, 4, 10, 54, 71, 74, 75, 76, 77, 78, 79, 80, 120, 124, 133, 134, 135, 140, 141, 165, 167, 173, 205, 206, 207, 208, 359, 432
Comingo, John.................................75, 207
Comingo, Lucy Jane (Morton) 74
Comingo, Sally *See* Griffith, Sally (Comingo)
Compton Family 266
Compton, George ....Apx G
Compton, James ....Apx G
Compton, John (child) ....Apx G
Conditional Unionist Party 81
Connor Family..266
Connor, John ....Apx G
Constitutional Unionist Party............. 75, 76
Cooper Family 266
Cooper, Miss Rose ....Apx G
Cooper, Z. G. (Mr.) ....Apx G
Cooper, Z. G. (Mrs.) ....Apx G
Copperhead movement 16
Cordell, Henry ....Apx G, 268
Corn Family 268
Corn, George (child) ....Apx G
Corn, John S. (child) ....Apx G
Corn, Louisa (child) ....Apx G
Corn, Martha (Cummings) (age 40) ...Apx G
Corn, Martin (age 50) ....Apx G
Corn, Nancy E. (child) ....Apx G
Corn, Samuel (child)..........................Apx G
Corn, Susan F. (child) ....Apx G
Cowherd Family 268
Cowherd, C. J. (age 36) ....Apx G
Cowherd, Edmond (age 47) ....Apx G
Cowherd, Francis K. ....Apx G
Cowherd, Francis K. (Mrs.) ....Apx G
Craig, Henry 206
Crawford Family................................... 269
Crawford, (Mr.) ....Apx G
Crawford, Riley 98
Crenshaw Family ....Apx G, 269
Crittendon, Thomas, Sr. 206
Crowther Family 270
Crowther, Benjamin ....Apx G
Crowther, children ....Apx G
Crowther, Minerva Ann Lilly ....Apx G
Cummins, Jim 107, 110
Curtis, Samuel R......67, 85, 86, 89, 166, 185
Cusenberry Family................................ 270
Cusenberry, (Mr.) and Family ....Apx G
Daily Journal of Commerce .. 63, 64, 86, 96, 97, 98, 129, 140, 164, 191, 192, 211, 212, 402, 413
Dalton Family 270
Dalton, Lewis (age 64) ....Apx G
Dalton, Matilda (age 50) ....Apx G
Daniels, John and Family ....Apx G, 270
Darby Family 271
Darby, John ....Apx G
Darby, Mary Jane (Austin) ....Apx G
Daugherty, Charles 171
Davenport, William ....Apx G, 271

Davidson, J. N. .... Apx G, 271
Davis Family 272
Davis, James .... Apx G
Davis, Jefferson 15, 65
Davis, John .... Apx G
Davis-Smith Cemetery 91, 98, 100, 297, 408
Dayton, Missouri 45, 89
Dean, Abner Holton (Reverend) 169
Dehoney Family 272, 273
Dehoney, Howard (child) .... Apx G
Dehoney, Leander M. .... Apx G
Dehoney, Lina (child) .... Apx G
Dehoney, Mary (Chism) .... Apx G
Dehoney, Simeon .... Apx G
Delevan Family .... Apx G, 273
Democrat Party 22, 25, 34, 75
Dempsy, Barney 160
Denver, John W. 32, 33
Devlin, Pat 43
DeWitt, Daniel B. (Mrs.) .... Apx G, 273
Dillingham Family 274
Dillingham, C. J. (age 15) .... Apx G
Dillingham, Family .... Apx G
Dillingham, Joshua (age 47) .... Apx G
Dillingham, Susan Jane
(Walker) (age 36) .... Apx G
District of Kansas 208
District of the Border 10, 66, 69, 71, 81, 86, 90, 104, 105, 109, 119, 122, 130, 131, 133, 134, 208, 211, 223, 226, 228, 244, 303, 398
Dolan, James .... Apx G, 274
Donahue Family .... Apx G, 274
Doneghy Family 274
Doneghy, Abraham J. (age 6) .... Apx G
Doneghy, Alexander (age 8) .... Apx G
Doneghy, H. Weitman (age 2) .... Apx G
Doneghy, Herbert (age 4) .... Apx G
Doneghy, James C. (age 13) .... Apx G
Doneghy, John T. (age 11) .... Apx G
Doneghy, Katherine (widow) .... Apx G
Dougal, Joseph M. .... Apx G, 276
Douglas, Stephen A. 23, 34
Douglass Family 276
Douglass, Milda J. (Sanders) (age 29) Apx G
Douglass, Thaddeus T. (child) .... Apx G
Douglass, William I. (age 38) .... Apx G
Dover, Ohio 59, 61, 199
Drake Constitution 168, 169, 170, 171, 202, 357
Drake, Charles D. 202
Dred Scott 31
Drumm Farm 149
Duggen, John 153
Duke, B. F. 99
Duncan Family 276, 277
Duncan, Charles Thomas .... Apx G
Duncan, John .... Apx G
Duncan, Louisa Lucretia (Alley)
(Mrs. Rev. Pryor) .... Apx G
Dunham, (Mr.) .... Apx G, 277
Dunn, Absolom .... Apx G, 277
Duvall Family 277
Duvall, Carrie (O'Bryan) .... Apx G
Duvall, Crit (child) .... Apx G
Duvall, Flora (child) .... Apx G
Duvall, Mattie (child) .... Apx G
Duvall, Thomas .... Apx G
Duvall, W. A. (child) .... Apx G
Dye, Isaac .... Apx G, 278
Easley Family 278, 279
Easley, Achilles .... Apx G
Easley, Miller W. .... Apx G
Easley, Stephen .... Apx G
Elder, Thomas "Bud" .... Apx G, 279
Eldridge House Hotel 106, 114, 116, 227
Ellez, A. C. (Mrs.) .... Apx G, 279
Elmwood Cemetery 75, 207, 344
Emancipation Proclamation 4, 9, 16, 17, 21, 42, 49, 91, 103, 104, 105, 123
Estes, William .... Apx G, 279
Ewing, Hugh 84, 209
Ewing, Maria (Boyle) 82
Ewing, Thomas, Jr. 3, 4, 6, 9, 10, 11, 33, 36, 38, 42, 46, 69, 81, 82, 83, 84, 85, 86, 87, 88, 89, 90, 91, 93, 95, 96, 100, 101, 102, 103, 104, 105, 106, 107, 109, 110, 118, 119, 120, 121, 122, 123, 124, 125, 128, 129, 130, 131, 132, 133, 135, 136, 137, 142, 156, 160, 162, 164, 165, 166, 168, 169, 170, 174, 175, 176, 178, 179, 180, 181, 182, 185, 187, 189, 190, 192, 203, 208, 209, 210, 211, 212, 219, 222, 223, 224, 227, 269, 272, 273, 277, 346, 347, 361, 401, 409, 412
Faley, Bill 171
Farmer, Henry .... Apx G, 280
Farnsworth, C. L. .... Apx G, 280

Feather beds....................151, 281, 354, 382
See also Buried belongings
Ferguson Family.....................................280
Ferguson, Felix ....Apx G
Ferguson, Richard and Family ....Apx G
Fields Family...280
Fields, Amanda (Chiles) ....Apx G
Fields, Martha Frances (Chrisman) ...Apx G
Fields, William ....Apx G
Filley Family 280
Flanery Family.......................................281
Flanery, George Z. ....153, Apx G
Flanery, Horace W. ....153, Apx G
Flanery, James S. (child) ....153, Apx G
Flanery, John W. ....153, Apx G
Flanery, Laura (Harris) (son, about 3) Apx G
Flanery, Laura Elizabeth (Harris) (age 25) (later, Bridges) ....3, 149, 152, Apx G, 296
Flanery, Lucinda (age 2)............153, Apx G
Flanery, Mac P. ....153, Apx G
Flanery, Martin Van Buren......150, 153, 281
Flanery, Melvina W. ....153, Apx G
Flanery, Rebecca J. ....153, Apx G
Flanery, William V. (age 3) ....150, Apx G
Flanery, Zion (age 49) 150, 152, Apx G, 281
Flemming, Robert W. 142
Fletcher, Thomas C.........................130, 173
Flood, Stephen ....Apx G, 282
Ford Family ....Apx G, 282
Ford, Emily (baby) ....Apx G
Ford, Lewis ....Apx G
Ford, Martha (baby) ....Apx G
Ford, Patsy (Holmes) ....Apx G
Fort Davidson, Missouri 185, 208
Fort Gibson, Oklahoma 69
Fort Leavenworth, Kansas.....34, 43, 64, 185
Fort Osage Township 276, 282, 304, 308, 311, 331, 334
Fort Scott, Kansas...................................41
Fort Sumter, South Carolina..........19, 37, 39
Fourteenth Senatorial District....................76
Franklin Family 282
Franklin, J. L. H. (age 21) ....Apx G
Franklin, John R. ....Apx G
Franklin, Lucy Ann (Walker) (age 35) Apx G
Franklin, Sarah ....Apx G
Franse Family ....Apx G, 283
Free State Hotel 29, 116, 227
Freeman Family ....Apx G, 283
Fremont, John C.18, 39, 40, 41, 42, 43, 45, 46, 249
Frick Family 283
Frick, Dr. W. E. ....Apx G
Frick (Mrs.) ....Apx G
Fristoe Family....................................... 284
Fristoe, Fannie *See* Twyman, Frances
Fristoe, John ....Apx G
Fristoe, Mary L. (Sullivan) ....Apx G
Frost Family.. 284
Frost, Daniel M. 35
Frost, Mary (Mrs.) ....Apx G
Gag Rule 23
Gage, John C... 125
Galloway (Mr.) 159
Gamble, Hamilton 45, 50, 51, 63, 67, 76, 129
Gammon Family 284
Gammon, Hattie ....Apx G
Gammon, James ....Apx G
Gammon, Sally (Stuart) and baby ......Apx G
Gammon, William ....Apx G
Garrett, Reuben ....Apx G, 284
Garrison Family ....Apx G, 284
Gault, James.. 153
Geam Family 284
Geam, Rile ....154, Apx G, 284
General Orders No. 9... 10, 103, 104, 105, 122, 131, 156, 219, 221, 347
General Orders No. 10. 10, 103, 104, 105, 122, 131, 156, 221, 223, 347
General Orders No. 11 1, 2, 3, 4, 6, 7, 8, 11, 12, 42, 51, 80, 88, 90, 96, 99, 100, 109, 117, 120, 121, 122, 123, 124, 125, 128, 129, 130, 131, 132, 133, 135, 136, 137, 139, 140, 141, 142, 143, 146, 148, 149, 150, 152, 153, 154, 157, 158, 161, 163, 164, 165, 168, 170, 174, 175, 176, 179, 180, 181, 182, 187, 190, 191, 192, 193, 200, 203, 205, 207, 209, 211, 212, 223, 224, 241, 242, 243, 244, 247, 248, 249, 250, 251, 252, 253, 254, 255, 256, 257, 258, 259, 260, 261, 262, 263, 264, 265, 266, 268, 269, 270, 271, 272, 273, 274, 275, 276, 277, 278, 279, 280, 281, 282, 283, 284, 285, 287, 288, 289, 290, 291, 292, 293, 294, 295, 296, 297, 298, 300, 303, 304, 305, 306, 307, 308, 309, 310, 311, 312, 313, 314, 315, 316, 317, 318, 319, 320, 321, 322, 323, 324, 325, 326,

327, 328, 329, 330, 331, 332, 333, 334, 335, 336, 337, 338, 339, 340, 341, 342, 343, 344, 345, 346, 347, 348, 349, 350, 351, 352, 353, 354, 355, 356, 357, 358, 359, 360, 361, 362, 363, 364, 365, 366, 367, 368, 369, 370, 371, 372, 373, 374, 375, 376, 377, 378, 379, 380, 381, 382, 383, 384, 385, 386, 397, 398, 399, 400, 401, 402, 405, 411, 413, 430
General Orders No. 140 72
Gentry, Mary (Henley) (Mrs. Joseph) Apx G, 285
George Family 287
George S. E. (Farmer) ....Apx G
George, Amanda J.
(later Mrs. John W. Tyler) ....Apx G
George, N. E. (Mrs.) and Family of 5 Apx G
George, Nathaniel "Nathan" Bass .....Apx G
Gibbons, Hank ....Apx G, 288
Gilky, (Mrs.) ....Apx G, 288
Gilliss, William..........................Apx G, 289
Gilmore, Joseph ....Apx G, 289
Gilpin, William ....Apx G, 289
Glass, James ....Apx G, 289
Graham Family 289
Graham, Ambers (age 26) ....Apx G
Graham, Mary ....Apx G
Graham, R. H. 65
Graham, Sarah J. (Stultz) (age 24) .....Apx G
Grand River 111
Granstaff, Molly 97, 98
Grant, Ulysses S. 10, 84, 202, 206
Gray, Lou (Munday) 98
Green Family ....Apx G, 290
Green, John R. 142
Gregg Family 290, 291, 292
Gregg, Cornelia (child) ....Apx G
Gregg, D. (age 31) ....Apx G
Gregg, George H. (child) ....Apx G
Gregg, Larius (child) ....Apx G
Gregg, Lera H. (child) ....Apx G
Gregg, Maria (Bryant) ....Apx G
Gregg, Martha (child) ....Apx G
Gregg, Mary (Harper) (age 36) ....Apx G
Gregg, Robert L. (child) ....Apx G
Gregg, Samuel ....Apx G
Gregg, Wallace (child) ....Apx G
Gregg, William (Mrs) ....Apx G
Gregg, William H. 110, 114
Gregg, William 194
Griffith Family 292
Griffith, G. W. 207
Griffith, George W. (age 34) ....Apx G
Griffith, Mary A. (about 60) ....Apx G
Griffith, Sally (Comingo) 75, 207
Groves, (Mrs.) ....Apx G, 292
Gudgell homestead 75, 167
Guerrilla warfare.... 91, 109, 122, 132, 169, 180, 346
Guerrillas . 6, 10, 12, 16, 19, 21, 35, 42, 43, 44, 45, 50, 58, 59, 63, 64, 65, 66, 68, 73, 81, 82, 87, 89, 90, 91, 93, 98, 101, 104, 106, 107, 108, 109, 110, 111, 114, 118, 121, 132, 142, 144, 147, 153, 154, 155, 165, 166, 171, 172, 194, 195, 196, 205, 206, 253, 272, 291, 331, 336, 343, 378, 384, 385
Hadley, J. M. 128
Hagan, William 150
Hahn, August Rosa "Rose" (Becker) 124
Hall (Miss) 98
Hall Family 294
Hall, Charles B. (age 10) ....Apx G
Hall, Ella Virginia (age 13) ....Apx G
Hall, Franklin (age 9) ....Apx G
Hall, James D. ....Apx G
Hall, Lemuel Edward (age 33) ....Apx G
Hall, Mary (about 6) ....Apx G
Hall, Mary E. (Thompson) (age 34) ...Apx G
Hall, Richard (about 5) ....Apx G
Hall, Willard Prebel 50
Halleck, Henry.....46, 47, 50, 63, 78, 90, 129
Haller Family ....Apx G, 294
Hambright Family 294
Hambright, John ....Apx G
Hamilton Family ....Apx G, 295
Hamilton, Bird ....Apx G
Hamilton, Sam ....Apx G
Hanley Family 295
Hanley, Sam ....Apx G
Hanley, Sam (Mrs.) ....Apx G
Hannahs, Harrison 95, 212, 223
Hannibal Bridge 187
Harbaugh Family 295
Harbaugh, Amanda (age 5) ....Apx G
Harbaugh, George (age 7) ....Apx G
Harbaugh, Mary Elizabeth (Pence)
(Mrs. Franklin E.) ....Apx G

Harbaugh, Minnie (age 1) ....Apx G
Hardeman, Peter 194, 411
Hardesty, Richard ....Apx G, 296
Hardin, Charles H. 187
Harding, Chester 52
Harper's Ferry, Virginia 32, 61
Harrelson Family ....Apx G, 296
Harris Family..296
Harris, Eliza 99
Harris, Jane (Hall) (Mrs. Samuel) 149, Apx G
Harris, Samuel 149
Harris, Fanny (Mrs.) ....Apx G
Harrison, William Henry 53, 83
Harrisonville, Missouri 76, 121, 122, 157, 170, 223, 227, 244, 251, 254, 268, 271, 283, 289, 304, 305, 307, 315, 318, 320, 321, 322, 323, 326, 336, 340, 349, 352, 353, 355, 363, 364, 370, 377, 378, 386, 397, 404
Hatton Family 297
Hatton, Henry ....Apx G
Hatton, J. W. ....Apx G
Hawkins, Robert L., III..........................199
Haynes, Elizabeth ....Apx G, 297
Hays Family..298
Hays, children ....Apx G
Hays, Elfleda (child) ....Apx G
Hays, Linville 88
Hays, Margaret Jane (Watts) ....88, Apx G
Hays, Sam...110
Hays, Upton..... 88, 89, 110, 251, 257, 298, 301, 303, 317, 384, 401, 402
Henderson, John B.................................202
Henley Family 303
Henley, Elizabeth (Belt) (Mrs. Alonzo) Apx G
Hereford Family 303
Hereford, H. F. (doctor) ....Apx G
Herold, David A. 208
Hickman Family 303, 304
Hickman, Amelia ....Apx G
Hickman, J. T. ....Apx G
Hickman's Mill, Missouri................121, 351
Higginsville, Missouri 192, 199
Hightower Family...................................304
Hightower, John (age 47) ....Apx G
Hightower, Sarah A. (Van Ardol) ......Apx G
Hightower, W. H. (about 5) ....Apx G
Hinkel, Charley (child) ....Apx G
Hinshaw, Nathan ....Apx G, 304
Hockensmith Family...................... 304, 305
Hockensmith, G. W. R. ....Apx G
Hockensmith, Thomas Lee ....Apx G
Hocker .... *See* Hoecker
Hoecker, Joe 75
Holcomb, A. ....Apx G, 305
Holloway, T. ....Apx G, 305
Holmes Family...................................... 305
Holmes, Hezekiah and Family ....Apx G
Holmes, Nancy (Tyler) ....Apx G
Holt, John D.. 111
Holt, Joseph 209
Hon Family 307
Hon, Jerry V. ....Apx G
Hon, Richard ....Apx G
Hood, John B. 201
Hooker, Joseph 9, 10, 15
Hoover, S. T. ....Apx G, 307
Hord, William ....Apx G, 307
Horn, Richard ....Apx G, 307
Houk, Solomon 125
Houston Family ....Apx G, 307
Howkinsmith, Clark............................... 195
Hudson Family................................ 308
Hudson, Geneva Elisabeth (Hamblen) Apx G
Hudson, James Russell ....Apx G
Hudspeth Family........................... 308, 312
Hudspeth, "Babe" 195
Hudspeth, Cyntha/Cynthia ....Apx G
Hudspeth, Elizabeth (Jesse) (age 33) .Apx G
Hudspeth, George (age 43) ....Apx G
Hudspeth, Paralee (child)...................Apx G
Hudspeth, Thomas B. (child) ....Apx G
Hudspeth, Tybithia (age 5) ....Apx G
Hughes Family...................................... 313
Hughes, Merritt ....Apx G
Hughes, Will ....Apx G
Hunt Family 313, 314
Hunt, E. F. (age 11) ....Apx G
Hunt, J. B. (age 3) ....Apx G
Hunt, J. M. (age 5) ....Apx G
Hunt, Mary A. (age 7) ....Apx G
Hunt, N. E. (Cave) (age 30) ....Apx G
Hunt, Noah (age 32) ....Apx G
Hunt, Noah (Mrs.) ....Apx G
Hunter Family....................................... 314
Hunter, Berrilla (age 22) ....155, Apx G
Hunter, David (about 33) ....38, 46, Apx G

Hunter, Elizabeth (Jeans) (age 71) 155, Apx G, 261
Hunter, John D. 155, 159
Hunter, John ... (age 75) 155, 159, 161, 162, Apx, G, 314, 344, 362
Hunter, Juda (Potter) Hunter (about 30) 155, Apx G
Hunter, Matilda *See* Ousley, Matilda (Hunter)
Hunter, Nancy *See* Cave, Nancy (Hunter)
Hunter, William (age 47) .... 155, Apx G
Hunter, William J. 159, 261
Hylton, J. D. 195
Independence Sentinel 171
Independence, Missouri 1, 2, 10, 15, 16, 23, 25, 44, 45, 54, 55, 59, 61, 62, 64, 71, 74, 75, 79, 86, 87, 91, 93, 97, 99, 101, 103, 106, 109, 111, 121, 122, 128, 130, 131, 140, 143, 144, 146, 147, 149, 150, 154, 156, 163, 167, 170, 171, 172, 173, 174, 181, 185, 187, 195, 205, 206, 219, 223, 227, 243, 244, 245, 246, 247, 250, 253, 256, 258, 262, 263, 264, 266, 267, 269, 273, 274, 276, 280, 281, 284, 285, 286, 287, 290, 294, 295, 303, 309, 312, 313, 315, 320, 321, 323, 324, 325, 326, 327, 330, 332, 333, 334, 335, 336, 337, 338, 339, 342, 343, 344, 352, 353, 354, 355, 356, 357, 358, 359, 360, 362, 365, 366, 368, 369, 371, 372, 373, 376, 379, 382, 384, 385, 397, 398, 400, 401, 403, 404, 405, 406, 407, 408, 410, 432
Ingram, Charles ....Apx G, 315
Ingrum Family 315
Ingrum, Hanna (Mrs. E. A.) ....Apx G
Iron Clad Oath ....*See* Loyalty Oath
Irwin Family 315
Irwin, Julia (Harris) (age 20) .....151, Apx G
Irwin, Mary Jane (Miss) ....Apx G
Jackman, Sidney Drake 36, 41, 408
Jackson County Jail .. 62, 91, 170, 171, 172, 206, 337, 402
Jackson Family ....Apx G, 315
Jackson, Andrew (portrait) 55
Jackson, Claiborne Fox......35, 39, 40, 50, 76
Jackson, Thomas "Stonewall"........9, 34, 293
James, Alexander Franklin "Frank" 110, 144, 171, 187, 192, 195, 196, 206, 207, 234, 373, 432, 433
James, Jesse Woodson 94, 171, 172, 173, 187, 188, 195, 198, 206, 373, 407, 409, 410, 432, 433
Jayhawkers 19, 35, 43, 44, 45, 47, 58, 62, 63, 65, 81, 82, 86, 105, 111, 124, 140, 142, 147, 148, 150, 168, 244, 267, 281, 293, 337, 347, 372, 380
Jefferson City, Missouri 19, 47, 54, 56, 58, 71, 76, 77, 78, 79, 80, 87, 93, 122, 125, 136, 171, 174, 188, 206, 268, 324, 343, 407, 409, 413, 414
Jefferson, Thomas (portrait) 55
Jefferson, Thomas 22
Jefferys Family 316
Jefferys/Jeffreys/Jeffries, Charles Peter Beauchamp ....Apx G
Jeffreys Family ....*See* Jefferys Family
Jeffries Family ....*See* Jefferys Family
Jenkins, Gaius 32
Jennison, Charles "Doc" 35, 36, 43, 44, 45, 46, 47, 58, 61, 63, 115, 155, 168, 174, 176, 181, 205, 206, 268, 372, 380
Jerrard Family ....Apx G, 316
Johnson Family 316
Johnson, Andrew 185, 201, 202, 325
Johnson, Coger W. (age 6) ....Apx G
Johnson, Elizabeth (Hall) (age 31) .....Apx G
Johnson, Greenville ....Apx G
Johnson, James R. (age 11) ....Apx G
Johnson, Joseph P. 201
Johnston Family ....Apx G, 316
Jones Family 317, 318
Jones, Billy ....Apx G
Jones, James ....Apx G
Jones, Jemima B. (Mrs. William Sanders Jones) ....Apx G
Jones, Samuel J. 27, 28
Judson, Ben ....Apx G, 318
Kansas City Police Commission............ 187
Kansas City Times 176
Kansas City, Missouri 3, 8, 10, 11, 43, 44, 45, 51, 56, 57, 58, 59, 63, 69, 75, 86, 91, 93, 94, 96, 97, 98, 99, 100, 101, 103, 104, 105, 106, 107, 109, 114, 119, 121, 122, 124, 125, 130, 131, 135, 136, 137, 142, 144, 148, 150, 154, 158, 159, 162, 168, 173, 179, 185, 187, 190, 191, 192, 198, 202, 206, 207, 208, 211, 212, 219, 221, 223, 224, 227, 243, 244, 245, 246,

248, 250, 252, 257, 259, 263, 264, 265, 268, 273, 279, 281, 282, 283, 293, 296, 297, 307, 316, 317, 318, 319, 321, 322, 325, 326, 329, 330, 331, 332, 334, 335, 337, 338, 344, 345, 346, 347, 349, 350, 351, 356, 357, 358, 359, 360, 362, 369, 373, 377, 378, 384, 385, 397, 398, 399, 400, 401, 402, 406, 408, 409, 410, 411, 412, 413, 432
Kansas-Nebraska Act 7, 16, 23, 24, 29, 34, 54, 60, 75, 83
Kaw Township..223
Keeney Family..319
Keeney, Elvira (Campbell) ....Apx G
Keeney, Isaac ....Apx G
Keeney, Thomas ....Apx G
Kemper Family..319
Kemper, Berilda (Bryant) ....Apx G
Kemper, James Harvey ....Apx G
Kendal, Charles (Mrs.) ....Apx G, 319
Kern Family..319
Kerr, Charity (McCorkle) 93, 95, 96, 98, 100
Kerr, Nathan 93
Keshlear Family.......................................319
Keshlear, Alice (age 20) ....Apx G
Keshlear, George B. (child) ....Apx G
Keshlear, James C. (child) ....Apx G
Keshlear, Joseph B. (age 3) ....Apx G
Keshlear, Mattie A. (child) ....Apx G
Keshlear, Sarah (Riegs) (widow age 45) Apx G
Keshlear, William G. (child) ....Apx G
King, A. A. 134
King, Austin A.................................71, 167
King, Clara (Bingham) ..*See* Bingham, Clara
King, Kate (aka Mrs. William Quantrill).195
King, Thomas Benton.............................167
Knight Family...320
Knight, Jim ....Apx G
Knight, Mary (France) ....Apx G
Koger, John...195
Kreeger Family..............................321, 322
Kreeger, Araminta (Daniels) ....Apx G
Kreeger, Fannie (age 10) (later Haller) Apx G
Kreeger, George W. ....Apx G
Kreeger, J. D. (age 8) ....Apx G
Kreeger, R. R. (under age 7) ....Apx G
Laffoon, Drury ....Apx G, 322
Lamar, John ....Apx G, 323
Lane, Amos...22
Lane, James Henry "Jim" 3, 21, 22, 32, 36, 37, 39, 43, 61, 63, 69, 84, 85, 89, 111, 115, 116, 118, 119, 123, 124, 128, 129, 135, 147, 165, 183, 208, 295, 404
Lane, Mary E. (Baldridge)........................ 22
Lane, Mary... 22
Lawrence, Kansas 4, 6, 10, 12, 22, 24, 27, 28, 29, 30, 31, 33, 44, 51, 61, 96, 106, 107, 108, 109, 110, 111, 114, 115, 116, 117, 118, 119, 120, 121, 122, 124, 129, 132, 134, 135, 155, 160, 164, 186, 194, 200, 201, 215, 226, 227, 228, 270, 291, 324, 331, 341, 404, 405, 406
Lazear, Bazel F. 130, 141
Lea, Frank 110
Lea, Joseph 110
Leavenworth, Kansas 23, 28, 30, 33, 44, 84, 88, 89, 96, 97, 105, 110, 116, 119, 123, 124, 186, 222, 227, 307, 346, 412
Lecompton, Kansas...........26, 27, 31, 32, 61
Lee Family............................74, Apx G, 323
Lee, Robert E. 9, 10, 15, 34, 109
Lee's Summit, Missouri 111, 146, 150, 251, 255, 267, 295, 315, 317, 339, 342, 344, 347, 348, 372, 381, 404
Lewis Family 323
Lewis, Robert ....Apx G
Lewis, Sarah ....Apx G
Lexington, Missouri 7, 30, 41, 42, 43, 45, 46, 57, 74, 87, 133, 140, 141, 154, 163, 176, 249, 279, 282, 295, 320, 324, 335, 346, 356, 358, 365, 368, 380, 412
Liberty Tribune 63, 97, 109, 110, 120, 129, 166, 172, 188, 414
Lincoln, Abraham 4, 9, 11, 15, 17, 18, 19, 21, 28, 33, 34, 35, 36, 37, 38, 39, 40, 41, 42, 46, 58, 67, 68, 76, 85, 86, 89, 90, 91, 103, 104, 105, 122, 123, 129, 130, 132, 133, 134, 135, 140, 146, 184, 185, 191, 201, 205, 208, 209, 210, 213, 214, 267, 359, 360, 363, 408, 411
Little Blue River 100
Little Blue, Missouri..99, 100, 273, 313, 347
Little, Jim... 195
Livesay Family 323
Livesay, (Mrs.) ....Apx G
Lone Jack, Missouri 3, 16, 64, 105, 111, 150, 151, 154, 156, 157, 159, 163, 247, 250, 253, 257, 261, 268, 270, 285, 289, 313, 314, 322, 341, 345, 349, 362, 378, 383, 400

Louisiana Purchase 22, 23
Loyalty Oath.... 168, 169, 170, 171, 312, 354
Ludwick Family .... Apx G, 323
Lundy, Creed .... Apx G, 324
Luttrell Family...324
Luttrell, Armilda (age 31) .... 144, Apx G
Luttrell, Granvile (age 33) .... 146, Apx G
Luttrell, Samuel (age 48) .... 144, Apx G
Luttrell, Sarah (age 72)...........................146
Luttrell, Sarah (daughter of Sam) 146, Apx G
Lykins Family.................................324, 325
Lykins, Johnston (doctor)........................190
Lykins, Martha A. (Livingston) (age 39) ......
See Bingham, Martha A. "Mattie"
Lynch Family..325
Lynch, Miranda (Snodgrass) ....Apx G
Lyon, Nathaniel 39, 68, 174
Maddox, Asa.............................. Apx G, 326
Majors, Isham .... Apx G, 326
Manifest Destiny.......................................22
Mann, Christopher .... Apx G, 326
Massachusetts Emigration Aid Society 24
Masters Family 326
Masters, M. C. (age 30) ....Apx G
Masters, Mary C. (Leader) (age 21) ..Apx G
Mattin, S. E. .... Apx G, 327
Maupin, John 110
Maupin, Tom 110
Maxwell Family......................................327
Maxwell, Martha A. ....Apx G
Maxwell, Tom (Mrs.) ....Apx G
Mayes, Sarah Louisa (Wilson)
(Mrs. Henry C., when a child) ....Apx G
McClellan, George B.........................72, 184
McCook, Alexander McDowell.................84
McCook, Daniel, Jr...................................84
McCool Family.......................................327
McCool, James ....Apx G
McCool, Peter V. ....Apx G
McCorkle, Jabez 93
McCorkle, John 93, 106, 110, 195
McCorkle, Nancy Elizabeth "Nannie" (Harris)
....93, 99, Apx G
McCready, (Mrs.) .... Apx G, 327
McCullough, Ben......................................40
McDowell, Irvin 72
McFarland, T. J. .... Apx G, 327
McFerrin, Elizabeth .... Apx G, 328
McGee, Elijah M. 101
McGill Family .... Apx G, 328
McHenry, James E. (Mrs.)..................... 125
McKinley, William 203
Meade, George... 10
Meador Family....................................... 328
Meador, Sara Angeline, a sister and
three children .... Apx G, 328
Mers Family....................... Apx G, 328, 329
Mers, Andrew 328
Mers, W. T. 329
Mexican War 22, 30, 32, 76, 343, 344
Milam, Thomas .... Apx G, 329
Miller Family 329
Miller, George (Rev.) ....135, Apx G
Mims, F. (professor) .... Apx G, 330
Missouri Compromise 16, 23, 31, 83
Missouri General Assembly 49, 54, 55, 56, 57, 188
Missouri Home Guard 57
Missouri River 39, 41, 51, 54, 61, 63, 69, 71, 77, 84, 86, 91, 111, 119, 132, 142, 150, 163, 167, 185, 187, 190, 244, 249, 278, 289, 291, 294, 296, 308, 326, 337, 338, 345, 354, 366, 367, 368, 376, 378
Missouri State Guard 7, 39, 40, 41, 49, 50, 57, 81, 147, 150, 249, 257, 327, 356
Montgomery Family 330
Montgomery, Elizabeth (Holloway) ..Apx G
Montgomery, James 32, 35, 41, 45, 61, 63, 84
Montgomery, Levi ....Apx G
Moore Family 330, 331
Moore, (Mr.) ....Apx G
Moore, Frances Fitzhugh (George)
(later Kabrick)..................................Apx G
Moore, Lawson and 6 children ....Apx G
Moore, Will ....Apx G
Moore, Zirilda ....Apx G
Morris, John Calvin .... Apx G, 331
Morrison, A. W....................................... 51
Morrow Family...................................... 331
Morrow, Jesse (age 53) ....Apx G
Morrow, Nathan (child) ....Apx G
Morrow, Rufus F. B. (child) ....Apx G
Morrow, Sylvia (Hudspeth) (age 56) .Apx G
Mudd, Samuel A.................................... 208
Muir Family 332, 333
Muir, Bettie (child) ....Apx G

Muir, Buck ....Apx G
Muir, Buckner (age 84) ...151, 247, 332, 374
Muir, John S. ....Apx G
Muir, Nancy (Parr) ....Apx G
Muir, Sam (age 56) ....151, Apx G
Muir, Samuel (age 16) ....Apx G
Mulligan, James A..........................7, 41, 57
Munday, Martha "Mattie" 98
Munday, Susan 98
Murray Family 333
Murray, Mary F. (Smith) ....Apx G
Murray, William F. (age 7) ....Apx G
Myers Family...333
Myers, Harry ....Apx G
Myers, William ....Apx G
Nailor Family...333
Nailor, Fountain ....Apx G
Neat Family ....Apx G, 333
Nelson Family 334
Nelson, Hector ....Apx G
New Franklin, Missouri 52
Newman Family 334
Newman, J. N. ....Apx G
Nicholson Family 334
Noel, James W. and wife ....Apx G
Nolan, R. H...............................Apx G, 334
Noland, Harriett (Newton) ....Apx G, 334
Norman, C. P. ....Apx G, 334
O'Bannon, E. ....Apx G, 335
O'Donnell, (Mrs.) ....Apx G, 335
O'Laughlin, Michael 210
Oak Grove, Missouri 106, 247, 271, 277, 283, 288, 308, 316, 401
Oldham Family...............................335, 336
Oldham, Albert G. ....Apx G
Oldham, Althea (child) ....Apx G
Oldham, Amelia ....Apx G
Oldham, John R. ....Apx G
Oldham, John R. (Mrs.) ....Apx G
Oldham, Mary Julia ....Apx G
Oldham, Thomas (child) ....Apx G
Osage River 42, 69, 120
Osceola, Missouri 42, 43, 114, 147, 225
Oulsey, Andrew......................................161
Ousley Family 336
Ousley, Andrew (age 17) 155, 159, Apx G, 261, 336, 344, 346
Ousley, Dana (Mrs.) ....Apx G
Ousley, Dana 161
Ousley, David 155
Ousley, Matilda (Hunter).... 155, 161, Apx G
Ousley, Miss (daughter) ....Apx G
Owings Family....................................... 336
Owings, Martha E. (George) (Mrs. Joshua) and Family of 7 ....Apx G
Pacific House 69, 86, 105, 124, 125, 136, 174
Paddock, George 109
Page Family 336
Page, Granville Dyson ....Apx G
Page, Venitia Mary "Nish" (Colcord) Apx G
Palmer Family............................... 337, 338
Palmer, (Dr.) ....Apx G
Palmer, Jacob Teaford ....Apx G
Pappinsville, Missouri 45
Pardee (Captain) 110
Parish Family ....Apx G, 338
Park Family ....Apx G, 338
Park, Anna Lou (Gault) 153
Park, Annie (age 2) ....Apx G
Park, G. L... 153
Parker Family................................ 338, 339
Parker, William H. ....Apx G
Parker, Zachariah ....Apx G
Parker, Zachariah (Mrs.) ....Apx G
Parmer, Allen 195
Parsons Family....................................... 339
Parsons, Barbara (Ely) ....Apx G
Parsons, Daniel White ....Apx G
Parsons, Melvina A. ....Apx G
Parsons, Samuel ....Apx G
Partisan Ranger 65
Paschall, Nathaniel 57
Patterson Family 340
Patterson, Jesse (child) ....Apx G
Patterson, Joel (age 47) ....Apx G
Patterson, Joel B. (child) ....Apx G
Patterson, Mollie (child) ....Apx G
Patterson, Sarah (age 43) ....Apx G
Paw Paws 82
Payne Family 340
Payne, E. S. ....Apx G
Payne, Lewis.. 210
Pearl Harbor, Hawaii 204
Pemberton, John C.................................. 10
Perry, Joab 171
Phillips and Comingo 206

Phillips, Comingo and Slover 206
Phillips Family..340
Phillips, James M. ....Apx G
Pierce Family..340
Pierce, A. J. (doctor) ....Apx G
Pierce, Franklin...................................22, 31
Pillard, Phillip D. ....Apx G
Pink Hill, Missouri 62, 263, 264, 283, 376, 406
Pinkerton Detective Agency..............36, 187
Pleasant Family 340
Pleasant Hill, Missouri 45, 121, 122, 154, 158, 159, 163, 223, 244, 253, 255, 267, 268, 279, 280, 283, 285, 304, 307, 320, 328, 329, 333, 339, 342, 345, 346, 348, 358, 359, 362
Pleasant, Edward ....Apx G
Pleasanton, Alfred 9
Plumb, Preston B. 128
Polk, James K. 22
Pollard Family 341
Pomeroy, Samuel C. 37, 202
Poole, David 110
Potter Family 341
Potter, Amanda (child) ....154, Apx G
Potter, Benjamin (son) ....Apx G
Potter, Benjamin .... (age 75) 154, 159, 161, Apx G, 261, 341, 344, 346, 360
Potter, David ....Apx G
Potter, John 155
Potter, Marian (child) ....154, Apx G
Potter, Martha ....Apx G
Potter, Mary "Polly" (Hays) 154
Potter, Melchert 155
Potter, Steven..155
Potter, Thomas (age 12) ....154, Apx G
Potter, William 154
Potts Family...341
Potts, Peter J. (Mrs. Fannie E.) (Parsons) Apx G
Powell Family..................................341, 342
Powell, Absalom.....................................156
Powell, Alvis ....151, Apx G
Powell, John Beeler ....Apx G
Powell, Jonathan R. ....Apx G
Powell, Parthen (Meador) ....Apx G
Poyntz, John ....Apx G, 342
Price, Sterling 7, 40, 43, 57, 76, 77, 98, 147, 150, 184, 195, 208, 259, 280, 332, 339, 344, 349, 376
Prince, W. E..43
Provost Marshal 3, 10, 65, 71, 72, 73, 79, 80, 104, 120, 124, 133, 135, 140, 141, 165, 205, 208, 254, 333, 359, 366, 405, 432
Quantrill, Caroline Cornelia (Clarke) . 59, 60
Quantrill, Thomas Henry 59, 60
Quantrill, William Clarke 3, 4, 6, 10, 11, 12, 16, 44, 59, 60, 61, 62, 63, 64, 65, 66, 82, 85, 87, 91, 93, 94, 98, 102, 106, 107, 110, 111, 114, 115, 116, 117, 118, 119, 121, 135, 141, 144, 150, 153, 154, 155, 163, 165, 183, 185, 192, 194, 195, 196, 197, 198, 199, 200, 227, 228, 243, 244, 262, 271, 291, 308, 310, 331, 341, 346, 371, 373, 378, 380, 403, 404, 405, 406, 408, 410, 411, 412, 413
Quick, Cornelius ....Apx G, 342
Ralston Family.. 342
Ralston, Adam (age 8) ....Apx G
Ralston, Anna (age 10)
(later, Mrs. Frank James) ....Apx G
Ralston, Fedora (age 15) ....Apx G
Ralston, Harry M. (age 1) ....Apx G
Ralston, Katherine (age 3) ....Apx G
Ralston, Mary Catherine (Hill) ....Apx G
Ralston, Samuel (age 13) ....Apx G
Ralston, William H. (age 5) ....Apx G
Ralston, Samuel "Sam" 25, Apx G, 342, 343
Ransom, Wyllys C. 130, 174
Raytown, Missouri.. 98, 100, 272, 273, 297, 332, 336, 347, 348, 364, 376, 377, 407, 430, 432, 433
Red Legs 19, 35, 43, 44, 58, 63, 66, 82, 85, 86, 100, 105, 109, 111, 142, 144, 148, 151, 168, 174, 176, 178, 205, 248, 267, 268, 270, 272, 293, 307, 351, 376, 379, 384
Reed, Farlow ....Apx G, 343
Reid, John W. (age 40) ....Apx G, 343
Republican Party 21, 33, 34, 36, 78, 82, 132, 157, 168, 169, 170, 171, 179, 185, 201, 202, 210
Reynolds, Thomas 50, 57
Rice Family.............................344, 346, 347
Rice, E. N..................................159, Apx G
Rice, Elizabeth....................................... 157
Rice, Hollow ....Apx G
Rice, Isaac (age 17) ....157, 159, Apx G
Rice, James 157
Rice, Martha ... *See* Tate, Martha Jane (Rice)

Rice, Martin (Age 49) 105, 142, 156, 157, 158, 159, 161, 163, Apx G, 252, 253, 314, 344, 345, 346, 347, 360, 361, 383, 399, 401, 412
Rice, Mary (Lynch) (age 70) ...... 157, Apx G
Rice, Mary (Martin's daughter) .... Apx G
Rice, Nancy 157
Rice, Spencer 157
Richardson Family.................................347
Richardson, Joseph .... Apx G
Richardson, Mary I. .... Apx G
Ridge, Isaac (doctor)...............................128
Ridgeway (Captain) 173
Ridlesberger, Jesse...................... Apx G, 347
Ritter Family...347
Ritter, Henry T. (age 40) .... Apx G
Ritter, Taylor (Mrs.) (age 36) .... Apx G
Robinson Family 348
Robinson, Charles..........................24, 31, 36
Robinson, John .... Apx G
Rockingham County, Virginia...................51
Rollins, James S.. 49, 50, 52, 54, 55, 56, 57, 58, 102, 106, 136, 168, 174, 178, 179, 187, 189, 190, 191
Ross, (Mrs.) .... Apx G, 348
Ross, Edmund G. 202
Ross, John..110
Roupe Family 348
Roupe, John .... Apx G
Roupe, John Henry (son of John) ...... Apx G
Roupe, John 154, 349
Rout, Henry .... Apx G, 349
Rowland Family 349
Rowland, Jesse .... Apx G
Rowland, William .... Apx G
Ruger, T. H. 204
Runyan, A. O. and Family .... Apx G, 349
Russel, Elizabeth Southworth (Williams) Apx G
Ryley, Archiey (age 15) .... Apx G
Ryley, Emma (child) .... Apx G
Saint Clair, Joseph .... Apx G, 349
Samuel, Reuben 188
Samuel, Zeralda (James)..........................188
Sanders, (Mr.) .... Apx G, 350
Sappington, Erasmus 54
Sartain, John 8, 54, 178, 179, 181
Saunders Family 350
Saunders, Drucilla (Stayton) .... Apx G
Saunders, John B. .... Apx G
Saunders, Susan (age 11) .... Apx G
Schofield, Georgia (Kilbourne) 203
Schofield, Harriet (Bartlett) 203
Schofield, John McAllister 3, 4, 9, 40, 67, 68, 69, 71, 82, 86, 87, 89, 90, 91, 103, 116, 117, 118, 123, 124, 129, 130, 131, 135, 136, 164, 165, 166, 180, 194, 201, 203, 204, 211, 227, 402
Scott, (Mr.) .... Apx G, 351
Scott, William W. 197
Scrivner Family 351
Scrivner, Joseph .... Apx G
Scrivner, Pameila .... Apx G
Scruggs, P. T. .... Apx G, 351
Secession 17, 18, 39, 49, 50, 56, 57, 76, 77, 78, 79, 81, 205, 344
Sechrest, Charles and Family ..... Apx G, 351
Seddon, James A. 65
Selvey, Armenia (Crawford).............. 96, 98
Settle, J. F. .... Apx G, 351
Sevier, Henry Clay .... Apx G , 351
Seward, William H. 214
Shaler, N. T. .... Apx G, 352
Sharp, Matt .... Apx G, 352
Shaw, Robert Gould......................... 10, 103
Shawnee Trail 157
Shawneetown, Kansas 64
Shelby, Joseph Orville "Jo" 98
Sheley Family 352
Shelley, James K. 76
Shelton Family....................................... 352
Shelton, Minerva .... Apx G
Shelton, William H. and father's family Apx G
Shepard, "Ole" 195
Shepherd, George 110
Sherman, Charles 83
Sherman, Mary 83
Sherman, William Tecumseh 11, 33, 83, 84, 210
Shipley Family...................................... 352
Shipley, Amanda .... Apx G
Shipley, Preston (youth) .... Apx G
Shipley, William .... Apx G
Singer, John 95
Sixth Congressional District 10, 71, 79, 165, 205
Slaughter Family................................... 353
Slaughter, Elizabeth .... Apx G
Slaughter, O. V. .... Apx G

Slavery 10, 17, 18, 27, 29, 32, 42, 49, 51, 53, 56, 61, 62, 67, 75, 91, 103, 104, 105, 122, 131, 147, 213, 214, 219, 221, 253, 265, 267, 268, 275, 278, 282, 283, 291, 305, 306, 307, 309, 310, 354, 356, 357, 362, 372, 374, 380, 406
Sloan, F. X. .... Apx G, 353
Slover, James 206
Smart, Susan (Mrs.) .... Apx G, 353
Smith Family 354, 355
Smith, B. F. and Family .... Apx G
Smith, Dicey .... Apx G
Smith, Eliza (Wallace) .... Apx G
Smith, Franklin (Jr.) .... Apx G
Smith, Franklin (Sr.) .... Apx G
Smith, Isaac .... Apx G
Smith, John .... Apx G
Smith, Minor T. and Family .... Apx G
Sni-A-Bar River 163
Sni-A-Bar Township 156, 247, 259, 261, 274, 316, 326
Snow, William F. 158
Southwest Expedition 36
Spangler, Edward 209
Square and Compass (Wayside Inn) 51
St. Louis, Missouri 9, 35, 39, 41, 45, 51, 57, 67, 68, 71, 76, 77, 78, 79, 80, 83, 91, 117, 122, 123, 131, 132, 166, 168, 174, 178, 180, 181, 182, 184, 185, 202, 208, 247, 255, 285, 286, 303, 313, 323, 325, 344, 359, 360, 381, 407, 412
Stanton, Edwin 201, 202, 203, 209, 210
Staples Family .... Apx G, 356
Staples, Abner Ford .... Apx G
Staples, Elsie .... Apx G
Stayton Family 355
Stayton, Lucy Ann (Hite) .... Apx G
Stayton, Moses .... Apx G
Stayton, Rebecca .... Apx G
Steele Family 356
Steele, Albert Gallatin .... 140, Apx G
Steele, Emily (Hampton) .... 140, Apx G
Steele, George 140
Steen, Elizabeth .... Apx G, 358
Stegall Family .... Apx G, 358
Stephens Family 358
Stephens, John L. L. .... Apx G
Stevens, Flora 98, 125
Stevens, William (Judge) 125
Stewart, John E. 61
Stewart, Robert Marcellus 35
Stine, E. (Mr.) 125
Stonestreet Family 358
Stonestreet, James .... Apx G
Stonestreet, James (Mrs.) .... Apx G
Storms Family .... Apx G, 358
Stuart, J. E. B. 9, 34
Summers, John W. and Family .... Apx G, 358
Sumner, William 28
Surratt, Mary E. 210
Sweeney, James and Family .... Apx G, 358
Symington, Robert Smith .... Apx G, 359
Talbot, Joshua .... Apx G, 359
Talley Family 360
Talley, David C. .... Apx G
Talley, Mary (Fristoe) .... Apx G
Tate Family 361
Tate, John (age 5) .... Apx G
Tate, John 157
Tate, Martha Jane (Rice) (age 27) . 157, 161, Apx G
Tate, Mary (age 3) .... 157, Apx G
Tate, Nancy Rosalie (baby) .... 157, Apx G
Tate, William Calvin(age 32) 157, 159, 161, Apx G, 345, 360
Tate, William 346, 360
Tatum Family 362
Tatum, Elizabeth (Clark) and Family Apx G
Tatum, John W. (youth) .... Apx G
Tatum, Thomas Jefferson .... Apx G
Taylor Family 363
Taylor, Elizabeth .... Apx G
Taylor, Fletcher 106, 110
Taylor, Zachary 83
Teager, Dick 153
Terrell, Edwin 195
Thatcher, J. M. (Mrs.), her father, and her four children .... Apx G, 363
Thayer, Eli 24
Thomas Family 363
Thomas Property.. *See* Union prison collapse
Thomas, Griffin .... Apx G
Thompson Family 363
Thompson, David (age 14) .... Apx G
Thompson, George M. (age 47) .... Apx G
Thompson, James (age 16) .... Apx G

Thompson, Lizzie H. (age 13) ....Apx G
Thompson, Lucinda (Webb) ....Apx G
Thompson, Susan A. (age 41)............Apx G
Thompson, William ....Apx G
Thorne, Joshua (doctor)..........100, 101, 102
Thornton, C. T. ....Apx G, 364
Thruston, W. Creighton ....Apx G, 364
Timberlake, James 206
Todd, George 110, 194
Triplett Family..365
Triplett, Dan (age 4) ....Apx G
Triplett, Eli and Family ....Apx G
Triplett, Margaret (Hart) ....Apx G
Trow, Harrison 107, 110, 132
Truman Family 366
Truman, Anderson Shipp "Andy" ......Apx G
Truman, Harry S... 147, 149, 262, 306, 324, 353, 384, 409
Truman, John Anderson (child) ....Apx G
Truman, Mary Jane ....Apx G
Trusten, Cappell (child) ....Apx G
Tucker, Tom ....Apx G
Tuggle, E. L. ....Apx G
Twyman Family.......................................366
Twyman, Frances C. "Fannie" (Fristoe) ....143, Apx G
Twyman, George T. (age 11) ....Apx G
Twyman, Julia (age 13) ....Apx G
Twyman, Lyddall W. (doctor) ... 143, Apx G
Unconditional Unionist Party 56, 81
Underwood, John ....Apx G, 370
Union Cemetery........................98, 192, 325
Union Hotel 91
Union prison collapse..........10, 98, 100, 102
Van Buren Township 156, 250, 252, 268, 289, 292, 346, 363, 364
Van Horn, Robert Thompson 57
Van Kirk, (Mr.) ....Apx G, 370
Van Ness, Alice ....Apx G, 370
Vandevere, Susan (Crawford) 96, 98
Vanness, Alice Fay 97
Vaughan, R. C. 134, 140
Vaughn Family 370
Vaughn, Rebecca (Harper) ....Apx G
Vaughn, Reuben ....Apx G
Vergin, William G. ....Apx G, 370
Vincent, Charles H. 101
Wade Family 370
Wade, W. ....Apx G
Wakarusa War 27
Wakefield, James H. 195
Walker Family 371
Walker, Andrew J. (age 27) ....Apx G
Walker, Collins (age 39) ....Apx G
Walker, Elizabeth (Braden) ....Apx G
Walker, James ....Apx G
Walker, Jefferson (age 35) ....Apx G
Walker, John Riley (age 33) ....Apx G
Walker, Morgan............61, 62, 63, 370, 371
Walker, Nancy C. (Montgomery) ......Apx G
Walker, Polly (Cox)................................ 61
Walker, Susan Jane ....Apx G
Walker, Thomas J. (age 5) ....Apx G
Walker, Zachaiah (age 25) ....Apx G
Wallace Family 371, 373
Wallace, Addison A. (infant) ....Apx G
Wallace, Addison H. ....Apx G
Wallace, Charles H. (child) ....Apx G
Wallace, Jefferson Davis ....Apx G
Wallace, Jessamine (Ryley) (age 40) .Apx G
Wallace, John (child) ....Apx G
Wallace, John D. ....Apx G
Wallace, Joseph (Reverend) ....Apx G
Wallace, Nellie (Duncan) (Mrs. John D.) Apx G
Wallace, William H. (age 14) ....Apx G
Wallace, William 206
Walton, (Mrs.) ....151, Apx G, 374
Walton, Thomas (age 3) ....151, Apx G
Ward Family 374
Ward, Cyrus ....Apx G
Warren Family 375
Warren, Lucinda Jane (Robertson) ....Apx G
Warren, Nelson (Bates County)..........Apx G
Warren, Nelson (Jackson County) ......Apx G
Warren, Sarah ....Apx G
Washington Township 147, 318, 382
Washington, George (portrait)................. 55
Wayne City Landing..........61, 150, 372, 376
Webster, Daniel 83
Wells, Washington................................. 150
West Family...............................Apx G, 376
West Point...........................67, 83, 203, 204
West, George Cornelius (age 7) ....Apx G
West, George Cornelius (Mrs.) ....Apx G
West, Lysander ....Apx G
Western Journal of Commerce 96

Weston, Missouri....................................84
Westport, Missouri 45, 64, 88, 89, 103, 121, 122, 124, 125, 150, 185, 206, 219, 223, 227, 256, 287, 296, 298, 300, 303, 316, 325, 331, 332, 359, 363, 377, 384, 400, 402
Whig Party...34, 52, 53, 55, 75, 82, 157, 191
White Family ....Apx G, 377
White Rags 150
White, William ....Apx G
Whitsitt, J. B. ....Apx G, 377
Wiley, William D. ....Apx G, 377
Wilhite, Jefferson ....Apx G, 377
Willett, J. N. ....Apx G, 377
William(s), David ....Apx G
Williams Family 378
Williams, J. R. ....Apx G
Williams, J. W. ....Apx G
Williams, John 150
Willis, Alpheus ....Apx G, 378
Wilson (Mrs.) 96
Wilson Family 378
Wilson, (Mr.) ....Apx G
Wilson, Louisa (McMurray) ....Apx G
Woodruff Family ....Apx G, 379
Woodruff, Julia ....Apx G
Woodson Family....................................380
Woodson, Chrisman and Comingo 74
Woodson, M. A. (Mrs.) (formerly Mrs. Ervin; originally Miss Bowers) ....Apx G
Woodson, Margaret J. (Ashley)................75
Woodson, Samuel "Sam"....................54, 75
Woodson, Samuel H. (Mrs.) ....Apx G
Woolery Family.....................................380
Woolery, Caroline ....Apx G
Woolery, J. J. ....Apx G
Worrell Family ....Apx G, 380
Wright Family 380
Wright, Emory (Mrs.) (her unnamed grandfather) ....Apx G
Wyatt Family 381
Wyatt, Loraine (Mrs. A. M.) ....Apx G
Wyatt, Wiley ....Apx G
Wyatt, Wiley (Mrs.) ....Apx G
Yancey, George ....Apx G, 382
Yankee Family...................................... 382
Yankee, Jacob ....Apx G
Yankee, Jacob (Mrs.) ....Apx G
Yankee, Orlena (Beeler) (Mrs. David) Apx G
Yankee, Wesley ....Apx G
Yeager, Dick 110
Yeager, Richard 152
Yoder, Solomon ....Apx G, 383
Young Family 383
Young, Ada (child) ....147, Apx G
Young, Harriet "Louisa" (Gregg)146, Apx G
Young, Harrison (age 17) ....147, Apx G
Young, Laura Jane (age 13) ....147, Apx G
Young, Martha "Mattie" Ellen (age 11) ....147, Apx G
Young, Rebecca Jane (Smith) ....Apx G
Young, Sarah Ann (age 20) ....Apx G
Young, Sarah Ann .... *See* Chiles, Sarah Ann (Young)
Young, Solomon 146, 149, 262, 264, 307, 353, 384
Young, Susan M. ....147, Apx G
Young, William A. 147
Younger, Caroline................................... 98
Younger, Coleman "Cole" 98, 107, 110, 111, 115, 144, 353, 383, 410
Younger, Josie 98
Younger, Sara "Sally"............................. 98
Youngs Family...........................Apx G, 385
Zion Family 38
Zion, Nancy ....Apx G, 385

ACKNOWLEDGEMENTS

No book of this magnitude could be written without the forethought and dedication of many individuals who have given diligently of their time, commitment and service to the cause of memorializing the history of the Border War, Civil War, and Order No. 11 in Jackson County.

To some, their recognition and acknowledgement for years of service comes posthumously, including Tim Cox, Donald R. Hale, Ethylene Ballard Thruston, Harold F. "Sonny" Wells, and Pearl Wilcox.

Another remarkable lady from Raytown, Missouri, that deserves recognition is Roberta Bonnewitz—the "Raytown Historian."

Several individuals who have dedicated their lives preserving, advocating, researching and/or writing about the rich history of Jackson County deserve the gratitude of the author and all others who have been touched and inspired by their efforts. With profound respect, special tribute is paid to local historians Carolyn Bartels, Bill and Annette Curtis, Harold Dellinger, Joanne Chiles Eakin, Nancy M. Erhlich, Ann Fanning, and Brian and Sharon Snyder. Without their endeavors, this book would never have been written.

Historian Tom Rafiner, Parkville, Missouri, kindly provided the bulk of the information and material on the ravages of Order No. 11 along the border in Cass and Bates County. His contribution and research on the identity of those evicted in these two counties under Order No. 11 has been an integral part of this book. His current research and future publication on Jackson County is anxiously anticipated, and it is hoped the content in this work will aid his endeavors.

There is simply no way this book could have been completed without the valued assistance of the Jackson County Historical Society ("JCHS"). Steve Noll, Executive Director, provided guidance and patience. David W. Jackson, Archivist, and his chief volunteer assistant in this project, Pati Kidwell, gave yeoman aid, assistance and advice. Without their tireless efforts and countless hours mining data over several months, Appendices G and H would not have been compiled, or completed. David, my esteemed friend, colleague, and author in his own right, also spent countless hours searching and digitizing images, and designing and indexing this book in preparation for publishing on a tight turn-around so we would have a product for the summer 2013 Order No. 11-related sesquicentennial events described below.

JCHS has a mission to preserve the rich history of Jackson County and a determination to share the story through educational and living history programming. To commemorate the 150th Anniversary of Order No. 11, JCHS partnered with several historically rich organizations and living historians to produce and present "Blood & Ashes," a trilogy of living history programs commemorating the Civil War Border conflict in Jackson County in August 1863, and Order No. 11. Their affiliations have been instrumental in making the history of these troubled times come to life. I want to personally thank the following partners of JCHS and living historians for their contributions.

> *The Friends of the Rice Tremonti Home* in Raytown, Missouri, and its dedicated Board: Barb Schlapia, president; Leigh Elmore, vice president; Dave Wurth, treasurer; Marilyn Fleming, recording secretary; Vi Guinn, corresponding secretary; Bob Gilham; Bob Grissom; Mike McDonough; Jessie Jan Martin; Phyllis Miller; Mick O'Neal; and Aaron Smith;
>
> *Wornall-Majors House Museums,* and in particular the tirelessness of Director Anna Marie Tutara, Herb Kline, Alex Rosser, and their gifted volunteers;
>
> *Wide Awake Films* and its marvelous team of Shane Seely, Ed Liedecker, and dedicated staff;
>
> *Living historians* David Bears, Dan Hadley, Gregg Higginbotham, Honorable Michael W. Manners, Shirley Wurth, Aaron J. Racine, Dan Enloe, and Bill Wehner deserve recognition for their hours of time and commitment to "Blood & Ashes."

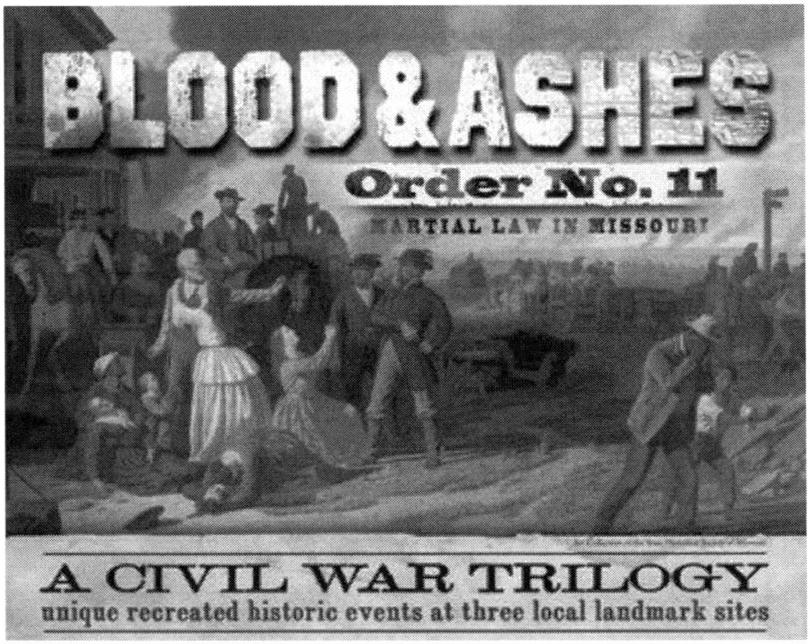

ABOUT THE AUTHOR

Ralph A. Monaco, II is a life-long resident of Raytown, Missouri. He graduated from Rockhurst University in 1978, summa cum laude, earning bachelor degrees in history and political science. In 1981, he received his law degree from the University of Missouri at Kansas City School of Law. He is a charter member of the Kansas City law firm, Monaco, Sanders, Gotfredson, Racine & Barber, L.C. He has been in the private practice of law for the last thirty-two years and he is a member of the Missouri Bar and the Western District of the Federal Court in Missouri. His peers in the legal community have recognized him for his legal work by bestowing upon him an AV attorney rating.

Monaco has been active in the political, historical and civic communities in Jackson County. He formally served on the Raytown School Board, and served as School Board President. He was elected to the Missouri House of Representatives, District #49 in 1996, and went on to serve four terms, including serving as Chairman of the House Judiciary Committee. He is past President of the Jackson County Historical Society and currently serves as Vice-President of Living History and Programming for the Society.

Monaco has received numerous public service awards. These have included the Acorn Lifetime Membership Award by the Missouri PTA, Missouri Judiciary Legislator of the Year Award, the Missouri Bar Legislator of the Year Award, the Association of Retarded Citizens of Missouri Friend of the ARC Legislator of the Year Award, Outstanding Service Award from Jackson County Parks & Recreation, Volunteer of the Year from Jackson County Parks & Recreation; Making History Come Alive Award from the Jackson County Historical Society, Lifetime Membership Award in the Raytown Historical Society, Pioneer of Harrodsburg Award and Kentucky Colonels.

Monaco continues to remain very active with historical organizations and in historical portrayals. He has a passion for living history and Civil War reenacting. For well over 20 years, he has portrayed Abram Comingo, a 19th century attorney and U.S. Provost Marshal for the Sixth Congressional District during the Civil War. In addition, he has provided first-person portrayals of other historical figures including Manuel de Lisa, a Spanish Fur Trader who helped outfit the Lewis & Clark Expedition; James Slover, Frank James' defense attorney; Edgar Allan Poe, the poet and master of the macabre;

George Caleb Bingham, the Missouri artist and politician; Col. Thomas H. Swope, the philanthropist and purported murder victim at the hands of Dr. B. Hyde; and Jesse James, Jr., the son of the bandit, Jesse James.

Monaco is the author of, "The Strange Story of Col. Swope & Dr. Hyde." (Independence, Mo.: Two Trails Publishing, 2010); "Son of a Bandit: Jesse James and the Leeds Gang" (Kansas City, Mo.: Monaco Publishing, LLC, 2012); and, "Blood on the Streets: The Civil War Comes to Jackson County, Missouri, August 1862." (Independence, Mo., Monaco Publishing LLC, 2012). His book, "Son of a Bandit," received the coveted John Newman Edwards Award from the Friends of the James Farm in 2012. Over two decades, he has also authored several manuscripts and historical living history programs, including the murder of Jackson County Jailor, Henry Bugler; and three significant "re-trials," of Frank James; Jesse James Jr.; and, Dr. Bennett Clark Hyde, for the alleged murder of Col. Swope.

Monaco and his wife, Karen, live in Raytown, and they have two daughters, Lindsay and Lisa.

Made in the USA
Monee, IL
13 September 2026

62981011R10240